EXPERIMENTAL PSYCHOLOGY
A Case Approach

FIFTH EDITION

ROBERT L. SOLSO
University of Nevada, Reno

HOMER H. JOHNSON
Loyola University of Chicago

 HarperCollins*CollegePublishers*

Acquisitions Editor: Catherine Woods
Project Coordination and Text Design: Proof Positive/Farrowlyne Associates, Inc.
Cover Design: Kay Petronio
Production Manager: Kewal Sharma
Compositor: ComCom Division of Haddon Craftsmen, Inc.
Printer and Binder: Malloy Lithographing, Inc.
Cover Printer: Malloy Lithographing, Inc.

EXPERIMENTAL PSYCHOLOGY: A Case Approach, Fifth Edition

Library of Congress Cataloging-in-Publication Data
Solso, Robert L., 1933–
    Experimental psychology : a case approach / Robert L. Solso, Homer
H. Johnson. — 5th ed.
        p.   cm.
    Rev. ed. of: An introduction to experimental design in psychology.
3rd ed. c1984.
    Includes bibliographical references and index.
    ISBN 0-06-501142-2
    1. Psychology, Experimental.   2. Experimental design.
3. Psychology—Case studies.   I. Johnson, Homer H., 1935–
II. Johnson, Homer H., 1935–   Introduction to experimental design
in psychology.   III. Title.
BF181.S68   1994
150'.724—dc20   93-4977
                                                          CIP

Table of Contents

To the Instructor

We are pleased to present this fifth edition of *Experimental Psychology: A Case Approach.* Experimental psychologists have covered a lot of territory during the past century when psychology became a formal discipline, and the many changes have necessitated changes in books and professional articles. This edition of *Experimental Psychology* reflects the current status of the major themes in psychology and shows how experiments are conducted as we approach the close of the 20th century.

Originally we wrote this book to identify the basic principles of experimental design as practiced by experimental psychologists. Too often the teaching of experimental psychology involves lengthy discussions of theoretical statistics, the philosophy of science, or a concentration on a highly specialized area of research. So after several years of experimenting with how to best present the material for a first course in experimental psychology, we have developed a method by which the student studies actual case examples in psychology and then generalizes the ideas from those examples to the principles of experimental design. The approach is "bottom-up," in the argot of cognitive psychology, in that we emphasize actual experiments from which principles are derived. Although this approach builds from the basic to the more abstract, it is our intention to give both aspects of experimental psychology due attention. Both examples *and* principles are important in the study of experimental psychology. It is our purpose to teach both.

The pedagogical method in this book uses actual experiments to help the student learn how design principles are applied in research. In this edition of *Experimental Psychology* the student will read, critique, or analyze some 75 cases or experiments that exemplify various design principles and problems. In addition to understanding design, the student will become comfortable with the research literature and learn much of the content material of psychology.

This edition is based on the feedback and suggestions of numerous instructors who used the first editions in a wide variety of courses, ranging from introductory psychology through graduate psychology, as well as courses in other disciplines.

Their recommended improvements, as well as our class testing, have strengthened the text considerably.

Using examples in the teaching of disciplines such as experimental psychology has been a traditional means of instruction and is still widely practiced—whether the subject is high-energy physics, carpentry, accounting, computer programming, psychotherapy, creative writing, or cellular biology. But in courses involving experimental design in the behavioral sciences (or, more plainly stated, "how to do research in . . .") the common practice is to meander through a series of philosophical/theoretical issues that, while important to any scholar's education, make little or no effort to establish a link between the principles and the real world of research. We believe this link is essential to the development of critical thought and the practice of research, and hope that this book will help you establish that link as you teach experimental design.

We wrote the book from a tutorial standpoint: as if we were a private tutor instructing a student as he or she was reading the material. First we present a principle or a problem in experimental design. Then we show how the principle or problem has been dealt with in the psychological literature. We provide annotated reviews of actual articles, much as a master teacher might do if he or she sat down with a student and critically read an article with him or her. Based on the comments we have received from students and instructors alike, this technique is remarkably successful. We hope this edition will be even more useful and instructive in that sense than the previous editions.

Part One of this edition, which deals with the basic principles of experimental design, has been expanded and is more inclusive. Also, while we have long advocated the view that experimental psychology is a laboratory science, in this edition we have expanded the view expressed in previous editions that a researcher needs to be prepared to ply his or her trade in a variety of settings, inside or outside the laboratory. In addition, we have expanded the section on ethics in experimental research to reflect changes in that area over the past few years. We have included the most recent version of the ethical standards as adopted by the American Psychological Association. We have also split the previous Chapter 2 on the anatomy of experimental design into two separate chapters and reorganized sections of other chapters so they progress more logically. We continue to stress the use of computers in library research and trust that the material we have presented will be helpful to your students in using electronic data banks. We also discuss the development of research ideas. In both Part One and Part Two we are mindful of the role psychological research plays in shaping thinking in society in general, and, as such, we have included examples (on gender differences, for example) that address these issues. We believe that such examples of "controversial" research in an introductory book on experimental psychology are entirely appropriate.

In Part Two we have added or deleted several articles. The scope of articles in Part Two was carefully selected to sample the major areas of psychology, including industrial and practical issues, cognitive psychology, social psychology, animal and ethological studies, practical problems, cross-cultural studies, psychotherapy, single-subject designs, educational psychology, behavioral modification, child psychology, and so on. We have found that by using this format students learn a great deal about the different fields of psychology in addition to learning about the vast diversity of experimental designs and techniques currently being employed in the behavioral sciences. In this edition the case analysis notes for the case studies in Part Two appear on (or near) the same page as the referenced section. Thus, the student does not have to flip back and forth between the case analysis and the article. Some professors have told us that Part Two of the book is the primary reason they select the book for class use. (Curiously, about the same number have told us they select the book for the contents of the first section!)

Finally, in Appendixes A and B we have responded to frequent requests to include a section on basic statistics. This section is modest, but allows the book to be used in a much wider range of courses where basic statistics is necessary. It also allows instructors to demonstrate the computational procedures for a large number of the statistical tests demonstrated in this book.

It would be inappropriate for us to tell the instructor how to use the text; however, we would like to note briefly how we have used these materials with students having only a minimal knowledge of psychology. Our procedure was to make short daily reading assignments and to discuss this material during the next class session. Some professors have told us that the material works well in large classes in which they cannot provide personal tutoring. Chapter 1 is introductory material and was usually handled in one or two class sessions. Chapter 2 is expanded for discussion in four or five class sessions. Students could usually understand the first three sections of this chapter without much help from the instructor, but some students needed help with the later sections. Chapter 3 usually required about four class sessions. With Chapter 4 we discussed four control-problem examples at length in class to ensure that the students understood the design and procedures involved.

We handled the design critiques in Chapter 5 in two ways. In one approach we had students redesign the experiments outside of class and then we asked a few students to present their designs in class. Then the rest of the class commented on and criticized the new design. Our other approach was similar to the first, except that we put the students in groups of three or four and had the groups present their new designs. This generated considerable discussion and debate.

In the class discussion of Chapter 6, we started by emphasizing the problems that could arise if a biased assignment of subjects occurred. We found that emphasizing the problem generated more appreciation of possible solutions. The procedure for teaching Chapter 6 was similar to that for Chapter 4.

The exercise at the end of Chapter 7 asks the students to develop a series of questions that they can use to critique experiments with respect to their design. Having students do this as a homework assignment and comparing their lists in class worked quite well.

Chapter 8 promises to create a lively class discussion on the ethics of experimental research. Although the examples are intended to illustrate clearly different ethical principles, there are, in many cases, other sides to the issues that will make for interesting debates. Chapter 9 explains the practical matters of using the library and developing research ideas. Chapter 10 provides a model experimental write-up, which we suggest students use in trying their hand at preparing a similar paper, using a hypothetical experiment and hypothetical results.

Part Two of the book contains 16 reprinted articles. We selected these articles to illustrate the design problems presented in Part One. We have also identified some "special issues" of experimental design that are embodied in some of the articles. These issues include control problems in field-based experiments, small n experiments, research with animals, clinical research, and so on. The articles are arranged in four groups with different types of experiments in each group. In general, the articles in each group are ordered from least complex to most complex. Some instructors may choose to assign the articles in the sequence presented, while others may select only those articles that reflect the nature of the course. We hope we have presented a bountiful smorgasbord that will satisfy the most selective diet.

For beginning psychology students, it is important to discuss some of the experiments thoroughly, even to the point of reading them aloud in class and discussing them point by point. Reading the psychological literature is a new and somewhat anxiety-provoking experience for most students. A thorough discussion of a few articles seems to relieve much of this anxiety and acquaint the student with the design and technique in a specific content area. We have also included a new section on "poster sessions" because an increasing number of students, both at the graduate and undergraduate levels, are using this format to present their work. We have also left several chapters for the student to analyze. This can be done individually or in groups and makes for good class discussion.

Our goal in this book is to make psychological research more understandable, more interesting, and perhaps even exciting to the student. We invite ideas from instructors on how to best achieve these goals.

An Instructor's Manual, with answers to questions and comments on the ethics section (Chapter 8), is available. The manual also contains test questions.

Finally, we would like to acknowledge the assistance of those who reviewed the book: Mary Anne Baker, Indiana University Southeast; W. Robert Batsell, Jr., Southern Methodist University; F. Samuel Bauer, Christopher Newport University; James R. Council, North Dakota State University; Michael Frank, Stockton State College; Paula Goolkasian, University of North Carolina, Charlotte; Maria S. McLean, Southwest Texas State University; and Theron Stimmel, Southwest Texas State University. Each read the entire manuscript and provided cogent remarks and valued advice. To all, we offer our sincere thanks.

Robert L. Solso
Homer H. Johnson

To the Student

You have chosen an excellent time to study experimental psychology. Exciting new developments are emerging in nearly every domain of the behavioral sciences, and a strong background in the methodology of experimental psychology will form a basis for understanding these new developments and provide you with the skills necessary for conducting research on your own.

This book examines the methods of experimental psychology. A great deal of the book is devoted to controlled psychological experiments and the collection of reliable data based on observations. We use a "case approach," meaning that we illustrate each of the principles of experimental design with an example, or case, drawn from the professional literature in psychology. Study these cases carefully as they represent excellent examples of skillfully conducted experiments from every major area in psychology—including animal studies, child psychology, social psychology, cognitive psychology, applied psychology, and so on. We also emphasize ethics in experimental research and give some guidelines on how to best develop research ideas and write a research paper.

We believe that a powerful lesson can be learned by studying the techniques of experts. In addition to helping you learn proper experimental design, we hope the book will teach you a little about the various topics of psychology and inspire some of you to continue your studies and the work that some of us have started. We would be interested to hear your reaction to the material in this book and how it affects your study and work in the field. Good reading!

RLS
HHJ

Basic Principles in Experimental Psychology

This book is divided into two distinct sections. Part One focuses on the basic principles of experimental psychology while Part Two analyzes actual experiments drawn from the psychological literature.

We begin with an introduction to scientific inquiry and methodology in psychology. Each major point is illustrated with an example from the experimental literature. In Part One we also discuss the fundamentals of research design as they apply to experimental psychology. A large portion of this material is devoted to the issue of experimental control, which is the means by which experimenters ensure the integrity of a psychological experiment.

Chapters 5 and 7 are called "Design Critiques" and offer brief descriptions of experiments that contain at least one conceptual or technical flaw. As you read these critiques, try to discover the error. Practice with these problems can strengthen your own ability to design experiments devoid of error.

Part One concludes with three important and practical chapters: "Ethics of Experimental Research," "The Psychological Literature: Reading for Understanding and as a Source for Research Ideas," and "Conducting Research and Writing a Research Paper." These chapters should guide you through the experimental process from the conceptualization of an experiment, to running an experiment, and finally to submitting a manuscript for publication.

Part One is intended to present, succinctly yet comprehensively, the rudiments of good experimental design and to show how to present research in a valid experimental paper. In Part Two we have selected for presentation actual experiments from the psychological literature. Each experimental case illustrates one or more specific design/experimental issues and, in most cases, is accompanied by our detailed analysis. (Several cases are without comment and await your remarks.) We suggest that you read these cases as you study the material in Part One to help enliven the textual material. Upon completion of Part One and Part Two you should be equipped to understand, analyze, and plan research in experimental psychology.

Chapter 1

An Introduction to Scientific Inquiry

Man and science are two concave mirrors continually reflecting each other.

—Aleksandr Herzen

Periodically a journal of the American Psychological Society, called *Psychological Science,* is delivered to my office. Not long ago, a chemist friend saw the journal in my office and asked, "Isn't that title an oxymoron? Do you believe that psychology is really a science?" My friend's questions raise the issue: Are *psychology* and *science* incongruent terms?

It seems that psychology has lived in the shadow of "real" science for so long that many still believe it is in a prescientific period—a kind of alchemic stage in which everything from the mercurial nature of human behavior to weird social practices is investigated using arcane techniques. When I pointed out to my chemist friend that "science" is characterized by its method (i.e., the "scientific method"), not by its subject matter (e.g., chemistry, the brain, or demographic trends), the chemist argued that psychology was not a "true" science because it lacked a laboratory tradition in which technical instruments played a role in the search for scientific principles. "But that is a false impression," I retorted. "From the very beginning (by which I meant the late 19th century), we have had laboratories, equipment, subjects, experimental techniques, statistical analysis of data, and so on, upon which reliable conclusions are based. And presently, experimental laboratories in psychology are equipped with a dazzling array of computers, physiological instruments, and all sorts of scientific gadgets sure to impress the hardest of the hard among scientists." "But still," he responded, "you study weird, inconsistent, even berserk behavior." "But, the more bizarre the subject," I replied, "the greater the necessity to apply scientific methodology."

Presently, the experimental investigation of thought and behavior by psychologists follows the basic principles of scientific inquiry prescribed in other more traditional fields of science. While the things psychologists study (thought and behavior) distinguish psychology from other scientific fields, the method of inquiry is similar.

CAN THOSE WHO STUDY THE TRANSCENDENTAL THOUGHTS OF PEOPLE IMMERSED IN HOT TUBS BE CALLED SCIENTISTS?

Experimental psychologists study subjects such as the interpersonal relations between people of different status, the predatory behavior of the barn owl, the eye movements of newborn children, personality traits, the etiology of schizophrenia, brain patterns of people solving logic problems, and the transcendental thoughts of people immersed in hot tubs. These topics, and many more, are legitimate issues a psychologist could examine using the experimental method.

The lay person frequently ponders these and other topics using speculation, subjective arguments, and phenomenological proofs. These nonscientific interpretations, although sometimes interesting, frequently lead to invalid conclusions.

But it is possible to examine these topics scientifically, by which we mean that they may be subjected to analysis that follows a prescribed, logical method of investigation in which the conclusions are reliable and valid. To do this, experimentation and rational analysis—tools of investigation that are less susceptible to subjective interpretation or idiosyncratic bias—are designed to yield valid conclusions about a wide range of behavior and thoughts. Someone interested in the thoughts of a person sitting in a hot tub is not necessarily committed to using kooky, nonscientific techniques to study that phenomenon. Indeed, the kookier the subject matter, the more rigorous psychologists must be in their application of experimental techniques.

WHAT IS SCIENCE?

If undergraduate students are asked, "What is **science**?" many answer by giving examples of science: Science is physics, chemistry, biology, and so on. This definition suggests that science is a generic term describing specific subject areas. For example, college students are required to take a number of "science courses," and these courses usually are drawn from physics, chemistry, astronomy, botany,

"ARE YOU A STUDENT?"

Experimental psychologists study a wide range of phenomena—from the physiological components of emotion, to the eating behavior of people in fast-food restaurants, to poetry learning—but few avenues of research are more interesting to college students than the study of interpersonal relations. How these relations are established in opening conversations has been the focus of research conducted by Kleinke, Meeker, and Staneski (1986), and Cunningham (1988). In one survey Kleinke et al. had subjects rate opening lines such as, "Isn't it cold? Let's make some body heat."; "Are you a student?"; and "I'm sort of shy, but I'd like to get to know you." These lines were conceptualized to represent three different types of social advances: cute-flippant, innocuous, and direct. Overall, respondents agreed that cute-flippant opening lines were the least desirable.

With these normative data, Cunningham (1988) field-tested the opening lines by having researchers approach members of the opposite sex and begin a conversation in a singles bar with different types of social advances. In general, women reacted negatively to the cute-flippant lines and were positive to direct or innocuous opening lines. The reaction of men was ambiguous. Perhaps female initiated conversation, no matter how innocuous or cute, is perceived by men as a come-on. The research illustrates that even the vagaries of social mingling may be analyzed experimentally.

biology, and the like. While some casual support for this definition can be found, it is less than specific and does not appear to be adequate. If you ask why chemistry is a science but history is not, or why physics is a science but music is not, the definition of science becomes more complex and usually confusing. People usually argue that science deals with facts (yet, so does history), or that science deals with theories (so does music), or that it involves laboratory experiments (but what about astronomy or the classification of plants in botany, both of which are usually characterized as sciences and yet frequently do not engage in laboratory experiments).

The difficulty in finding an adequate definition of science is also shared by scientists. Many scientists specify the collection of facts, the use of experimentation as a method of "proof," the use of theories as tentative explanations, and so on as prerequisites for a field to be labeled a science. But other definitions emphasize the ongoing or dynamic nature of science—the search for new "facts" and theories to replace the old, much as some of Einstein's theories replaced those of Newton. James Conant (1951) expressed this quality of science when he defined science as "an interconnected series of concepts and conceptual schemes that have developed as a result of experimentation and observation and are fruitful of further experimentation and observation." We can see this dynamic process when a scientist, through observation and experimentation, attempts to ascertain what is related to what, or what cause is related to what effect. These new facts are put into conceptual schemes. These schemes (usually called theories or models) are tentative; they try to explain the relationship among the information we have on hand. As new information becomes available, that information, and then new conceptual schemes, will arise to replace the old ones.

Modern scientists—be they physicists, geologists, astronomers, anthropologists, or psychologists—approach their specialized fields with a few basic assumptions about the structure of the universe. Common to **scientific thought** is the assumption that nature is somehow structured by laws that govern its operation. For example, a simple but fundamental law of physics is that objects, such as a lead ball, fall toward the earth if dropped from a high structure, such as the Leaning Tower of Pisa. Furthermore, this observation can be repeated with essentially identical results. Natural laws are reliable and reflect an order in the universe.

It is in building upon these assumptions through experimentation and observation that we reach the core of science. For example, we know that the speed at which an object falls can change. An object such as a lead ball falls slowly at first and then accelerates. Through observation, scientists can discover the general "lawful" properties of acceleration, which can lead to general principles or models that incorporate more and more features of the universe in ever-expanding theories. It is this kind of process that demonstrates a clear distinction between nonscientific and scientific thought.

► **CASE STUDY**

The Greek philosopher Aristotle, writing on the acceleration of falling objects, stated that, based on "logic," heavier bodies should fall at a faster rate than lighter bodies. A boulder would fall to the earth from atop a building faster than a rock because the boulder is heavier. Many people still believe that a large cannonball falls faster than a small lead ball—all other factors that might affect the fall being held constant. The "commonsense logic" seems valid, but science is suspicious of commonsense logic.

The scientist Galileo questioned the validity of this logical conclusion and, as all schoolchildren know, decided to observe the relative speed of falling objects. His "laboratory" was the conveniently tilted Leaning Tower of Pisa. (Although any high building would have done just as well, the tower does make a more interesting story and has undoubtedly drawn hordes of tourists to northern Italy over the centuries.)

In retrospect, we can see that the experimental procedure used by Galileo consisted of four stages, which are remarkably consistent with modern scientific inquiry. They are as follows:

1. Statement of a hypothesis. Objects of different weight will fall to the earth at the same speed.
2. Observation. Measure the relative rates of descent of objects of different weight.

3. Replicability. Multiple observations with objects of different weights can be made.
4. Development of a law (or model). If the observations confirm a relationship between the weights of objects and the speed with which they fall a generalized conclusion may be drawn.

Of course, this early experiment was beset with problems, which we call *control problems.* We will discuss these in some detail later, but for now consider a sample of these problems. First, Galileo had to make sure that both objects were dropped at the same time. For example, if he chose to drop them by hand, the tendency might be to drop the larger, heavier object first. Or, if he wanted to support his hypothesis, he may have unconsciously dropped the lighter object first to give it a brief head start. (Even psychological factors enter into scientific observations in physics!) To control for these problems, Galileo may have constructed a box with a trap door to allow both objects to be released at the same time.

Aristotle: "Logic shows that heavier objects fall to the ground faster than lighter objects."

Galileo: "Experiment shows that all objects of differing weights fall to the ground at the same speed!"

Figure 1.1   Aristotle's method of rational logic versus Galileo's method of experiment.

Then there was the matter of judging the speed of descent, as measured by which object hit the ground first. Some objective criteria needed to be established so an observer, or observers, could reliably judge the arrival of the objects.

There is another important variable that might affect the rate of descent, and that is the influence of atmospheric conditions, such as air resistance, on the fall of the objects. Observation shows that a feather descends more slowly than a brass ball of the same weight. One means of controlling the variable of air resistance would be to remove all air from the laboratory. But since Galileo's laboratory consisted of the Leaning Tower of Pisa and the space immediately surrounding it, construction of such a vacuum chamber was well beyond the technology of the time. (Subsequent experiments, however, did measure falling objects in a vacuum and such observations confirmed Galileo's observations.)

Since Galileo's time, his crude experiment has been replaced with progressively sophisticated observations that have confirmed that objects, be they feathers or cannonballs, fall at the same rate given a constant force of gravity. The principle upon which this law is based is called the principle of equivalence and is considered to be one of the most important laws that govern physical bodies throughout the universe.

The law of gravity and the experiments upon which it is based serve as examples of two levels of scientific inquiry that are critical to science. The first level is basic observations; the second level is the realization that these observations are part of a larger system. Throughout this book, and hopefully throughout your career as an experimental psychologist, you will be cognizant of these two levels. While it may be interesting as a first step to observe that children from upper socioeconomic levels attend college in greater numbers than children from lower socioeconomic levels, the integration of that observation into a larger model of human behavior and society is an important and necessary second step. Without this second step, Galileo's observations would surely be uninteresting to the history of science and, at best, be a tiny footnote about an eccentric Italian who dumped things from the Leaning Tower of Pisa.

SCIENCE AND PSYCHOLOGY

Behavioral scientists, including psychologists, go about their work with the same kind of assumptions as did Galileo about the lawful nature of the universe. Only the subject matter and technical equipment, not the scientific method, are different. Psychologists tend to study thought and behavior, not accelerating objects, and while the subject matter and the specialized tools seem to separate these disciplines, the scientific method and the assumption of a lawful universe whose secrets may be revealed through observation and experimentation unites these and all scientists in a common adventure.

The scientific inquiry of psychology is based on the same assumption as other scientific fields: There is an underlying reality to behavior and thought that may be revealed through scientific analysis. Our inquiry is based on a further assumption:

that the human is basically a system—a very complicated system—that may be understood and explained through scientific experimentation and rational analysis of the results of these experiments. While the musings of someone in a hot tub may seem beyond the scope of scientific analysis, they are not! And it is because such slippery topics are susceptible to such varied interpretation that experimental psychologists must apply scientific methodology with the greatest of precision to their subject matter, be it skin conductivity during periods of stress, eye movements of a dyslexic child, or even the thoughts of someone in a hot tub. Join us as we explore this point further.

WHY A SCIENCE OF PSYCHOLOGY?

You may ask, "Why are psychologists so concerned with developing a science of psychology?" The simple answer is this: Psychologists attempt to understand the laws of behavior, and to understand, predict, or control behavior with any precision is a very difficult task. Scientific inquiry seems to offer the best possibility of handling such a task.

Laypeople also understand behavior; or, they at least understand it to the extent that they can coexist with other people. They have principles of behavior, such as, "Absence makes the heart grow fonder." But there is also a contradictory principle that says, "Out of sight, out of mind." These contradictions occur because people's interpretations of behavior are highly colored by each person's perspective. Demonstrations at abortion clinics, for example, are called everything from Christian fundamentalism to atheistic radicalism, depending on one's beliefs and concerns. Psychologists, however, try to avoid this subjectivism by engaging in systematic studies of behavior, formulating hypotheses about behavior, and then testing them in an organized manner. The techniques that psychologists use to test their hypotheses are described as scientific inquiry.

ON THE SCIENTIFIC METHOD

From the earlier discussion it is apparent that an exact definition of science is somewhat elusive. The same problem arises when one attempts to define the **scientific method.** Some authors seem to imply that the scientific method consists of a few simple steps that, if followed, will inevitably lead to amazing and dramatic discoveries about nature. But if there is one such method, why does science progress in such a slow, fumbling, bumbling manner? A well-publicized example is the search for a cure for cancer that has lasted at least half a century. In spite of the use of some highly talented researchers and millions of dollars, the progress has been slow. Likewise, the search for the cause of abnormal personalities, such as childhood autism, has also been tortuously slow. Meanwhile, advances in some areas of psychological research have produced brilliant results, some of which have won the accolade of "breakthrough" research. Here we refer to studies of the components of memory, the serological basis of gender behavior, the perception of geometric forms, the treatment of obesity, and the reinforcement contingencies that lead to productive responses, to offer only a brief sample of the considerable list of success stories in modern experimental psychology. But still, the slow pace of progress casts doubt on the idea that the scientific method is simple or infallible.

Scientific methodology consists of a variety of techniques, approaches, strategies, designs, and rules of logic. These vary from problem to problem and discipline to discipline. This book gives the student an indication of the experimental techniques used in psychology. In keeping with this overall philosophy, we will now consider a case study showing a type of experiment done by psychologists and how it illustrates certain aspects of **experimental design.**

> ### ➤ *Case Study*
>
> The tendency for boys to play with dump trucks, tractors, race cars, and erector sets, and for girls to play with dolls, doll furnishings, and kitchen equipment has been known for a very long time. What causes this difference in play between the sexes? How might it be studied and analyzed? What would the results of such a study tell us about gender behavior?
>
>
>
> These questions and others were developed by Sheri Berenbaum and Melissa Hines, who work as research psychologists in medical settings.[1] In an article that appeared in *Psychological Science* (Berenbaum & Hines, 1992), entitled "Early Androgens Are Related to Childhood Sex-Typed Toy Preferences,"[2] the basis of gender-specific preferences for certain types of toys was examined.

It is well known, not only from the psychological literature but also from common knowledge,[3] that young boys and young girls are encouraged to play with certain toys and are discouraged from playing with others. Boys who play with dolls, for example, soon learn that such behavior may be seen as unacceptable and they may be labeled as "sissies," while girls who eschew kitchen toys for dump trucks may be labeled "tomboys." Although social learning and social pressure surely influence what toys children play with, might other forces be operating, such as hormones and/or genetics?

To test this idea, Berenbaum and Hines selected girls who experienced a genetic disorder known as congenital adrenal hyperplasia (CAH), a condition that produces a high volume of the hormone androgen, normally found in large concentrations in boys. (Boys with similar conditions were also reviewed in this study.) The researchers then evaluated the amount of time girls with CAH spent playing with "boys' toys" versus the amount of time with "girls' toys" and "neutral toys." They discovered that girls who had CAH spent more time playing with cars, fire engines, and Lincoln Logs than did girls with similar environmental backgrounds but without CAH. The authors concluded that "early hormone exposure in females has a masculinizing effect on sex-tied toy preferences." Thus, one more part of the determinants of sex-linked behavior was solved.

1. Dr. Berenbaum is at the Chicago Medical School, Department of Psychology and Department of Psychiatry, and Dr. Hines is at the UCLA School of Medicine, Department of Psychiatry and Biobehavioral Sciences.

2. This article is analyzed in detail in Chapter 16.

3. In general, experimental psychologists do not trust common knowledge but seek scientific validation for beliefs.

Applying Assumptions of the Scientific Method to Experimental Psychology

The scientific method, as applied to psychology and other sciences, is based on several assumptions about the nature of the physical and psychological universe. While philosophers of science may debate these assumptions, they form the foundation upon which scientific experiments are based.

Order.   Psychological and physical nature is **ordered,** not random or haphazard. Furthermore, it is assumed that phenomena are related systematically, and that events follow an orderly sequence that may be observed, described, and predicted. For example, we can see that children crawl before they walk, the sensation of an object precedes its recognition, and hungry animals seek food rewards more aggressively than satiated animals. In the case of the sex-linked toy preferences, there appears to be an orderly relationship between the presence of certain hormones (androgens) and a certain behavior (toy selection).

Determinism.   All events have a cause. Toy preference in young girls seems to be caused by high levels of a male hormone (although this may not be the universal cause of such behavior). This assumption is called **determinism,** meaning that psychological phenomena (as well as physical phenomena) have antecedents or preceding circumstances that cause an event, although in many instances it may be difficult to specify exactly what the cause is. Schizophrenia develops; its cause may be childhood stress, hormonal imbalance, genetic factors, disease, or, perhaps, some other factor that has not yet been discovered. The important feature, though, is that schizophrenia, preference for certain types of toys, nightmares, bed wetting, intellectual keenness, musical ability, belief in the ideology of the Republican Party, good grades, "bad" behavior—indeed all human actions and thoughts—do not spontaneously erupt from nothing. All human behavior and thought are caused. Finding the cause or causes of behavior is frequently a very difficult problem for experimental psychologists, but it is also exhilarating, much as it must have been exciting for Galileo to contemplate the forces of gravity that caused balls of unequal weight to fall to the earth at the same rate. Central to these inquiries is the assumption that behind each thought or action a cause exists. That assumption is basic to the scientific investigation of the human condition.

Empiricism.   **Empiricism** is defined as the practice of relying on observation and experimentation, especially in the natural sciences, which include experimental psychology. Empirical research relies on measurements (frequently in the form of numerically expressed measurements), such as evaluating the amount of learning, the strength of a belief, or the number of minutes spent playing with boys' toys. These measurements then are expressed as **data,** which give factual information in numerical form. For example, we measure learning by test scores, beliefs by preference poles, and preference for toys by observing which toys a child plays with.

Not all empirically gathered data are valid. Data can be invalid if they are selectively collected to support a preconceived notion, or if they overlook important aspects of a problem. As for the first case, one can prove almost anything by selecting supporting data and concealing contradictory data for one's pet theory. Unfortunately, overzealous politicians frequently practice this form of deceit by using polls that support their positions while disregarding data that don't. Such a practice is sometimes called "How to lie with statistics," and while this practice may be helpful in selling candidates, it has no place in scientific psychology.

In the second case, in which relevant information is overlooked, the culprit is ignorance—the scientist is unaware of the many possible causes of behavior and reports only a portion of the available data. Thus, the conclusion may be flawed due to the unintentional disregard of critical information. The cure for this problem is closer attention to the complexity of scientific issues and a broader scope of inquiry.

Parsimony.   In general, scientists prefer simple, or parsimonious, explanations of natural phenomena to complex ones. The assumption of parsimony is important in making generalizations about the broader nature of human behavior because it allows scientists to extrapolate from specific findings to more general statements. In the experiment of hormones and toy preference, the representation of the data is very simple—an antecedent condition for toy preference is the hormone androgen. From this parsimonious statement it may be possible to make judicious generalizations, such as, effeminate behavior in men and masculine behavior in women *may* be related to hormone levels. Such a generalization, in order to be validated, needs further empirical research and the collection of appropriate data.

The **law of parsimony,** as it is sometimes called, is also important in finding common explanations for research gathered within the same species and even for observations made across species. In the first instance, for example, human vision is similar among all people with normal eyes. When your eye and visual cortex initially process light, the procedure is the same for you as it is for someone from Maryland, Michigan, or Madagascar. Although the interpretation of the things the eye sees varies enormously among humans, the way the eye works in detecting light, converting light energy to neural energy, and passing that information on to the visual cortex is essentially the same for all people. The law of parsimony allows us to make such a generalization without testing each person. In addition, the law also may work across species. For example, data collected on the visual properties of nonhumans, such as rats, chimpanzees, dogs, pigeons, and so on, have some common features with data collected on human vision. For example, at the detailed level of the physiology of the eye, initial processing of visual stimuli is very similar among mammals. Scientists often seek parsimonious explanations for their observations that they may generalize over a wide class of people and species.

Psychological Experiment

We use the term **psychological experiment** to refer to investigations in which at least one variable is manipulated in order to study cause-and-effect relationships. We will emphasize experimental research in which the researcher manipulates some factors (variables), controls others, and ascertains the effects of the manipulated variable on another variable. In some experiments the researcher may not manipulate a variable physically, but can manipulate it through selection, as was the case in the toy selection example. The researchers did not inject androgen in the bloodstream of the girls (although such a procedure might be an even more direct test of the hypothesis), but they did select children whom the researchers had good reason to suspect had a high level of androgen in their blood. In this case, the manipulation of one variable was done through selection rather than with the imposition of a factor. It is this search for relationships between specific events and their consequences (cause and effect) that is so characteristic of scientific research.

Other types of scientific inquiry are commonly made in psychology. These are valid techniques and have led to important discoveries. One is naturalistic observation.

Naturalistic Observation

When Charles Darwin planned an expedition to the Galapagos Islands during the early part of the 19th century, he did so with the intention of observing the natural life he found there. He did not plan a series of experiments in which factors would be manipulated and their effects measured. Given Darwin's interest and the subject matter, **naturalistic observations** (i.e., observations made in natural or native settings) led to a theory that has been one of the most influential developments in science.

Naturalistic observation involves the systematic recording of perceived information. The setting may be as undefiled by humans as parts of the Galapagos Islands were in Darwin's time, or it may be as "civilized" as the interior of a health club in the heart of Los Angeles. Scrutiny of the minute details of the mating rituals of giant tortoises on the Galapagos Islands or careful examination of the subtle precopulatory maneuvers of patrons of a health club are both examples of naturalistic observations.

For a period, naturalistic observations in American psychology seemed as taboo as some of the things observed. Recently, though, they have gained greater popularity and are once again considered an important method of gathering data. But scientists must remember that when making naturalistic observations it is important to make objective and systematic observations to guard against distorting information through personal prejudices, feelings, and biases.

SCIENTIFIC METHODOLOGY

Our approach to the use of scientific methodology in the study of psychology is established on two principles. The first is that scientific observations are based on sensory experiences. We see, hear, touch, taste, and smell the world we live in. These observations, which are made under certain defined circumstances, or **controlled conditions** as they are called in experimental psychology, should correspond to the observations made by another scientist under comparable conditions. This feature of replicating the results of an experiment is called **reliability of results** and is a major requisite of scientific credibility.

However, because our sensory systems are limited in capability as well as scope, many signals outside the range of normal sensitivity remain unnoticed, meaning that those things that are detected take on a disproportionately greater significance. We call this the **tyranny of the senses,** recognizing that it is difficult to consider the importance of some of the real phenomena in the universe that lie outside the range of unassisted human perception.

Consider the electromagnetic spectrum, whose presence is ubiquitous. At one end of the spectrum are cosmic rays, gamma rays, and X rays, and on the other end are radio waves and TV waves. In the middle, between about 380 and 760 nanometers, are waves that are detectable by the human eye. For most of human history, the energy that fell within the visual spectrum was considered to be "reality." The same constricted view of reality applies to all the other senses. Even though we have become aware of the presence of forms of energy that we cannot experience, we continue to emphasize the sensations that we can detect through the ordinary senses.

Some augmentation of the senses has been achieved through the development of technology. Many instruments and techniques in science are designed primarily to make "visible" those things that are "invisible" to the unaided sensory system. These instruments—the microscope, radio telescope, and spectroscope, for example—translate energy outside the normal range of human detection into signals that can be understood by humans. In psychology, many sophisticated instruments have been developed that allow us to see deep within the psyche of a species and reveal secrets of human and animal life that were invisible and left to conjecture and speculation only a few years ago. We will encounter many of these techniques in this book, and we remind the reader that if a technique does not yet exist for a topic of interest, there is no prohibition against the invention of one.

The second principle upon which scientific methodology relies is that observations from our senses are organized logically into a structure of knowledge. Frequently in experimental psychology these structures of knowledge are called **models.** Cognitive psychologists may, for example, develop a model of memory that may be based on their observations of two types of memory and the laws that govern their relationship and storage of information. The structural web that creates models from observations and turns these models into theories is based on the principles of logic, which in the current context has developed out of the rich history of Western empirical thought. It is not our purpose here to deal in detail with this structural web, but rather to discuss some topics in the philosophy of science that relate to experimental psychology. The interested student can find extensive writings on this topic.

Development of Thoughts and Hypotheses in Experimental Psychology

One of the most difficult tasks confronting beginning students in experimental psychology is to organize their thoughts and develop a testable hypothesis for a given topic. For many reasons, this is difficult not only for the novice but also for the seasoned researcher. A major reason for the difficulty is a lack of knowledge. New research ideas rarely, if ever, erupt spontaneously from an intellectual void. Rather, new ideas and hypotheses usually are built on existing knowledge and past research. Therefore, the best advice on how to develop new thoughts and hypotheses is to immerse yourself in the literature of a branch of psychology that holds some real interest for you.[4] Read, discuss, investigate, and become well-versed in the subject

4. Recently, the growth of knowledge in nearly every branch of scientific inquiry, including psychology, has been so rapid that it is difficult for students and professional scientists to keep up with current facts and theories in their field. More and more we are seeing scientists use data banks, which store vast amounts of information in computer memories. (See Solso, 1987, for a more complete discussion of this topic.) As a consequence of the explosion of scientific information, a first step in the experimental process is frequently a computer-assisted search of the literature. Chapter 9 will provide a more complete description of these techniques.

matter. But passive knowledge is not enough. As you acquire knowledge about a topic, question the premise, the conclusions, and the technique and relate it to your knowledge of other matters. The development of new ideas in psychology, as well as in other disciplines, rests on the acquisition of the fundamental elements of a subject and flexibility in thinking about them, which allows one to combine and recombine the elements of thought in increasingly novel and meaningful patterns.

The final ingredient in the development of creative and meaningful research in psychology is the **dedication of scientists.** As we review the lives of great scientists—Pavlov, Curie, Bartlett, Skinner, Tinbergen, Darwin, Wundt, James—we see their devotion to the pursuit of knowledge. An inquiring, creative mind develops with knowledge, resources, flexibility, and dedication.

New ideas are based on old ideas, new inventions on old inventions, and new hypotheses on old hypotheses. Contrary to popular lore and media fiction, scientific advancements come from small increments of progress, rather than from a single brilliant discovery. Of course, we all aspire to that major scientific breakthrough, and your budding enthusiasm for achieving scientific eminence should not be discouraged. But such profound achievements are rare, and while most research projects fall short of seminal programs, they can nonetheless contribute mightily to the overall growth of scientific knowledge.

To illustrate the point of the **accumulation of knowledge,** consider an innocent question asked a few years ago by a son of one of the authors: "Who invented the automobile?" Trying to be instructive, the author told him that in about 1886 Karl Benz invented the automobile. "Wow, he must have been a real genius to figure out the engine, the brakes, the spark plugs, the wheels, and how everything worked together!" "Well, there were others, such as Henry Ford, R. E. Olds, and Daimler, and someone else invented the tires; I think it was Firestone. And then there was even the person who invented the wheel. . . ." But then the author experienced a moment of realization. "I think I may have misled you. No one person invented all of the components of the automobile any more than a single person invented the television, the theories of memory, or the symphony. Many people made significant discoveries that led to the invention of the automobile and other scientific discoveries."

It seems to us that the development of knowledge in psychology progresses along similar lines. Given an inquiring mind and a determined heart, many important scientific truths lie waiting to be uncovered by future scientists. Past discoveries beget future discoveries, past knowledge begets future knowledge, and even past wisdom may beget future wisdom.

In this chapter, we attempted to answer several questions frequently raised by students: What is science? What is the scientific method? Why are psychologists so insistent on being scientific? What is a psychological experiment? What is naturalistic observation? What is the source of hypotheses in psychology? Although the answers have been brief, we hope that they will provide some general understanding and justification for what follows in this book. The next chapter will offer a more specific attempt to examine some of the basic concepts of experimental design.

DEFINITIONS

The following terms and concepts were used in this chapter. We have found it instructive for students to go through the material a second time to define each of the following:

accumulation of knowledge   naturalistic observations
controlled conditions   order
data   psychological experiment
dedication of scientists   reliability of results
determinism   science
empiricism   scientific method
experimental design   scientific thought
law of parsimony   tyranny of the senses
models

Anatomy of Experimental Psychology: Design Strategies

Rules of Science: First was never to accept anything as true if I had not evident knowledge of its being so. . . . The second, to divide each problem I examined into as many parts as was feasible. . . . The third, to direct my thoughts in an orderly way. . . . And the last, to make . . . general surveys that I might be sure of leaving nothing out.

—*René Descartes*

THE LOGIC OF EXPERIMENTAL PSYCHOLOGY

Experimental psychology is a relatively simple matter; however, the subject matter, analysis of data by statistical means, control measures, theoretical considerations, and implications of research are among the most complicated matters known to science. Before we can intelligently address some of these more weighty topics, however, it is important to understand the simple and exacting methodology used in experimental science.

Fundamentally, experimental psychologists use two basic techniques: The first involves introducing a factor, or variable, and then measuring its effect. The second involves selecting a group that has certain well-defined characteristics (such as Olympian athletes) and contrasting the behavior of that group with a control group.

17

We begin with an example that illustrates the rudiments of the first technique.

> ➤ **CASE STUDY**
>
> Over 200 years ago, the Italian scientist Spallanzani attempted to determine what part of the male semen fertilized the female egg. On the basis of some earlier studies, Spallanzani hypothesized that it was the sperm cell. To test this hypothesis, Spallanzani filtered the semen from dogs to remove the sperm cells, and then artificially inseminated female dogs with either normal semen or sperm-free filtrate. The dogs inseminated with the normal fluid became pregnant, whereas those inseminated with the sperm-free filtrate did not. Thus, Spallanzani demonstrated that it was the sperm that fertilized the egg.

This experiment, although it was conducted in 1785 and was a biological experiment, still illustrates many of the basic principles of experimental design that are presently used in psychology and other sciences. Spallanzani started with a **hypothesis** that was based on previous research. In order to test this hypothesis, he used an artificial insemination technique through which he could *control* the insemination process and *manipulate* the content of the fluid prior to its injection into the dogs. Having this control, he set up a two-group experiment. One group of dogs (the experimental group) was inseminated with the sperm-free filtrate, and the other group (the control group) was inseminated with the normal semen. The two groups were treated alike in all other ways. Since the only difference between the two groups was whether sperm cells were present in the seminal fluid, any difference in the pregnancy rates between the two groups must have been due to this manipulation. It was this type of logic and this type of design that enabled Spallanzani to arrive at his valid conclusion.

BASIC STEPS IN EXPERIMENTAL PROCESS

Hypothesis → Design Experiment → Perform Experiment →
Make Observations → Conclusion

As shown in the box, the basic steps in the experimental process are amazingly simple, and in fact, even more complicated psychological research uses the same simple framework. The principle difference between simple experiments, like the Spallanzani experiment, and complex ones is not in the basic experimental paradigm, but in the inclusion of sometimes complicated factors that need to be properly controlled. The central idea is that experimental design is based on just a few elementary principles. It is wise to keep these basic principles in mind when you study experimental design and plan experiments.

Psychological research often uses the same logic as did the Spallanzani study. In the simplest case, one factor (variable) is manipulated by the experimenter, who holds all other factors constant. The manipulated variable is called the **independent variable** (IV), and the variable whose reaction is being observed is called the **dependent variable** (DV). The rudiments of the experimental design used in the Spallanzani experiment are shown below. Usually, one group of subjects receives one *level* (type or amount) of the independent variable, and another group receives a different level. Since both groups are treated exactly alike except for the level of the independent variable, then any difference observed in the dependent variable is due in all probability to the change in the independent variable. Through this process, the psychologist hopes to ascertain precisely the effects of one variable on another and to build knowledge about cause-and-effect, or functional, relationships in behavior. Some writers have talked about this process in terms of a "theory of control." This label expresses well the idea behind research. An attempt is made to control variables either by manipulation or by holding them constant. Once this control is obtained, the determinants of behavior can more likely be discovered.

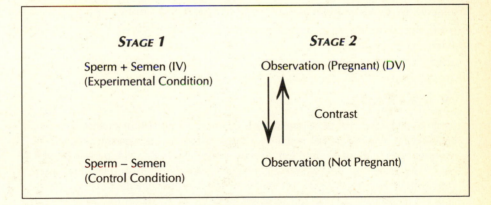

STAGE 1	STAGE 2
Sperm + Semen (IV) (Experimental Condition)	Observation (Pregnant) (DV)
	Contrast
Sperm − Semen (Control Condition)	Observation (Not Pregnant)

With this overview of experimental logic, the following material will illustrate some of the concepts and designs in psychological research.

INDEPENDENT AND DEPENDENT VARIABLES

It has already been noted that in simple experimental design only one variable is manipulated, and its effects on another variable are observed and measured. The manipulated variable is called the independent variable; the variable being observed is called the dependent variable.

In experimental psychology, the dependent variable often takes the form of direct answers to an experimenter's questions. These *dependent responses* have been part of experimental psychology for a very long time and continue to be of use today. Meanwhile, there are two types of independent variables used in psychology: The first type occurs when experimenters systematically manipulate a variable, such as when experimenters change the amount of reward given to an animal upon making a correct response. The second type of independent variable is based on the selection of subjects who possess a certain trait or characteristic of interest. We will consider each of these types in turn and see how they can influence a dependent variable.

Experimenter-Manipulated Independent Variables

We can illustrate the use of dependent variables and experimenter-manipulated independent variables in a study of what is known in the psychological literature as "massed practice" versus "distributed practice." The question addressed is: In learning a task, such as tracing an object while looking at it in a mirror (which reverses the image), is it better to practice the task continuously or practice for brief periods over a longer time period? Such questions have practical implications for students of calculus, or poetry, or even golf. Consider the following case study, which clearly shows an independent variable and a dependent variable.

➤ CASE STUDY

In a study of the effects of practice on learning, Lorge (1930) investigated the question of whether performance on a task is better when a person practices it continuously without interruption (massed practice) or when that person distributes the practice sessions with rest intervals between them (distributed practice). Lorge chose for his task a problem in which the subject traces a pattern (e.g., a star) but can only see the pattern (and his or her hand) in a mirror. Lorge had three groups of subjects and had each group trace the pattern 20 times. For one group the 20 trials were completed consecutively with no rest between trials. For the second group each trial was followed by a 1-minute rest period. The third group did one trial a day for 20 days; thus there was a 24-hour interval between trials. The measure of performance was the length of time it took the subjects to trace the pattern—the shorter the time, the better the performance.

Lorge's results indicated that, except for the first trial, the average performance of the 24-hour-interval group was better than that of the 1-minute-interval group, and the performance of the 1-minute-interval group was better than that of the continuous-practice group. The conclusion was that on this type of task, spaced practice leads to better performance than does massed practice.

In Lorge's experiment the independent variable was quantitative and controllable: The time interval varied between no interval, 1 minute, and 24 hours. (In contrast, recall the example from Chapter 1 in which hormonal level in children and play objects were not actively manipulated, but inferred through the presence of a genetic condition.) The dependent variable was the time it took the subjects to trace the pattern. Lorge attempted to keep all other factors constant; for example, the subjects all performed the same task, and they all had the same number of trials. This example is drawn from a traditional area of experimental psychology, and is reasonably straightforward. Only slightly more complicated is an experiment drawn from social psychology to which we now turn. In this experiment, the investigator is interested in whether we give more weight to information we hear first about someone, such as a potential friend, than we do to later information. As you read this example, try to identify the independent and dependent variables.

➤ *CASE STUDY*

Asch (1952) conducted an experiment to determine whether the first information (primacy information) you hear about another person is more important in forming an impression of that person than later information (recency information). Asch used two groups of subjects. A series of adjectives that were said to describe a certain person were read independently to both groups; however, one group received positive information first and negative information last, while the other group received negative information first and positive information last. The adjectives read to the group receiving positive information first were (in this order):

intelligent
industrious
impulsive
critical
stubborn
envious

The group receiving the negative information first heard the same list but in reverse order. Asch then asked the subjects to write down their general impression of the person. The group receiving the positive information first described him as an able person who had certain shortcomings. But the group who received the negative

THE EXPERIMENTAL RUDIMENTS OF THE ASCH EXAMPLE

Experimental Condition	Dependent Variable
presentation order of adjectives (forward) → impression of person	
presentation order of adjectives (backward) → impression of person	

information first described him as a "problem" whose abilities were hampered by serious difficulties. Since the group that received the positive information first tended to have a positive impression of the person, and the group that heard the negative information first tended to have a negative evaluation, the researcher concluded that first impressions are important in forming opinions about a person.

The independent variable in the Asch experiment was the order of presentation of the information—either positive to negative or negative to positive. The dependent variable was the subjects' descriptions of the person. All other variables were held constant; for example, both groups received exactly the same adjectives. Since the only difference between the treatment of the groups was the order of presentation, Asch concluded that the different impressions must be due to this manipulation. Thus, Asch was able to demonstrate a "law of primacy" in the formation of impressions.

Experimenter-Selected Independent Variables

An experimenter-selected independent variable is usually a **subject variable** such as IQ, authoritarianism, gender, race, presence of male hormone, or some other feature or characteristic that is not easily controllable. For example, an experimenter may want to ascertain the influence of authoritarianism on concept learning, so he or she can *select* two groups of subjects. One group consists of people who receive high scores on a standard test for authoritarianism and the other group consists of people who receive low scores on the same test. Then both groups can perform the same concept learning task, and the amount or speed of learning is the dependent variable.

Note that the experimenter has not actively manipulated authoritarianism but has selected for it. Some experimental psychologists would view this type of study as being nonexperimental because the independent variable (in this case a subject variable) is selected. Strictly speaking, an experiment in psychology is a piece of research that measures the effects of a *manipulated* independent variable. However, the use of subject variables as legitimate independent variables has proved to be a very useful method in psychological research. From a practical standpoint, some important questions in psychology, such as research on gender, personality types, diagnostic classification, race, age, social status, and other like concepts, can only be addressed through the use of subject variables. We now turn to an example of one such case: Are there gender differences in mathematical ability?

> ► *CASE STUDY*

A study by Benbow and Stanley (1980) used the subject variable of gender in trying to differentiate mathematical ability between boys and girls. The researchers gathered test scores for 9,927 seventh- and eighth-graders who were matched on (i.e., had equal amounts of) mathematics courses. Then they were given the Scholastic Aptitude Test. On the mathematics portion of the test, the boys' average score was significantly higher than the girls' average score. Also, more than 50 percent of the boys scored above 600 (out of a possible 800), while not one of the girls scored above 600. The researchers also reported extreme scores. The highest score for a boy was 190 points higher than the highest score for a girl. As might be expected, the results were controversial, but the researchers stand by their data and suggest that one of the purposes of scientific inquiry is to search for the causes of such data.

There are numerous subject variables that have been used as independent variables. For example, children of wealthy and poor parents have been asked to draw pictures of dimes or quarters in order to examine the relationship between children's economic background and their estimates of the size of money. There are also many experiments that compare the responses of males and females on a variety of tasks. In addition, comparisons have been made between the incidence of lung cancer among people who smoke cigarettes and people who do not. Sometimes such investigations make use of correlational analysis in which, for example, the incidence of lung cancer is compared with cigarette use. But be aware that some critics of this technique point out that the occurrence of two phenomena, for example increased incidence of lung cancer as related to increased use of cigarettes, does not, in and of itself, evince a causal connection between the two, and that in order to establish a cause-and-effect relationship other evidence must be presented. The long-standing battle in the United States over the health hazards associated with cigarette smoking is just one example of the issues that scientists explore using subject variables and correlational research.

EXPERIMENTAL AND CONTROL GROUPS

Whereas in many experiments treatment groups are exposed to different levels of the independent variable, on other occasions an experimental group and a control group are used. Although these experiments can be described using our definition of an independent variable, these concepts are discussed here because they present some unique problems in experimental design.

The **experimental group** is the group that receives the experimental treatment—that is, some manipulation by the experimenter. The **control group** is treated exactly like the experimental group except that the control group does not receive the experimental treatment. The Spallanzani experiment is a good example of this. The group of female dogs receiving the sperm-free filtrate was the experimental group, and the group receiving the normal semen was the control group. In the next example we look at control and experimental groups who are clearly different from most people in one area.

➤ CASE STUDY

Blind people are very adept at avoiding obstacles; however, little was known about how they do this. One hypothesis was that blind people developed "facial vision"; that is, they react to air pressure on exposed parts of the skin. A second theory was that avoidance of obstacles comes through the use of auditory cues. Supa, Cotzin, and Dallenbach (1944) set out to test these theories. They had blind people walk around in a large room in which obstacles (screens) had been set up. Two experimental treatments were used. In the first treatment, blind subjects wore a felt veil over their face and gloves on their hands (thus eliminating "skin perception"). In the second treatment, blind subjects wore earplugs (thus eliminating auditory cues). A third treatment was the control treatment, in which blind subjects walked around the room as they would normally. The results indicated that subjects in the control group and in the felt-veil group avoided the obstacles every time, but the subjects in the earplug group bumped into the obstacles every time. Based on these results, the authors concluded that the adeptness of the blind in avoiding obstacles is due primarily to their use of auditory cues and not to any facial vision.

The previous experiment is an abbreviated version of a series of experiments on the perception of sighted and blind subjects. In this example, it is somewhat difficult to specify an independent variable. The experiment is most easily described as having two treatment groups—one in which facial vision is eliminated and one in which auditory cues are eliminated. The control group is treated the same as the other treatment groups except they do not receive the veil or earplug treatment. The control group provides a baseline to help determine whether the treatments improve or hamper the avoidance of obstacles. The dependent variable in this study—the ability to respond to sensory-deprived cues—was measured by the number of times the subjects walked into the obstacles.

Sometimes more than one control group is needed. For example, in pharmacology a **placebo control group** is frequently used. A placebo group is best described as a group who is made to believe that it is getting a treatment that will improve its performance or cure some symptom, when in fact no treatment is provided. This type of control group is also used in testing the effectiveness of therapy. Consider the following example drawn from the psychological literature.

➤ CASE STUDY

Paul (1966) conducted an experiment to test the effectiveness of two types of therapy in treating speech phobia. His subjects were students enrolled in public speaking classes at a large university. Paul took 67 students who had serious performance problems in the course and assigned them to one of four conditions. One group of 15 subjects received a form of behavior therapy. A second group of 15 subjects received an insight therapy. A third group of 15 subjects received placebos in the form of harmless and ineffective pills, and were told that this would cure them of their problems. A fourth group of 22 subjects was informed that they would not be given any treatment and simply answered questionnaires that were also given to the other three groups. All subjects had to give one speech before the treatment began and one after the treatment had been completed. The dependent variable was the amount of improvement shown by the subjects from the first to the second speech, based on ratings made by four clinical psychologists. The four psychologists were not involved in the treatment the subjects received nor did they know which subjects were in which treatment group. The results indicated that 100 percent of the behavior therapy subjects improved, 60 percent of the insight therapy subjects improved, 73 percent of the placebo subjects improved, and 32 percent of the no-treatment control subjects improved.

The Paul experiment illustrates the need, in some experiments, for different types of control groups. The interpretation of the results of the experiment would have been quite different if Paul had not used a placebo control group. Without it, insight therapy would have appeared effective as a therapy in improving speech difficulties. But with the placebo group included in the design it appears that the insight therapy was ineffective as a therapy and may have only acted as a placebo. In fact, the placebo group's performance improved more than the insight therapy group's performance. The experiment also points out the need for a no-treatment control group, as over 30 percent of the subjects in this group improved in spite of the fact that they received no treatment. This can form a baseline to measure the effectiveness of a treatment compared to no treatment at all.

Different types of control groups are used in different areas of research. Researchers who remove a part of the brain of animals and use the animals as an experimental group sometimes use a control group that undergoes all of the surgical procedures except that the brain is left alone. This would control for a factor such as postoperative shock causing the effect found in the experimental group. The point to remember is that the control group must be treated exactly like the experimental group except for the specific experimental treatment.

The Paul experiment also illustrates an important control procedure used to avoid **experimenter bias.** The psychologists who rated the subjects' speaking performances were not the same people who treated the subjects in therapy, nor did they know which subjects were in which experimental group. It is reasonable to assume that therapists might be biased when it comes to evaluating the improvement of their own patients. Furthermore, the four judges might prefer a particular therapy, and if they knew which subjects had received this therapy, they might be prone to see more improvement in these subjects than in subjects in the other experimental conditions. Or perhaps the judges would have assumed that the subjects in the no-treatment control group could not have improved and would therefore rate those people's performances accordingly. Paul controlled for these potential biasing effects by using independent judges and keeping the judges blind as to what experimental group a particular subject belonged.

The term *blind* is used in a special sense in experimental research. **Single blind** usually means that the subjects of an experiment are not informed as to which treatment group they are in and might not be informed as to the nature of the experiment. **Double blind** is frequently used in drug research or any research that involves observers who are judging the performance or progress of the subjects of an experiment. Both the judges and the subjects are kept blind as to the type of experimental treatment that is being used as well as the type of effect that might be expected. The case study we have selected illustrates the influence of experimental bias in experimental psychology and the serious implications it can have for psychology and other scientific studies.

> ## CASE STUDY

An obvious demonstration of experimental bias was shown by Rosenthal and Fode (1963). A group of student experimenters who had some background in experimental psychology was asked to evaluate maze performance of two groups of rats. One group, so the experimenters were told, was selected from a long strain of "maze-bright" rats, while the second group was supposedly selected from a long strain of "maze-dull" rats. The experimenters conducted a study of maze performance and, as expected, the maze-bright rats did significantly better than the maze-dull rats. The odd thing was that the rats were randomly selected from a standard sample of rats—there was no bright or dull distinction. Had the maze-bright rats actually performed better than the maze-dull rats? Probably not, but the experimenters who observed the bright rats expected them to perform better, and this expectation seemed to cloud their observations.

Within-Subject Control

Basically, there are two simple types of control used in psychological experiments. In the first type, as illustrated in the previous examples, two or more sets of subjects are treated to different conditions, one of which may serve as a control condition. Contrasts are made between the results of the treatments. This design is called a **between-subject design** (see Chapter 6, Models 1 and 2). We can state the relationship as follows:

Group A	Experimental Condition 1	Measure Effects
Group B	Absence of Experimental Condition 1	Measure Effects

In the second type of design each subject undergoes two or more experimental conditions. The experimenter observes the results obtained after one treatment as contrasted with the results obtained after another treatment or treatments. This type of design is called **within-subject design** (see Models 3 and 4 in Chapter 3). We can state the relationship as follows:

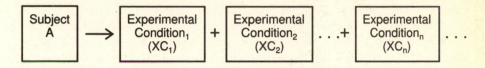

Within-subject designs offer several advantages over between-subject designs. They require fewer subjects, as each subject is treated to several conditions rather than requiring two sets of subjects for each experimental condition as in the between-subject design. Also, there is no need to match experimental and control subjects as each subject serves as his or her own control. Finally, from a statistical point of view, the variability (the amount scores vary around an average score) may be less for within-subject experiments than for between-subject experiments.

Nevertheless, some real problems may be encountered with within-subject design that may render it inappropriate for some experiments. For example, experiments in which one treatment may affect the results of a subsequent treatment are not suited for within-subject designs (unless, of course, that is the focus of the experiment). As an obvious example, consider the ease of learning two computer word-processing programs: Starry Words and Perfect Words. In a typical within-subject design a subject might learn Starry Words and then Perfect Words. The contrast would be between the ease of learning each. However, the experience with Starry Words might make it easier (or harder) to learn Perfect Words. In either case, the experimenter does not know what effects a previous experimental experience might have on the subsequent experience. The results would be spurious.

COMPUTERS AND EXPERIMENTAL PSYCHOLOGY

In recent years a significant change in laboratory technology has taken place largely due to the use of small personal computers (PCs). These desktop versions of their ancestral behemoths now find a place in many experimental laboratories and are sure to play an even more important role in the future. Originally, computers were used in the statistical analysis of data—number crunching. However, psychologists soon discovered that PCs were excellent devices for the presentation of stimuli, especially timed stimuli. Programs are available that make the PC one of the most useful and necessary pieces of equipment in the laboratory.

An additional problem with these designs relates to the **demand character-istics** of experimental design, which means that the subject can figure out what the experimenter wants and then either try to supply the desired response or intention-ally withhold it. This can occur in the within-subject design because a subject experi-ences several experimental conditions, which may give the subject an idea as to what the experiment is about. He or she can then bias the responses to satisfy (or frus-trate) the experimenter.[5] Demand characteristics deserve serious consideration in experimental psychology, especially in studies of social psychology, consumer psy-chology, child psychology, cognitive psychology, and abnormal psychology. In Chapter 3 some sophisticated means for getting around some of these problems are discussed.

The Two Meanings of Control

The preceding discussion of control groups and single- and double-blind ex-periments points out that there are two uses of the word *control,* and both are important. In the first, and literal, sense it means the experimenter can make things happen when he or she wants them to happen. This is what was meant by control when it was stated that Spallanzani had control over the insemination pro-cess—he could control the contents of the seminal fluid. This kind of manipula-tion and/or selection of independent variables is a prime example of this type of control. Another example can be seen when studying how experimenters observe nest building by canaries. Under natural conditions, these birds do small amounts of nest building scattered over rather long periods of time. So in order to study this behavior experimentally, it is necessary to get the birds to build when the ex-perimenter is prepared to make his or her observations. Experimenters solve this problem by keeping canaries in cages and giving them the nest material only when the researchers are there to observe. Here the presentation of nest material has been controlled, and precise observations of nest-building behavior can be made.

The second way researchers *control* an experiment is by arranging condi-tions so that they can attribute the experimental results to the independent varia-ble and not to some other variable. Paul controlled for the judges' bias by keeping the judges blind. The use of control groups is another attempt to ensure that ex-perimental results are not due to some other variable. In the Paul experiment the placebo group controls for placebo effects, and the no-treatment group controls for spontaneous remission. This second use of control will be discussed more fully in Chapter 3.

---

5. Good students often satisfy their professors by giving in to the demand characteristics set forth in a class, but sometimes they discriminate between what the professor wants and what they think is correct. One bright student once wrote on his examination, "The correct answer is. . . . The answer the professor wants is. . . ." The professor wrote back, "Answer 1 = F, answer 2 = A. Overall grade = C."

CORRELATIONAL STUDIES

The term *studies* rather than *experiments* is used in this section because correlational studies do not involve cause-and-effect relationships (even though they may infer such relationships). **Correlational studies** are studies of the relationship between two variables. They can be valuable if the interpretation of the results of these studies does not go beyond the limited scope of the measure.

An example of a psychological variable that could be examined in a correlational study is intelligence—some people have more and some have less, and a quotient of intelligence can be expressed quantifiably. Running speed is another human variable that can be quantified. A correlational study could be done between these two variables. First, members of a sophomore class at Stanvard University could run a 50-meter race. Some would run fast, some slow, and quite a lot somewhere in between. The same group then could take a standard intelligence test, whose results would also vary. Then the two sets of data (intelligence and running speed) could be correlated by means of a simple statistical test. The results would indicate the degree of relationship between the two variables, which in this case would probably be weak but positive (i.e., intelligence and running speed would tend, slightly, to go up and down with each other).

➤ *CASE STUDY*

Or consider a correlational study that examined liquor consumption and birthrates in the San Francisco Bay area between 1970 and 1980. Both liquor consumption and birthrates are variables—women practice both activities at varying levels. The study yielded a strong positive correlation between birthrates and the consumption of alcoholic beverages. Did one cause the other? Well, you could make a pretty good case that drinking played a role in more children being born, but then another interpretation could be that the birth of children caused an increased consumption of alcohol. Or it could be that the coincidental data were caused by other factors. Any causal statement based solely on correlational analysis is not possible.

Does intelligence cause running speed (or does running speed cause intelligence), and does drinking lead to more children being born (or vice versa)? While any of these could be true, from a scientific position merely establishing a relationship between two variables *does not mean causality*. We stress this point because lay thinkers often take the co-occurrence of two events as proof of a causal relationship. Many times superstition follows from this thinking, and students can think of many examples of people carrying good luck charms because on one or more occasions they had good luck while carrying such amulets.

The following example provides a correlational analysis of humor and body type:

➤ *CASE STUDY*

The first test of a hypothesis involving two sets of data may simply be correlational analysis. If it is hypothesized that A causes B, then failure to find a correlation between A and B may be reason to question the hypothesis.

Wilson, Nias, and Brazendale (1975) hypothesized that a woman's body shape would influence her attitude toward chauvinistic cartoons focusing on female anatomy. Sixty-two female student teachers rated the "funniness" of a set of British seaside postcards that featured lecherous male interest in attractive women. Results of a correlational analysis between the subjects' "shapeliness" as measured by a bust/waist ratio and their ratings of the sexist cartoons indicated a positive correlation. Some of the results follow:

Table 2.1   CARTOONS SIGNIFICANTLY PREFERRED BY "SHAPELY" WOMEN ($N = 62$)

DESCRIPTION OF CARTOON	CORRELATION WITH "SHAPELINESS"
1. Two men lecherously surveying bikini-clad females	.34°°
2. Two male swimmers admiring a female sunbather	.31°
3. Man and woman viewing painting of a pirate standing behind a very phallic cannon	.26°

°$p < .05$
°°$p < .01$

The authors' interpretation was that women who are themselves physically attractive are more accepting of this kind of humor; an alternate hypothesis is that physically unattractive people are sensitive to humor in which the theme is to make light of anyone who is less than statuesque.

Of course, correlations do not determine the direction of cause and effect. They merely indicate that a relationship between two variables may occur with a probability greater than one would expect by chance (although it may be more logical to search for cause-and-effect relationships between body types and appreciation of risqué postcards than to conclude that appreciation of risque postcards causes body changes). The judicious use of correlational techniques is a valid approach to searching for causative relationships.

NONEXPERIMENTAL (BUT EMPIRICAL) RESEARCH

Rigid thinking and dogmatism have been the two enemies of creativity and flexibility in research. For years experimental psychology in America eschewed research that did not conform to the paradigm in which a variable is introduced (or inferred) and its consequence observed. The traditional experimental design followed the pattern of finding cause-and-effect relationships between antecedent events and their consequences.

But there are many interesting psychological issues that do not lend themselves to this neat experimental paradigm and we need to investigate these issues with reliable measures. Some of these issues include the buying habits of steel workers in Pittsburgh, the difference between the number of bipolar personalities in Miami and Seattle, and the trends in fashion over the past century. These topics, and hundreds of other like topics, are interesting, worthwhile, and important to psychologists and they may be investigated scientifically, studied empirically, and can yield reliable data.

Collecting Data in Nonexperimental Research

It is important to have some standard means of collecting data in **nonexperimental research,** such as research based on observation of a subject (or subjects) over a period of time. We turn to an example of a nontraditional research paradigm.

➤ *CASE STUDY*

One of the most engaging and important research projects conducted in recent times was the "Washoe" venture carried out by our colleagues Allen and Beatrix Gardner at the University of Nevada at Reno. Some years ago, they developed the idea that it might be possible to communicate with chimpanzees by use of American Sign Language—a sign language commonly used by hearing impaired people. The implication of their research was enormous in that it told us something about the fundamental properties of language, comparative psychology, and even phylogenetic considerations regarding apes and humans. The effort was truly "big science." But teaching sign language to chimps did not lend itself to the presentation of an independent variable and the measurement of a dependent response in a conventional research paradigm, so the Gardners devised an observational procedure that produced reliable data with regard to the processes of language acquisition. One technique that they used to evaluate the recognition vocabulary of the chimpanzees was based on the presentation of a hand sign by one member of the team (such as the sign for the word "doll"), a reaction by the chimp (such as pointing to the "correct" object), and the scoring of the chimp's response by another experimenter who was blind to the original sign. Only after the chimp's reaction was recorded was it compared with the original sign. Thus, the possibility of an experimenter bias, so common in like experiments, was minimized. By collecting a great volume of data over an extended period of time, it was possible to show reliable progress in language acquisition of their most famous chimpanzee, Washoe, and other members of the colony. Additionally, subsequent observations

Photo courtesy of R. A. and B. Gardner.

were made of spontaneous language patterns that were initiated by the chimpan-zees rather than in response to experimenter cues.

As shown in the example, the elements of time and frequency of events are conventional components of observational data. Three different means of quantifying behavior in observational studies are in use. They are the frequency method, duration method, and intervals method.

Frequency Method.   The **frequency method** is based on the recording of a specific behavior within a certain time period. Thus, if you were interested in aggressive behavior by children on a playground, you might establish an **operational definition** of aggressive behavior and record the occurrence of that behavior over a 30-minute time period, for example.

Duration Method.   With the **duration method** you would record the length of time that elapses during each episode of a behavior. In the example of aggressive behavior for example, you might record how long each aggressive behavior lasts.

Intervals Method.   The **intervals method** is a type of observation in which time is divided into discreet intervals, for example 3 minutes each. An observer might then note if a specific behavior, such as aggressive behavior, occurs within that time period. Such information might give a clue as to the continuity of behavior, as in the case of a chimpanzee beginning a behavior and continuing it over a series of time intervals.

Empirical Observations of Trends

As we saw, it is possible to make reliable, empirical observations of behavior even when a specific independent variable is not manipulated. The techniques described may also be applied to even more general types of information, such as cultural trends. For example, it may be useful to measure the frequency of suicides in Baltimore throughout the year, or over several years, in an effort to establish trends that may predict future rates of suicide. Or, such descriptive trends may be linked to specific dates, such as Christmas, or seasons, such as winter. These data may form the basis for more general theories relating mood and seasons and the interactive effect between them.

Establish a Base Rate

In observational research that is based on trends, it is important to establish a base rate or baseline for the critical variable so that a meaningful contrast may be made. As an example, consider the hypothesis that marital infidelity increases as the sale of alcoholic beverages increases. Correlational research calculated over time may be the first step in establishing a cause-and-effect relationship between these two variables. However, it is important to establish a base rate, or the normal rate of occurrence of marital infidelity, as well as liquor sales over time. Since these figures may not be known, it behooves the investigator to collect data on these variables so that meaningful observations may be made.

In general, using observations of historical trends as evidence for cause-and-effect behavior is less reliable than are experimental studies in which a specific independent variable is manipulated. For example, one historical trend showed a strong correlation between the hardness of asphalt and the number of infant deaths. But it is difficult to see how one variable might affect the other, and in fact the phenomenon may be due to a third variable or could be the result of chance. For these reasons, empirical observations of historical trends must be undertaken with strict adherence to prescribed standards of observation. We now turn to an example of empirical observations of a historical trend.

➤ *CASE STUDY*

Within the past few years, anorexia and bulimia have become more common, especially among young American women. Why? Might such irregular eating habits be related to the idealized image of "the beautiful woman"? And, if so, might it be possible to investigate the trends in "idealized" images in the common culture?

Rita Hayworth: the prettiest pinup in World War II.
Courtesy of *Life*.

In a study that investigated these questions, Mazur (1986) traced the history of the changing concepts of feminine beauty and the overadaptation by some women in their effort to conform to idealized forms. This research project posed some very special problems in that the quantification of idealized forms of feminine beauty is not obvious. Some investigators might assert that feminine beauty is something they could recognize but not define operationally. (Such a remark is reminiscent of a statement made by a Supreme Court justice a few years ago about pornography: "I known it when I see it.") Nevertheless, Mazur attempted to quantify some of the attributes of feminine beauty by illustrating some of the physical characteristics of female icons, as demonstrated by Miss America contestants and Playmates from *Playboy* magazine over an extended time period. Some of the data reported in the article are shown in Figures 2.1 and 2.2.

In general, these data suggest that the ideal beauty has grown more slender and taller. While people can do little to affect their height, they can do much to regulate their waist, hips, and bust through dieting. According to Mazur's interpretation of these data, some women have taken dieting to an extreme level with the result being abnormal eating practices.

As illustrated, the gathering of information on trends does not conform to the traditional experimental method in which the effect of some variable is measured. Nevertheless, in the previous example the researcher has attempted to study the causes of eating disorders in a scholarly way. A word of caution: Eating disorders, and other psychological phenomena, are likely caused by a multitude of factors, of

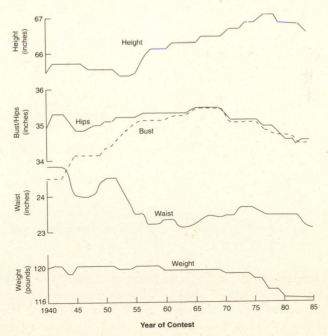

Figure 2.1   Body measurements of Miss America contestants, by year of contest (smoothed).

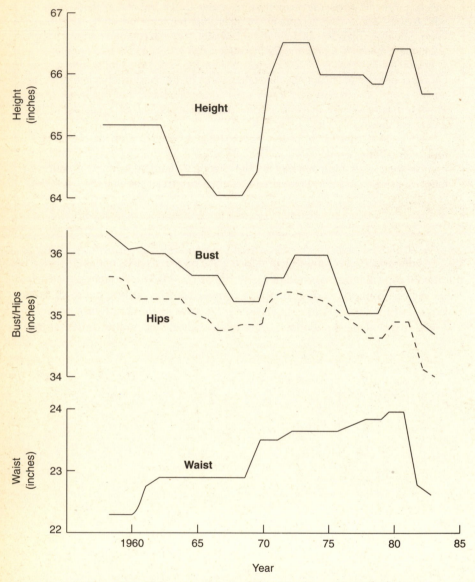

Figure 2.2   Body measurements of Playmates, by year (smoothed).

which idealized body shapes is but one. In order to specify all of the causes of behavior in this case, or in similar cases, additional studies are required—studies that may or may not conform to traditional experimental techniques. The conclusion is that complex behaviors, such as eating aberrations, are likely to be the result of complex causes and that the conscientious investigator is obligated to use a wide range of investigatory techniques in seeking these causes. These techniques may include traditional experimental procedures and nontraditional procedures. We close this chapter by discussing one other type of procedure.

Naturalistic Observations

As suggested by its name, naturalistic observations are studies based on observations of subjects in a native setting or "in the field." (See chapter 1.) In general, there is no attempt to manipulate the environment and measure the effect of a specific independent variable, but instead the social milieu and subjects provide the stimuli for actions that are the source of data. In some sense, all humans are naturalistic observers—that is, we all observe others in their natural setting, be it in an airport, a supermarket, a singles bar, a classroom, or a theater. But it should be emphasized here that because the subjects are in a natural setting, rather than in a laboratory, the scientific observations are inherently less well defined.

Suppose you were interested in a problem that has fascinated people for centuries: nocturnal behavior during a full moon. Legend tells us (experimental psychologists are inherently suspicious of legends) that people are more restless and do strange things when the moon is full; in fact, the word *lunatic* is derived from the same root word as *lunar*. Casual information, which is sometimes a source for hypotheses, suggests that people sleep more poorly, have more dreams, drink more heavily, and carouse more frequently during a full moon than otherwise. Police, hospital attendants, ambulance attendants, and others who deal with emergency conditions attest to the frequency of bizarre behavior during a full moon, and some research has tended to confirm these results. So several years ago an enterprising undergraduate student of ours, who was a night attendant in a mental hospital, collected data on the number of times patients got up during the night, and he looked for a correlation with the phase of the moon. The relationship was positive, but the cause was undetermined. It might have been that lunar light simply provided illumination for patients to make their way to the bathroom without stubbing their toes.

In order to gather reliable data on nocturnal activity and lunar phases, it is important to establish operationally defined criteria for the behavior in question, namely, nocturnal activity. Since the number of times a person uses the toilet during the night may simply be caused by lighting conditions, it is necessary to make more subtle observations to establish nocturnal behavior such as the position of a sleeper, the number of times a person moves during sleep, and the number of dreams one experiences—observations that can then be compared to the phase of the moon. All of these observations must be made as unobtrusively as possible, less the observer become an unwanted or contaminating stimulus. Furthermore, each factor should be quantified, which may require the use of some sophisticated equipment (although it should be pointed out that many naturalistic observations need no specialized equipment). To record the position of a sleeper, an experimenter might use the number of times a person moves from one position (e.g., faceup) to another (e.g., facedown) during a night. Activity could be empirically measured by mounting a bed on four micro-switches that would record "jiggles." The frequency of dreams could be measured by attaching tiny electrodes to the sleeper's eyelids to record rapid eye movements (REM), which are associated with dream activity. To quantify these three variables (the dependent variables in the study) the following coding sheet is provided:

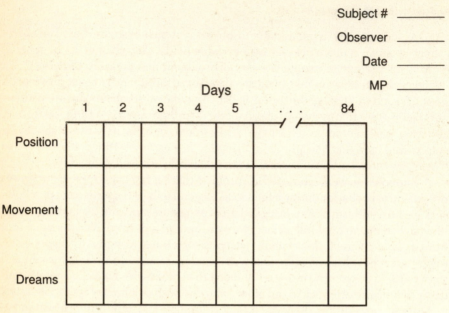

Figure 2.3   Data form for sleep activity study.

Note that in the upper right-hand corner of this form spaces are provided to identify the subject, the observer (in case more than one observer collects data), the date, and the MP (moon phase), which should be added later. In a single-blind experiment the subject is not told that the salient experimental variable is the moon phase. In a double-blind experiment both the person collecting the data and the subjects are not informed of the critical independent variables. This double-blind control would guard against experimenter bias and subject bias.

In this experiment, care must be taken to specify the characteristics of the sample being tested. For example, the activity of elderly single subjects might be less perceptible than that of newly married 18-year-olds, REM measures for subjects who use elevating psychoactive drugs might be higher than for non–drug users, and position activity may be greater among coffee drinkers than among non–coffee drinkers. All of these subject variables need to be considered because they may affect nocturnal activity in profound ways. Thus, the influence of lunar phases on activity (if it really exists) may be obscured by other, extraneous causes.

Thus far such an ambitious project has not been done,[6] but many experiments in which empirical measurements were made in a naturalistic setting have been done. The next example illustrates the case.

6. Perhaps some ambitious student will do this (or something like this project). Please keep us informed.

> ### CASE STUDY

The research scientists Dean, Willis, and Hewitt (1975) were interested in the interpersonal distance between people of unequal status. The hypothesis of their research was based on the assumption that people of lower status stand farther away from people of higher status than they do from people of equal status. A student, for example, may stand a bit farther away from his or her professor than from his or her classmate. The student may stand even farther away from the dean or president of the university. To test this hypothesis Dean et al. unobtrusively measured how far away enlisted men would stand when they approached naval officers. The results of this study showed that the distance between people of different rank increased as the difference in rank increased. Such a study, which tells us something about the dynamics of personal space and perceived status, does not lend itself to traditional experimental methodology but is well suited for naturalistic observation.[7]

7. A complete analysis of this article is found in Chapter 12.

Field-Based Studies

In yet another example of these types of studies, we now illustrate a **field-based study,** which is a research investigation conducted in a natural setting. A field study was conducted using the "lost-letter technique," which involves the distribution of bogus letters to see if people would mail them to the addressee. The return rate (i.e., the number of lost letters mailed) is measured for different neighborhoods, for example.

In the following example, Bryson and Hamblin (1988) use the technique to evaluate the return rate of postcards that contained either neutral news or bad news. Note the return rate by type of news and by gender.

LOST LETTER TECHNIQUE AND THE MUM EFFECT
BY J. B. BRYSON AND K. HAMBLIN

A variant of the lost-letter technique, the lost postcard, was employed to examine attitudes toward informing people of their romantic partner's apparent infidelity. Stamped and addressed postcards were left on the windshields of 180 cars parked near mailboxes, with an accompanying handwritten note reading "Found this by your car— is it yours?"

One-third ($N = 60$) of the cards had a neutral/good news (control) message ("Glad to hear you've worked things out. We're getting along better too. Keep in touch . . ."); 30 of these were addressed to a male, 30 to a female.

The other 120 cards, equally divided by sex, informed the addressee of his (her) girlfriend's (boyfriend's) apparent infidelity, in the following message: Dear Bob (Judy), I hate to be the one to tell you this, but I think I saw your girlfriend Ann

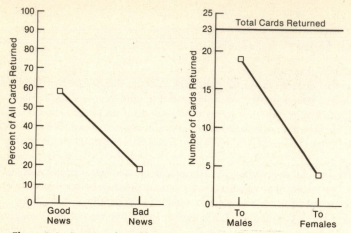

Figure 2.4 Percent of good/bad news cards mailed and number of bad news cards mailed to males and females.

(boyfriend Bob) coming out of the TraveLodge off El Cajon Blvd. with another guy (woman) on Thursday. It might not be important, but I didn't know how to tell you in person—Barry (Beth).

Consistent with a MUM effect hypothesis that bad news is not transmitted as often as good news: 35 (58.3 percent) of the neutral/good news postcards were mailed versus 23 (19.2 percent) of the bad news postcards ($\chi^2 = 28.10$, $p < .001$). There was also a substantial double standard effect caused by sex of the recipient of the bad news postcards: of the 23 cards returned, 19 were addressed to males and only 4 to females ($\chi^2 = 12.02$, $p < .01$). Figure 2.4 shows the results.

These findings indicate that, while there is a general unwillingness to transmit bad news (the MUM effect), there is a definite double standard in the willingness to transmit information regarding infidelities—the partner is always the last to know, especially if she is female.

Why should people be less inclined to report a male's infidelities? Two possibilities seem likely: It may be that females' infidelities are less accepted and hence more likely to be reported, or it may be that women are considered less capable of dealing with their partner's infidelities and need to be protected from learning of them.

Some field-based studies raise serious ethical questions because they may involve deceit, and the permission of the subject is rarely obtained. Bear this in mind and return to this problem when you read Chapter 9.

DEFINITIONS

The following terms and concepts were used in this chapter. We have found it instructive for students to go through the material a second time to define each of the following:

between-subject design

control group

correlational studies

demand characteristics

dependent variable

double blind

duration method

experimental group

experimenter bias

field-based study

frequency method

hypothesis

independent variable

intervals method

nonexperimental research

operational definition

placebo control group

single blind

subject variable

within-subject design

Chapter 3

Factorial Designs, Quasi-Experimental Designs, and Other Advanced Design Techniques

It is a capital mistake to theorize before one has data. Insensibly one begins to twist facts to suit theories, instead of theories to suit facts.

—Sherlock Holmes (Arthur Conan Doyle)

All experiments in psychology are designed to produce statements about the nature of behavior and thought from which theories and models can be fashioned. Psychological experiments are mostly designed to indicate cause-and-effect relationships, while other valid studies may be purely descriptive, as in the instance of case studies. In attempting to show cause-and-effect relationships in science in general, and experimental psychology in particular, researchers have shown that an effect (or effects) frequently have multiple causes. Thus, your ability to react to highway signs is not only a function of the ambient luminosity (you couldn't see signs in total darkness) but also of your ability to understand the meaning of the signs and your physiological reaction time.

In simple, one-factor designs only a single factor is manipulated and other factors are held constant or controlled. In this chapter we will discuss more complex types of experimental procedures, including those in which several factors are manipulated, quasi-experimental designs, and other topics.

FACTORIAL DESIGNS

Single-factor experiments are important tools in psychology, but it is evident that behavior is rarely a function of a single variable. For example, evidence from developmental psychology indicates that if a child's parents are hostile and excessively controlling, then the child may become withdrawn. On the other hand, if the hostile parental behavior is combined with a tendency to ignore the child and exert no control over his or her behavior, the child tends toward antisocial behaviors. Thus, these two variables—degree of kindness or hostility and degree of control—are both important factors in determining a child's behavior.

In order to determine the influence of two or more independent variables on a dependent variable, researchers employ a **factorial design.** In a factorial design two or more variables may be manipulated at the same time. To illustrate the use of a somewhat simple factorial design we have selected a case from comparative psychology in which the experimenters studied how fast rats run a straight course under varying levels of food deprivation and food reward.

> ➤ CASE STUDY

Ehrenfreund and Badia (1962) examined the performance of rats under varying food-deprivation and incentive conditions. The apparatus used in the experiment was a 5-foot-long, straight alley with a start box on one end and a goal box on the other end. The dependent measure was the speed at which the rats ran down the alley. Twenty rats were used in this experiment—half were maintained at 95% of their **ad lib,** or free-feeding, weight and the other half were maintained at 85% of their ad lib weight. For the purposes of this experiment, the high-deprivation treatment was defined as those rats maintained at 85% of ad lib weight, and the low-deprivation treatment was defined as those rats maintained at 95% of their ad lib weight.

Then experimenters put a 45-mg food pellet (low incentive) in the food box for half of the rats in the high-deprivation treatment, and put a 260-mg food pellet (high incentive) in the goal box for the other half. Using the same incentive treatments with the low-deprivation treatment yielded the four experimental treatments of the experiment:

1. high deprivation–high incentive
2. high deprivation–low incentive
3. low deprivation–high incentive
4. low deprivation–low incentive

There were five rats in each of the four conditions.

Performance was measured in terms of the speed at which the rats traversed the middle 2-foot section of the alley. The experimenters recorded the median running speed for each rat on the last 10 trials of the experiment. Each rat's median score then was divided into 10 so that a higher score reflected a faster running time. The means of these scores for each of the four treatment groups are presented in Table 3.1.[8]

Table 3.1   MEAN RUNNING SCORES FOR DEPRIVATION AND INCENTIVE GROUPS

		DEPRIVATION TREATMENT	
		95%	85%
Incentive treatment	45 mg	10.26	13.92
	260 mg	13.86	15.15

It is evident from the table that both deprivation and incentive are influential in determining performance. The mean for the high-incentive groups is higher than the mean for the low-incentive groups within each deprivation level. Furthermore, both of the means for the high-deprivation groups are higher than the means for the low-deprivation groups.

8. For the computation of factorial analysis see "Analysis of Variance" in Appendix A.

The preceding example is called a 2×2 factorial design. Two levels of one variable (deprivation) are combined factorially with two levels of a second variable (incentive) to yield 2×2, or four, separate treatments. Each level of the first variable occurs within each level of the second variable—that is, high and low deprivation are tested both under high- and low-incentive conditions. The design also could be extended to add other independent variables; for example, a $2 \times 2 \times 3$ design would consist of two levels of the first variable, two levels of the second variable, and three levels of the third variable. In this design there would be 12 separate treatments. One of the advantages of the factorial design is that it allows the researcher to ascertain how independent variables combine with one another to determine the values of the dependent variable.

MEASURING ATTITUDES

Social psychologists who measure public attitudes toward a variety of topics (e.g., foreign aid, the value of space exploration, abortion, or the death penalty) face several unique problems in their research. Among the most difficult is deciding the way a questionnaire is phrased in a public opinion poll. Take a recent poll conducted during the 1992 presidential election in which the views of people about government welfare projects were sampled. In one instance the question was phrased: "Would you favor more welfare to the poor?" In a second instance the question was phrased: "Would you favor more assistance to the poor?" In the first case 19% of the respondents agreed while in the second case 63% agreed.

The following experiment is a 2 × 2 factorial design applied to attitude formation—a common theme in social psychological research—that illustrates the power of such a design to pick out interactions between variables.

➤ CASE STUDY

Several studies have shown that if people are somehow induced to argue in favor of something to which they are opposed, the more they are paid for the task, the more they change their own attitude toward that subject. This result is consistent with a reinforcement explanation—money is a reward, and the more money that is paid for arguing in favor of a position that a person opposes, the more that position is reinforced. On the other hand, several other studies have found the exact opposite effect. These studies support a **dissonance theory** explanation (Festinger, 1957). This explanation assumes that dissonance (or a state of discomfort) is aroused when a person argues in favor of a position he or she actually opposes. Furthermore, the smaller the amount of money given for doing this task, the greater should be the amount of dissonance, since low sums of money provide inadequate justification for defending something to which a person is opposed. In order to relieve the high level of dissonance, the people must convince themselves that they really are in favor of the argued position. Therefore, more attitude change would occur in the high-dissonance (low-money) condition than in the low-dissonance (high-money) condition.

In an attempt to resolve these conflicting results, Linder, Cooper, and Jones (1967) suggested that the reinforcement hypothesis was applicable when subjects had no choice in arguing for the position they opposed. They further assumed that the dissonance hypothesis was appropriate under conditions in which the subject somehow chose to argue for the opposed position. The logic for their hypothesis was based on the assumption that dissonance can be created only when people choose to do something they oppose, but not when they are forced to do it.

The design used in this experiment was a 2 × 2 factorial design. Subjects were college students who each wrote an essay supporting a speaker-ban law for colleges (a position they were against). The subjects either were told to write the essay (no-choice condition) or were given a choice as to whether or not they would write the essay (free-choice condition). In addition, half of the subjects in each of the two conditions were paid 50 cents for writing the essay, while the other half of the subjects were paid $2.50. Thus, the design consisted of the four treatments, and 10 subjects were randomly assigned to each treatment. The dependent variable was the amount of change in the subjects' attitudes toward the speaker-ban law. This was measured by having the subjects check the point on a scale that indicated the degree to which they approved of the speaker-ban law. The mean change for the four treatments is given in Table 3.2. These data are also clearly shown in graphic form in Figure 3.1.

Table 3.2 MEAN ATTITUDE CHANGE FOR CHOICE AND INCENTIVE CONDITIONS

		INCENTIVE	
		50 CENT INCENTIVE	$2.50 INCENTIVE
Choice	No-choice treatment	−0.05	+0.63
	Free-choice treatment	+1.25	−0.07

Positive scores indicate change toward favoring the speaker-ban law, while negative scores indicate a greater aversion toward the position (a boomerang effect). The results supported the authors' hypothesis. In the no-choice treatment, the $2.50 incentive elicited more attitude change than did the 50 cent incentive. In the free-choice treatment, the opposite result was found.

The appropriate statistical test to analyze the previous results is **analysis of variance.** The actual calculations of this statistic will not be explained here. However, some of the logic of this test will be described because it gives some insight as to the logic of a factorial design, as well as pointing out how the results are analyzed. The experiment used a 2 × 2 factorial design with two levels of one variable (no choice or free choice) and two levels of a second variable (50 cent incentive or $2.50 incentive). In the analysis of variance, each variable (level of choice and level of incentive) is analyzed separately, and then the interactions between the variables are analyzed. In the previous experiment there were two factors; therefore, the analysis of variance will contain (1) an analysis of the main effect of the first variable (choice), (2) an analysis of the main effect of the second variable (incentive), and (3) an analysis of the interaction between the two variables.

One can conceptualize a 2 × 2 analysis in the following form:

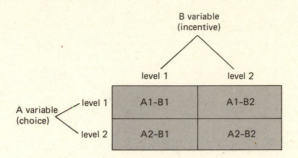

The main effect of the choice variable is analyzed by comparing the mean attitude change of subjects in the no-choice treatment with the mean attitude change of subjects in the free-choice treatment, ignoring the incentive manipulations. The mean of the no-choice treatment is calculated by adding the means for both incentive conditions and dividing by 2 (−0.05 + 0.63 = 0.58; 0.58 ÷ 2 = 0.29). The same process is used in the free-choice treatment with a resulting mean value of 0.59. These two means are then compared to analyze for the main effect of choice. The same process is repeated to analyze the main effect of incentive. The necessary addition and division reveal that the mean attitude change for the 50 cent treatment is 0.60 and the mean for the $2.50 condition is 0.28. These two means are then compared to ascertain the main effect of incentive. The next step is to analyze the interaction between the two variables. The test for an interaction effect determines if two variables are independent of each other with respect to their influence on the dependent variable. The variables are dependent if it is demonstrated that one variable affects the dependent variable differently under each of the two levels of the second variable.

Linder et al. (1967) analyzed their results in this manner. Their computations indicated that although the mean attitude did show a change both for the main effect of choice and for the main effect of incentive, these changes were not **statistically significant** (see Box), meaning the magnitude of these changes was not greater than would be expected by chance fluctuation in the means. **Chance** is the variation in results that is due to uncontrolled factors such as guessing, experimental error, and failure to achieve a perfect matching of subjects in each treatment group. Because of a variety of uncontrolled factors, we would expect the means to show some change even if the treatments had no effect. The interaction effect, however, was statistically significant in the direction hypothesized by the authors—high incentive facili-

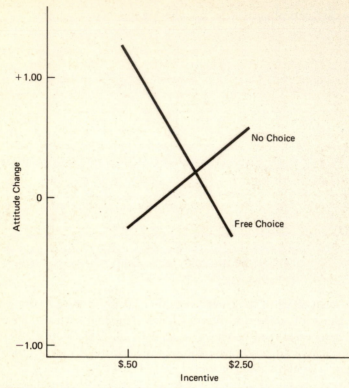

Figure 3.2 Mean attitude change as a function of choice and incentive conditions.

tated attitude change in the no-choice treatment, but low incentive facilitated attitude change in the free-choice treatment. Thus, the two variables interacted to determine the amount of attitude change.

➤ STATISTICALLY SIGNIFICANT

Prior to the collection and analysis of data, experimental psychologists commonly anticipate how the data will be analyzed statistically. A researcher often sets a **level of significance,** defined as the statistical point above (or below) which one can infer the operation of nonchance factors. Researchers usually set this level of significance in terms of a probabilistic statement. In many psychological experiments the level may be expressed as "the 0.05 level" or "the 0.01 level," which means that the results will occur by chance only 0.05 (1 in 20) or 0.01 (1 in 100) times by chance alone. Data that have been statistically analyzed and meet the preestab-

lished criterion (be it 0.05, 0.01, or even 0.001) then are called statistically significant.

It should be pointed out that a result may not be statistically significant but still
can be of interest. Often, especially during pilot studies or experiments with few
subjects or observations, the results may fail to reach the level of statistical significance but can still suggest that further investigation with perhaps better controls
and/or more observations may be worthwhile. At the same time, an experimenter
must be cautious not to go on a fishing trip for results by running and fine-tuning an
experiment until he or she gets the desired results.

In another example of a 2 × 2 design, Chi (1978) used two types of subjects and
two types of tasks to study the importance of specialized knowledge on memory.

➤ CASE STUDY

In this study, Chi (1978) examined the recall of digits and of chess pieces (A variable) by children and adults (B variable), thus completing the 2 × 2 design. The
children were 10-year-olds who were skilled chess players, while the adults were
novice chess players. The first task involved looking at chess pieces as they might
appear in a normal game and recalling the arrangement after they were removed. In
the digit portion of the task, recall was measured using a standard series of digits,
such as might appear on an IQ test.

The design can be conceptualized as follows:

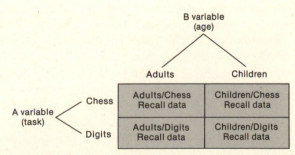

Figure 3.3 Design of Chi (1978) experiment examining digit/chess recall by adults and
children.

Figure 3.3 shows a strong interaction. It appears that specialized knowledge, such
as chess knowledge, facilitates recall of information in that domain but has little
effect on digit memory. Adults who are unsophisticated in chess recall fewer pieces
than do children but perform better on digit recall.

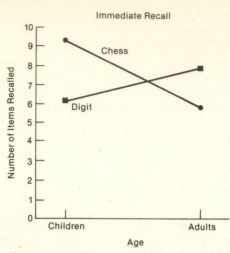

Immediate Recall

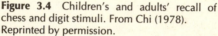

Figure 3.4 Children's and adults' recall of chess and digit stimuli. From Chi (1978). Reprinted by permission.

We can see the strong interaction graphically in Figure 3.4. In the text of the study, Chi (1978) presents a statistical analysis that confirms mathematically the results shown in the figure.

The 2 × 2 design is the simplest of factorial designs. These designs become increasingly complex as more factors are added, or when several levels of a factor are used. The following example illustrates this point. In it, only two factors are used, but while one factor still has two levels, there are four levels in the other factor. This example was taken from research in educational psychology.

Much of the evaluation of classroom learning takes place through tests, quizzes, or papers. Quite frequently a student takes a test or turns in a paper on one day, and it is returned several days later. This situation is called **delayed information feedback,** where a time delay occurs between when the student takes the test and when the student learns where he or she made mistakes. This delayed feedback has been criticized as poor educational procedure, leading to some new methods, such as programmed textbooks and teaching machines, that make use of an immediate feedback technique in that the student gives an answer to a particular question and finds out immediately whether the answer is right or wrong.

➤ *Case Study*

More (1969) designed an experiment to compare the effects of immediate and delayed information feedback using learning materials similar to those used in a school classroom. The subjects were eighth-grade students from four junior high schools who were asked to read a 1,200-word article on glaciers. After the students read the article they were immediately tested on its contents in a 20-question multiple-choice test.

To test for the effects of delayed information feedback, a 4 × 2 factorial design was used. The first variable was the length of delay between taking the test and learning the results. There were four time intervals: One group received the feedback immediately as the students answered each question, while the other three groups received the feedback either 2½ hours, 1 day, or 4 days after taking the test. After the students received this feedback, they took the same test again. At this point the experimenter introduced the second variable: The students either took the test immediately after receiving the feedback (acquisition treatment) or 3 days after receiving the feedback (retention treatment). The dependent variable was the number of questions the students answered correctly the second time.

Based on the symbolic description of factorial studies discussed previously, this procedure has two levels of an A variable (treatment) and four levels of a B variable (delay of feedback). This can be conceptualized as follows:

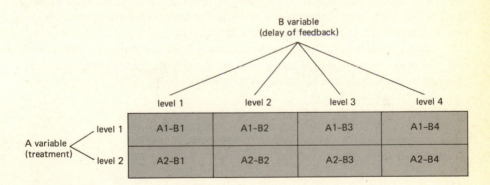

To summarize the procedure, the students started by reading an article on glaciers and were immediately tested on the material. Then they received feedback as to the correct answers after one of four time delays. They then were retested on the same test either immediately after feedback or three days after feedback. There were three different classrooms of students in each of the eight treatment groups. The mean number of questions answered correctly on the second test for each of the eight groups is given in Table 3.3. The results indicate that performance was higher in the acquisition group than in the retention group (16.4 compared to 14.8). This was expected as some forgetting would be likely after three days. For the acquisition treatment, the no-delay feedback group scored significantly lower than the other three feedback groups (15.1 compared to 16.7, 16.9, and 16.7), whose scores did not

Table 3.3   MEAN SECOND TEST PERFORMANCE OF TREATMENT GROUPS

| | DELAY OF FEEDBACK | | | | |
	NO DELAY	2½-HOUR DELAY	1-DAY DELAY	4-DAY DELAY	TREATMENT MEANS
Acquisition treatment	15.1	16.7	16.9	16.7	16.4
Retention treatment	13.0	15.4	16.0	14.6	14.8
Feedback means	14.1	16.1	16.5	15.7	

differ significantly from each other. For the retention treatment, the 2½-hour and the 1-day delay groups scored significantly higher than either the no-delay or the 4-day delay groups (15.4 and 16.0 compared to 13.0 and 14.6). No significant difference was found between the 2½-hour and the 1-day groups or between the no-delay and the 4-day groups.

More (1969) concluded that the results offer no support for the assumption that immediate feedback maximizes learning of the specific materials used in this study. Quite the contrary—delayed feedback of 2½ hours or more produced superior learning in the acquisition treatment. For the retention treatment, some delay (but not too much) maximized the students' retention of the materials at a later date. More felt that this latter finding was especially important since one primary objective of instruction is to maximize retention of material.

Most researchers find it valuable to draw a figure showing the results of a factorial design because it enables them to see the relationship between the variables more clearly. Figure 3.5 illustrates the results of More's (1969) experiment.

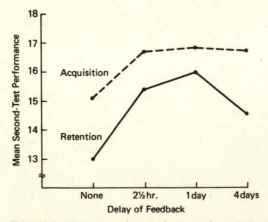

Figure 3.5   Mean second test performance for acquisition and retention conditions.

The dependent variable is on the vertical axis, and one independent variable is on the horizontal axis. Separate curves show each level of the second independent variable. The effects of delay in the experiment are more clearly illustrated by this figure. This experiment points out the possible complexity of the factorial design as well as the value of having several levels of factors. If only two levels of the delay factor had been used, the interpretation of the results for the retention treatment would have been quite different, regardless of which two delay levels had been picked. To illustrate a more complex factorial design, consider the color matching experiment by Solso and Short (1979) below that illustrates a $3 \times 3 \times 2$ factorial design.

Complex Factorial Designs

The following example illustrates several important principles of experimental psychology. The first is the use of a within-subject control. Note that subjects are treated to several conditions in the following experiment and thus serve as their own control. Also, note that measurement of mental processing is introduced in the form of reaction time. The assumption of this measurement is that intellectual processes, such as matching two signals, requires time, albeit a very small amount of time. Subtle differences between matching signals may reveal important psychological processes. Finally, the experiment introduces the use of the tachistoscope (or T-scope).

➤ CASE STUDY

Psychologists have been interested in the way information is "coded" or recorded in memory after it is perceived. In a series of important experiments done by Posner and his associates (see Posner, 1969; Posner, Boies, Eichelman, & Taylor, 1969; Posner & Keele, 1968), it was found that subjects initially formed visual codes for letters and then formed name codes. Following these experiments Solso and Short (1979) performed an experiment on color codes. This experiment allows us to introduce several relevant features of contemporary experimental design, including within-subject design, reaction times, tachistoscopic procedures, and visual representation of data.

In the experiment subjects were shown a color square and either an associate to the color, the name of the color, or another color square. The second stimulus appeared either simultaneously with the first color, after 500 msec (1/2 sec), or after 1,500 msec (1 1/2 sec). Subjects were asked to press a reaction-time key as quickly as possible if the two stimuli (the color square and the secondary stimulus) matched. An equal number of nonmatched secondary stimuli were shown to assure that the subjects were really reacting to the matching task. For example, in a given series of trials a subject might see a red square and then 500 msec later the word *blood*, which, being a correct associate, should produce a match response. In a second trial the same subject might see a green square presented simultaneously with the word *green*, which is also a match. In a third trial the same subject might see a blue square followed 1,500 msec later by a green square, which is not a match, and so on. The design could be thought of as a 3 × 3 × 2 design. The first variable represents the type of relationship being tested: color to associate, color to word, or color to color. The second variable represents the delay between stimuli: 0, 500, or 1,500 msec. The third variable represents either a match (e.g., red–blood) or a mismatch (e.g., red–blue).

  The design is a within-subject design. Each subject was treated with each of the experimental variables. The results are shown in Figure 3.6 and indicate that subjects initially respond fastest to color-color pairings (e.g., red-red) and slowest to color-association pairings (e.g., red-blood). But as the interval between the two stimuli is lengthened to 1,500 msec, the reaction time for matching the associate (e.g., blood) decreases significantly, so that the reaction times for all three groups are similar. The authors interpreted the results as a parallel development of codes to colors (see Figure 3.7).

The previous experiment also introduces the topic of **mental chronometry**—measuring the speed of mental events. This topic has been a favorite tool of

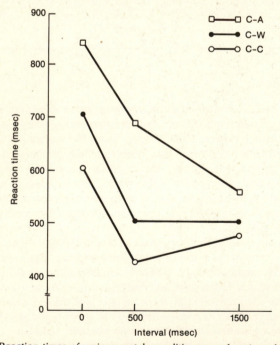

Figure 3.6 Reaction times of various match conditions as a function of priming interval: C–A = color to associate; C–W = color to word; and C–C = color to color. From Solso and Short (1979).

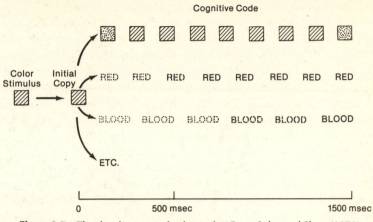

Figure 3.7   The development of color codes. From Solso and Short (1979).

many recent cognitive psychologists, but its history can be traced to the 19th century. The basic principle involved is that the processing of information requires time, and complex processing requires more time than simple processing.

> ➤ *BEHAVIORISM AND COGNITION PSYCHOLOGY*

Experimental psychology in the United States has undergone several major changes throughout the 20th century. Perhaps the most significant has been the change from it being a purely behavioristic study of human psychology to a study of the internal representations as expressed by cognitive psychologists.

In the behavioristic mode, overt behavior—what the organism in question did—was considered to be the main theme of psychology. But rigid adherence to the behavioral idea was challenged by cognitive psychologists who contended that important topics such as thought, language production, intelligence, memory, and other related topics required the development of models that reflected the complex internal thought processes. The characteristics of these models could be carefully ascertained from studying behavior. Both behaviorists and cognitive psychologists rely on overt behavior, but the cognitive psychologist is likely to use such observations as the basis for theories of the basic patterns of thought.

In the case of a cognitive experiment, when measuring reaction time for matching colors (see Solso & Short, 1979) the experimenters based their inference about internal representations (e.g., the associative strength between colors and color associates) on behavior (viz., reaction time). Other examples of behavioral studies may also be found throughout this text.

Reaction-time experiments are usually based on a forced-choice procedure in which the subject is required to make a decision between two (or sometimes more) alternatives. Because reaction-time experiments are highly susceptible to practice effects, subjects are normally given a series of warm-up trials to, as some researchers say, "limber up their index fingers." But much more is involved in warm-up, or practice, trials. Familiarization with the apparatus, the experimental environment, the experimenter, the procedure, and the stimuli are all important factors in warm-up and should be considered in all reaction-time experiments, as well as other experiments in which practice effects might be significant.

The experiment also introduces us to a standard piece of equipment: the **tachistoscope,** or T-scope. This device is not new to experimental psychology, but has recently become very sophisticated. It is primarily used in experiments that require great precision in presenting visual stimuli for brief time periods and accuracy in measuring reaction times. The typical T-scope is an enclosed box that has three channels, or viewing chambers, capable of exposing three different stimuli. The exposure time is controlled by small fluorescent tubes that show the stimuli when illuminated. Computers have replaced many of the functions of the T-scope, but for very precise presentation of stimuli, the T-scope remains an essential tool.

QUASI-EXPERIMENTAL DESIGNS

The concept of **quasi-experimental design** was introduced by Campbell and Stanley (1963)[9] to overcome some of the problems faced by psychologists who wish to study behavior that occurs in less formally structured environments than a laboratory. Quasi-experimental designs are studies in which the independent variables (the elements whose effects are to be measured) are selected from the natural environment. Sometimes quasi-experimental designs are termed **ex post facto research,** or "as if" designs, as the collection and analysis of data take place after an event has happened. These designs are similar to naturalistic observations, which were discussed previously. One part of the logic of this design is that if the experimenter had been able to introduce an independent variable into a situation, then that variable would have been the same as the variable that was naturally introduced. It is *as if* the experimenter was responsible for the introduction of the experimental variable.

In typical experimental work in psychology, subjects are selected for participation. Selection of subjects is either random or on the basis of certain well-defined characteristics (e.g., sailors between the ages of 19 and 25). The subjects are then presented with an experimental variable, frequently in a laboratory setting, which is

9. In order to give the reader a sense of the importance of these types of designs in psychology, surveys of graduate departments of psychology have been conducted that have asked which books are most frequently recommended to their graduate students who are preparing for departmental examinations. *Experimental and Quasi-Experimental Designs for Research* by Campbell and Stanley has been one of the most frequently recommended books throughout several surveys. (See Solso [1987b] for details.)

a controlled environment. The results of the experiment then are frequently generalized to a larger population or in some way applied to real life (**generalization of results**). But real life occurs naturally, and sometimes it is impossible to select subjects for an experiment and bring them to a well-controlled laboratory setting for precise recording of data. How could someone study a riot in the laboratory? And if someone could, would the conclusions be generalizable to a real riot? These topics are of interest to experimental and applied psychologists and need to be addressed. Essentially, the problem boils down to the fact that observations based on microcosmic life (as might be observed in the laboratory) may not be valid for macrocosmic life.

In laboratory work in experimental psychology we are dealing with a highly controlled environment, which has been called a **closed system.** The laboratory has many virtues, and many psychology experiments require this rigid control over stimuli. An **open system** is an environment over which we have little or no control, such as what is vaguely called the real world.

We have selected an example of a quasi-experimental design that deals with automobile accidents as related to the severity of traffic laws.

> **CASE STUDY**

After a record number of automobile fatalities in the state of Connecticut several years ago, harsh action was taken against speeders. After these measures were introduced, a decline in traffic fatalities was noted. Campbell (1969) made a detailed study of this phenomenon in a quasi-experimental design. The data for Connecticut traffic fatalities are shown in Figure 3.8. In this figure we see a decline in fatalities from 1955 to 1956 after the introduction of harsh treatment of speeders. However, one may correctly argue that many other causes may have entered into this picture (e.g., road conditions may have been improved, better weather may have occurred in 1956 than in 1955, or better driver education may have been available). One might note, however, that the decline in fatalities continued after 1955.

Another way to look at the data is to compare the automobile fatality rates among comparable states. Those data are shown in Figure 3.9. In this figure we see that although the fatality rates in these four states fell slightly from 1951–1959, the rate of decline for Connecticut was much greater, especially after the 1955 crackdown on speeders. The results therefore suggest that the 1955 treatment did have an effect on fatality rates.

In this case, no reasonable laboratory experiment could adequately test the effectiveness of the 1955 crackdown. Yet the issue is important, and many similar

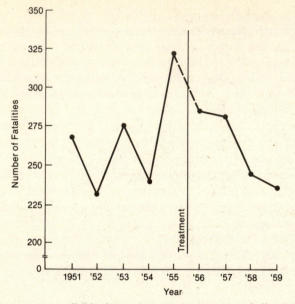

Figure 3.8   Connecticut traffic fatalities, 1951–1959. From Campbell (1969). Copyright 1969 by the American Psychological Association. Reprinted by permission.

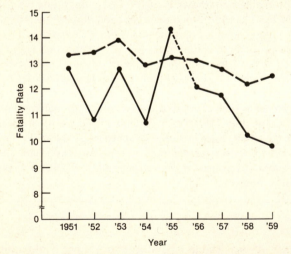

Figure 3.9   Connecticut traffic fatality rate (lower line) with the fatality rate of four comparable states. From Campbell (1969). Copyright 1969 by the American Psychological Association. Reprinted by permission.

real world problems need solutions. Given the circumstances, and despite the inherent lack of precision in these types of studies, quasi-experimental designs seem to render as close an approximation to valid conclusions as one could expect.

FUNCTIONAL DESIGNS

In the previously discussed factorial designs, the usual procedure was to assign several subjects to each experimental treatment. Then the mean (or percent) of the subjects' scores was compared using the appropriate statistical test. While this procedure is common, a different research strategy is frequently used by researchers interested in what they have called "the experimental analysis of behavior." This area of research originated with B. F. Skinner, and people in this field of study are sometimes called "Skinnerians." The design has been called a **functional design** because it uses functional definitions of terms and concepts. Developing a functional definition of a concept (e.g., punishment) is accomplished by specifying the relationship between a set of determining conditions and their effects on behavior, both of which can be precisely measured. The requirement that functional definitions be used to some extent leads one to adopt the type of research strategy described here.

Several differences can be noted in contrasting this research strategy with those previously discussed in this book. First, researchers in this area tend to be atheoretical in that they are more concerned with examining variables that control behavior than with testing some theory. Instead of viewing an experiment as a means of theory testing, experimenters systematically explore variables that control behavior with the assumption that theory will emerge inductively from the data.

Small *n* Designs

A second difference is that researchers in this area will sometimes use only one or two subjects, called a **small *n* design,** rather than large groups of subjects in each experimental treatment. These researchers tend to report their data not in the form of means and variances of several treatment groups, but in the form of a typical response curve. This curve is a segment of the subject's behavior that is deemed typical of his or her performance under the particular experimental conditions. Another difference is that a statistical analysis of the data sometimes is not used, but the typical curve (or curves) is presented for visual inspection of the response regularities that are representative of that particular stimulus condition.

A commonly used technique in small *n* studies is an **ABA design** in which a subject's untreated behavior (A) is observed first. This measure is sometimes called **baseline data** as it serves as a point of departure from which to contrast the effects of experimental treatment. In the second phase the experimental variable is introduced and its effect is measured (B). In the final phase the experimental variable is absent and behavior is observed (A). Schematically, the design appears like this:

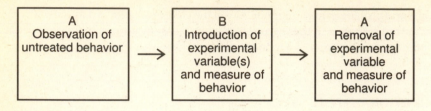

A behavioral therapist may, for example, be interested in treating a patient who overeats. In this case, the dependent variable could simply be the weight of the client. (In more elaborate designs the dependent variable could involve a range of physiological and psychological measures—metabolic rate, strength, feelings of well-being, absenteeism, and so on.) Thus, the observation of untreated behavior in the A phase could be the subject's weight before treatment. In the B phase the treatment or independent variable would be introduced, which could be a type of behavioral psychotherapy. Note that in some studies of this sort two or more independent variables may be used (e.g., the therapists may use positive reinforcement and exercise). It is critical to understand that the results of studies involving two or more independent variables do not permit unequivocal cause-and-effect conclusions. It is possible, however, to make a valid concluding statement to the effect that treatment by means of conditions 1 and 2 has led to the following results. In the case mentioned, if weight loss occurred, then a statement to the effect that behavioral therapy and exercise led to weight loss would seem justified, but not that behavioral therapy or exercise alone led to the weight loss. It may be that each of the variables alone would lead to weight loss, but the design does not permit such a conclusion. On the other hand, the combination of treatments (therapy and exercise) may effect weight loss through a **synergistic effect,** meaning the cooperative action of the treatments could result in more effective reactions. Astute experimental psychologists are always watchful for possible synergistic effects, which may have profound behavioral consequences.

Many of the previous designs are based on an internal validity in which the baseline data for a single subject serves as a control for subsequent observations. If an external control is used, the experiment uses the following setup, sometimes called an **AAA design:**

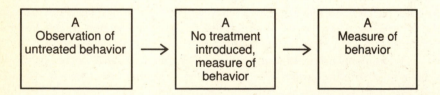

The ABA design can be extended to an ABABA design in which the experimental variable is reintroduced, or to an AB_1AB_2A or $AB_1AB_2AB_3A$ design, in which two or more different experimental variables are used. More elaborate designs are restricted only by the ingenuity of the psychologist.

Researchers performing these types of experiments may use a highly controlled experimental situation. For example, an animal may be placed in a Skinner box—a chamber that typically contains only a bar to be pressed by the animal, a food dispenser, and some signal lights. In this simple situation, numerous independent variables could be manipulated, such as the particular **schedule of reinforcement** under which the food is dispensed. For example, the reinforcement may be dispensed on a fixed-ratio schedule, in which the animal is reinforced with a food pellet after it presses the bar a fixed number of times (e.g., 1 pellet for 1, 16, 47, or 100 presses). In a fixed-interval schedule the animal receives a food pellet for pressing the bar at least once in every fixed time period (e.g., every 30 seconds or 4 minutes). The dependent variable in these cases is the frequency of the bar-pressing response and is usually presented in the form of a cumulative frequency curve. The following example from Ferster and Perrott (1968) illustrates research using a schedule of reinforcement.

➤ *CASE STUDY*

The apparatus used in this experiment was a Skinner box. This box was approximately 14 square inches and contained only a small Plexiglas plate and a food magazine on the same wall. The plate and food magazine were connected such that if the plate was pushed, a food pellet would be released from the food magazine. Both the plate and the releaser mechanism were attached to a cumulative recorder that consisted of a pen mounted on a sliding arm. The pen point rested on a strip of paper that passed slowly over a cylinder with the passage of time. If no responses were made by the animal, the pen point merely left a horizontal line as the paper passed over the cylinder. Each time the pigeon pecked the Plexiglas plate, the pen moved a small step in one direction on the paper and did not return to its original position. When the paper was examined, it was rather easy to see the animal's rate of response by noting the rate at which the pen moved upward in a given time period.

The subject used in this experiment was a pigeon that had had previous experience with this apparatus. The pigeon was kept at 80% of its free-feeding body weight throughout the experiment, and was placed in the Skinner box for 1 hour per day over a 6-week period. In the box the pigeon was reinforced for pecking the plate on a fixed-ratio schedule, meaning the pigeon received a food pellet after pecking the plate a fixed number of times. Several fixed-ratio values were used. During the first week the pigeon received a food pellet after 70 pecks (FR 70); during the second week a food pellet was received after 185 pecks (FR 185); and during the third week the food pellet was received after 325 pecks (FR 325). The order was then reversed for the next 3 weeks.

Figure 3.10 shows the performance during each of the three fixed-ratio schedules. Each segment is an excerpt that is typical of the pigeon's daily performance on each schedule. The dots indicate the point at which reinforcement was delivered. The rate of performance in each segment can be estimated by comparing the overall slope of each segment with the slopes given in the grid in the lower right-hand corner of the figure. The slope of FR 70 indicates that when the pigeon was responding, it pecked approximately three or four times per second. When 70 pecks were required for reinforcement, the bird's pecking was almost continuous, with a very slight pause after each reinforcement. For FR 185 there was a longer pause after each reinforcement; however, when the pigeon began pecking again, it started at a very rapid rate, which it maintained until the next reinforcement. The pause became much longer for FR 325; however, once the pigeon began pecking again, it did so at the same rate that was noted for FR 70 and FR 185. The experiment indicates that the number of pecks necessary for reinforcement does not affect the rate of pecking but does influence the length of pause between the dispensing of a reinforcement and the resumption of pecking.

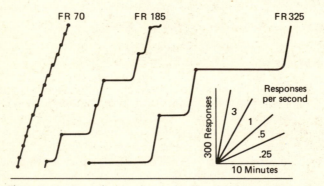

Figure 3.10 Rate of responding under three fixed-ratio schedules.

This example illustrates the design strategy used in this area. A single subject was used to participate in three experimental conditions (i.e., three different fixed-ratio schedules). The data are presented in terms of typical response curves that indicate the regularities of response under the three conditions. No high-level statistics are used. The experimental situation itself is highly controlled with precise measurement of the reinforcement conditions and the pecking responses.

This same design has been applied to a variety of subjects and experimental situations. The following example is an application of the same strategy in an experiment designed to control abnormal behavior in a chronic schizophrenic.

➤ CASE STUDY

The strength of a response may decrease as a function of continued reinforcement. This phenomenon is called satiation and can be easily demonstrated in the laboratory. If an animal is given continuous reinforcement over a long period of time, the animal will stop emitting the reinforced response. Ayllon (1963) used the satiation procedure to control hoarding behavior in a psychiatric patient. The subject was a 47-year-old patient in a mental hospital who collected towels and stored them in her room. Although the nurses repeatedly retrieved the towels, the subject collected more and had an average of 20 in her room on any given day. Ayllon's procedure was first to establish a baseline, representing the average number of towels in the subject's room under normal conditions. After a seven-week observation period, a satiation period began. The nurses no longer removed towels from the subject's room; instead they began bringing towels into the room and simply handing them to the patient without comment. During this period the number of towels brought into the room by the nurses increased from 7 per day during the first week to 60 per day during the third week. The satiation period lasted for 5 weeks until the subject had accumulated 625 towels and had begun to remove the towels.

Figure 3.11 shows the mean number of towels in the patient's room per week over the life of the experiment. Note that this is not a cumulative record. After the

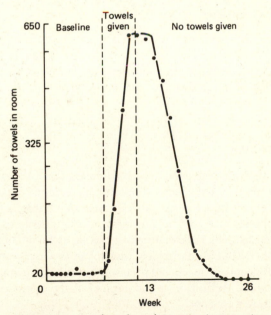

Figure 3.11   Number of towels in patient's room prior to, during, and after treatment.

satiation period the subject continued to remove towels from her room until by the 22nd week there was an average of 1.5 towels in the room. This average was maintained through the 26th week. Ayllon (1963) made periodic observations throughout the next year and found that this average continued. The subject never returned to her towel-hoarding behavior, and no other problem behavior replaced it.

While the Ayllon (1963) experiment is an interesting demonstration of a "cure" for the hoarding behavior, one should not lose sight of the fact that it is also a nicely controlled experiment using a functional design. Ayllon: (1) establishes a baseline for the frequency of a behavior, (2) institutes a well-designed experimental treatment, (3) terminates the experimental treatment, and (4) continues observation of the frequency of the behavior over an extended time period. A single subject is used, and no high-level statistics are used. A point to be emphasized is that sound experimental design is necessary to ascertain the regularity of behavior, whether the subject be a pigeon or a mental patient and whether the behavior be pecking or towel hoarding.

ADDITIONAL CONSIDERATIONS

This chapter has been concerned with some of the different types of experimental design used in psychology. While the basic logic of design is rather simple, the design and procedure themselves may become fairly complex, as has already been seen in the case of factorial designs and as will be seen in other types of design in the following chapters.

Operational Definitions

There are two points that need to be introduced at this time. First, one decision that experimenters make is how to operationally define their variables. Researchers usually already have conceptual definitions of variables they wish to investigate—anxiety, intelligence, ego involvement, drive, distributed practice, and reinforcement, for example. Their theories are based on the relationships between concepts such as these. But to do research, psychologists must somehow operationally define these concepts by specifying precisely how each is manipulated or measured. An **operational definition** is a statement of the operations necessary to produce and measure a concept.

There is considerable variability as to the extent to which variables can be operationally defined in a precise manner that retains the full meaning of the concept. On one hand, variables such as the spacing of practice, as used in the Lorge (1930) experiment (see p. 20), or the delay of feedback, as used in More's (1969)

". . . And then he raises the issue of how many angels can dance on the head of a pin, and I say you haven't operationalized the question sufficiently—are you talking about classical ballet, jazz, the two-step, country swing. . . ."

experiment (see p. 50), are fairly easy to operationally define. But psychologists also use abstract concepts such as intelligence or anxiety, which may be somewhat difficult to operationally define in a manner that includes the full complexity of the concept. Anxiety is a good example of such a variable. Almost everyone has some idea of what anxiety is. There are several dictionary definitions of anxiety, most of which agree that it is a complex emotional state with apprehension as its most prominent component. In attempting to operationally define this variable, researchers have used pencil-and-paper tests, a Palmar sweat technique, the galvanic skin response, heart rate tests, and eye movement tests. Each of these operational definitions probably measures some part of anxiety, although none of them measures its total complexity. A researcher somehow must pick or develop an operational definition that is suited for the specific situation.

It is absolutely necessary that the variables used in research be operationally defined. A series of abstract concepts used in psychology follows, and we suggest that the reader find operational definitions for these as part of a class exercise.

anxiety
creativity
aggression
intelligence
frustration

ego involvement
memory
learning
reinforcement
self-esteem
attitude
punishment
pornography
insight
leadership
effort
death
behavior

The second point that needs to be introduced is that experimenters are always confronted with the question of how far they can generalize the results and conclusions of their experiments. In the Paul (1966) experiment (see p. 24), which showed the superiority of behavior therapy over insight therapy, there are several questions that can be raised concerning the generality of the findings:

1. How far can the results be generalized across "illnesses"? Is behavior therapy superior only for speech phobia or for phobias in general? Does it work only for mild mental problems, including neurosis, or is it superior for all mental problems?
2. How far can the results be generalized across subjects? Are the findings applicable only to college students? Would the same results be found with children or with middle-aged people? Would the same results be found for less intelligent people?
3. Would the same results be found if the therapy time had been longer? Is behavior therapy effective as a quick therapy, while insight therapy needs more time to work?
4. Did improvement generalize to other speaking situations? Was the therapy only effective for speech class situations, for all classroom situations, or for all speaking situations?

There are additional questions that could be raised; for example, was the improvement temporary or permanent? But the previous numbered questions are sufficient to illustrate the point. The question of generalization of results enters into all research and is not a criticism of an experiment. Rather it points out the limitations of a single experiment. In a single experiment, the researcher strives for tight control over the variables so that he or she can be certain as to the validity of the results. Control is often achieved by limiting an experiment to a specific behavior, to a specific sample of subjects, to a specific measurement technique, and to a specific time period. While these techniques of control are helpful, perhaps essential, in ensuring the validity of the results, they also raise questions as to whether the results can be generalized. This is why one experiment often leads to several other experiments, and thus makes research an ongoing process with new problems to investigate and new knowledge to acquire.

DEFINITIONS

The following terms and concepts were used in this chapter. We have found it instructive for students to go through the material a second time to define each of the following:

AAA design
ABA design
ad lib
analysis of variance
baseline data
chance
closed system
delayed information feedback
dissonance theory
ex post facto research
factorial design
functional design

generalization of results
level of significance
mental chronometry
open system
operational definition
quasi-experimental design
schedule of reinforcement
small *n* design
statistically significant
synergistic effect
tachistoscope

Chapter 4

Anatomy of Experimental Design: Control

Information is any difference that makes a difference.
— *Gregory Bateson*

CONTROLLED CONTRASTS

Scientific experiments are based on controlled observations of **contrasts.** From these controlled observations **inferences** are drawn that are designed to generalize to a larger class of phenomena. This principle applies equally to physics, chemistry, botany, geology, astronomy, and psychology. In some scientific studies contrasts may be more obvious than in other endeavors. An example drawn from physics might be the contrasting of objects' masses. Consider someone's reaction, for example, to hitting a regular tennis ball and hitting one made of lead. The contrast between how well the two balls bounce off (or through) a tennis racket is pretty obvious. Then, from the study of the contrast it is possible to make inferences about such general topics as mass, inertia, and tennis rackets. Most scientific observations of contrasts are far more subtle.

One example of a subtle contrast drawn from physics is the measurement of electromagnetic fields created by charged electrodes. The existence of these "invisible" fields is not immediately obvious, but their influence can be measured by comparing the distribution of finely machined iron particles in the absence and in the presence of an electromagnetic field. In psychological terms, the electromagnetic field, created by a charged electrode, is an independent variable, while the distribution of iron particles is a dependent measure from which a dependent variable (the characteristics of an electromagnetic field) is deduced. Nearly all high school students have seen this demonstration. Iron filings are sprinkled on a thin board, and then a magnet is placed under the board. The distribution of the iron filings immediately changes, and a pattern emerges that clearly shows the presence of an electromagnetic field and its characteristics. From these observations theoretical characteristics can be inferred and mathematical models that generalize to a wide class of similar physical situations can be developed. See Figure 4.1 for a diagrammatic inference of such a field.

Figure 4.1 Hypothetical distribution of an electromagnetic field created by two positive charges.

The purpose of this example is not to teach a principle of electrodynamics but to illustrate a basic tenet of experimentation: Science is based on observing contrasts. The observations may be in a controlled environment, as in a laboratory, or in nature. These observations then form a system of knowledge, such as theoretical physics or psychology.

In the electromagnetic field example, the contrast was between the presence or absence of an electromagnetic field. In psychological research the contrast is commonly between the influence of several levels of an independent variable on a group of subjects while another group of subjects is observed in the absence of the independent variable. The first group is called an experimental group; the second group is a control group.

SCIENTIFIC INFERENCES

Scientific theories are based on inferences. The distribution of iron filings changes as a result of an electromagnetic field, much like human behavior changes under a variety of conditions (e.g., drugs, social setting, motivation). Just as the physicist makes inferences about the characteristics of electromagnetic fields, experimental psychologists observe behavior caused by some variable. Then psychologists can sometimes develop models that generalize to a wide class of similar situations that lead to reliable laws about human behavior.

Because the validity of our observations is based on the characteristics of the subjects (as well as on the influence of the independent variable), it is extremely important that the experimental and control groups be as similar as possible. Think, for example, of the erroneous conclusions about electromagnetic fields that would emerge if one scientist used iron filings in the experiment and another scientist used aluminum filings—the observations would be different, not because the laws of electromagnetic flux changed, but because different materials were used. And even if observations are valid, if the experiment lacks control, the results may be meaningless or may misdirect future research. These issues are important to the experimental psychologist, and techniques have evolved to safeguard against errors resulting from lack of proper control. We now turn to some of these possible error problems and procedures for their control.

TYPES OF CONTROL

The previous chapters have focused on design strategies. In previous examples, the experimenter manipulated some variable (the independent variable) and observed the effect on the dependent variable. The experimenter controlled the independent variable by determining how much or what kind of variable was presented. This is one form of experimental control. It was also emphasized that good design requires that the only variable being manipulated be the independent variable and that all other conditions be held constant for the various treatment groups. **Holding conditions constant** is a second type of experimental control. If the various treatment and control groups are treated exactly alike except for the independent variable, then any differences in the dependent variable must be due to the independent variable. If some other variable is affecting the results, this variable is usually called an **extraneous variable** or a **confounding variable.** The variable is extraneous in that it is an "extra" variable that has entered into the experiment. It is a confounding variable because the experimenter cannot be sure if the results are due to the independent variable or the extraneous variable. Thus, the results are inconclusive, and the experiment must be repeated using a design that eliminates the influence of the extraneous variable.

A variable is extraneous only when it can be assumed to influence the dependent variable. In an experiment studying the perception of visual illusions, for example, variables such as the color of the subject's eyes, height, athletic ability, and knowledge of Swiss mountain cheese probably have no effect on the subject's perception, and the experimenter would be foolish to attempt to control all these individual differences. (Also, these differences tend to balance out when a representative sample of subjects is used in an experiment.) On the other hand, variables such as eyesight and illumination of illusions are likely to affect the results, and these should be controlled.

Whether a variable should be considered extraneous (and therefore in need of control) is largely a function of the dependent variable and will vary from experiment to experiment. Intelligence may be an extraneous variable for someone interested in learning but probably is insignificant in an experiment to determine the threshold of tones of varying frequencies. Knowledge of Swiss mountain cheese probably is unrelated to attitudes toward abortion but may be related to sensory discrimination of dairy products. Of course, anyone can make a case for possible relationships that are improbable (such as a limited level of intelligence being necessary to participate in an experiment on sensory thresholds, or knowledge of cheese being remotely related to abortion views). But don't chase ghosts. There are many important problems to be solved in psychology, and too often neophyte researchers get bogged down in petty problems. Good experimental design is characterized by careful attention to the control of *real* extraneous variables and the investigation of *important* topics. Even the most exquisitely crafted experiment cannot save a stupid idea, and a brilliant idea is wasted in an experiment with a serious design problem. Good judgment combined with knowledge can improve an experimental design so that potential extraneous variables are manageable, thus allowing the experimenter to control only those variables that can be reasonably assumed to influence the dependent variable.

This chapter is primarily concerned with controlling extraneous variables that occur when experimenters manipulate the independent variable. A related problem deals with ensuring that the various treatment groups have the same subject characteristics. This will be discussed in Chapter 6 because a set of specific techniques has evolved to handle this problem.

So far, considerable emphasis has been placed on holding all other conditions constant as a means of controlling for the effects of extraneous variables. Actually, this is only one of two very general techniques of controlling extraneous variables. A second method involves the use of treatment or control groups. All experiments use the first technique of control in that experimenters attempt to manipulate only the independent variable. But some experimenters also add treatment or control groups to further control extraneous variables. This latter technique is frequently used when the experimental manipulation may contain an extraneous variable as well as the independent variable. By adding treatment or control groups it then may be possible to separate the effects of the extraneous variable from those of the independent variable.

Holding Conditions Constant

In the Lorge (1930) experiment comparing massed and distributed practice (see p. 20), the independent variable was the length of time between practice sessions. This was the only variable that was manipulated. All other variables were held constant: All treatment groups performed the same task; all treatment groups had the same amount of practice; the task was such that it would be difficult for the treatment groups with spaced practice to rehearse between practice sessions; the abilities of the subjects in each treatment were equal; and so on. All these factors potentially could be extraneous variables. For instance, if one treatment group received more practice than another group, the results could be due to the amount of practice rather than to the spacing of practice. If the different treatment groups had performed different tasks, then the results could be due to a difference in task variables, such as difficulty, rather than being due to the spacing of practice. If the subjects in one treatment had better task-related abilities than those in another treatment, then the results could be due to differences in the abilities rather than to the spacing of practice. By ensuring that these variables were the same for all treatment groups, Lorge eliminated them as explanations for his results. This is the logic behind holding all conditions except the independent variable constant.

Subject variables.   Holding conditions constant is obviously essential to good experimental design and is easily understood by the beginning psychology student. However, as will be seen in the following examples, even the most competent researchers may unknowingly violate this principle. While it is not possible to construct a checklist of extraneous variables (since they vary from situation to situation), there are some areas in which problems are especially prominent. For example, when the independent variable is a subject variable, there is always a danger that it is related in some systematic manner to another subject variable. If this is true, then any experimental results might be due to the second subject variable, introduced inadvertently, rather than the one the researcher introduced. This problem is called a **subject variable–subject variable confound.** To illustrate this problem, consider a study in which the researcher hypothesized that an authoritarian person would have more difficulty learning complex material than a person who was not authoritarian. This hypothesis was based on the assumption that high-authoritarian people think in a rather simplistic manner and therefore would have difficulty learning complex material. To test this hypothesis the investigator had high-authoritarian subjects and low-authoritarian subjects learn some complex material. When tested on the material, the low-authoritarian group recalled considerably more than the high-authoritarian group. But one criticism of several leveled against this study pointed out that it is well known that there is an inverse relationship between authoritarianism and intelligence—authoritarian people tend to be less intelligent, and vice versa. Therefore, the fact that the high-authoritarian group learned less could be explained by the fact that they were less intelligent, and authoritarianism may have had nothing to do with the results.

If a subject variable is not being manipulated, then a subject variable–subject variable confound is of little danger. However, the experimenter must be aware of other possible extraneous variables. Some of these problems will be discussed later in this chapter.

Experimental Paradigms and Statistics

A **paradigm** in experimental psychology is a model or pattern an investigator uses to organize research.

In this section, we introduce five experimental paradigms and the methods of control in each:

1. randomized group design
2. matched subject design
3. repeated measure design
4. repeated measure design with sequence counterbalanced
5. factorial design

Some of these paradigms and controls will be discussed further in later sections.

Each of these paradigms is supported by a specific statistical test, the details of which may be found in most statistics books. Some rudiments of statistical analysis are found in Appendix A, but the computational procedures for these design problems are beyond the intent of this book. The relevant statistical tests are commonly available on packaged computer programs. Thus, it is now possible to design an experiment, conduct the research, collect and statistically analyze the data, and draw valid conclusions with impressive efficiency.

Only a few years ago, the statistical analysis of data frequently required more time than all other phases of experimentation. With the appearance of high capacity desktop computers, the ubiquitous PCs and Macs, it is possible to encode data in a standard statistical package that calculates values and prints results in seconds. Even more extensive calculations are possible using mainframe computers. As impressive as some of these modern behemoths are, sufficient power usually exists in smaller, generally more accessible and inexpensive computers. It is even possible to collect data in a computer data base that can be easily transferred to a statistical package for analysis.

Certain statistical packages are considered too simpleminded by some psychologists and statisticians, while other people welcome such number-crunching devices as a boon to experimental psychology by emancipating the creative thinker to devote more time to seminal issues and to experimental design matters. Others suggest that the use of such packages, without understanding the basic rationale of statistics, is a serious mistake that may lead to grievous misapplications of statistical results.

We believe students of experimental design should understand the basic assumptions of statistics and the applications of specific tests. Access to this knowledge is readily available through statistics courses and texts. While some students may choose to pursue a specialized course of study in statistics, it is possible to plan and execute psychology experiments with only a rudimentary knowledge of statistics. At the same time we readily embrace the labor-saving, efficient, and accurate computer packages that are available presently, but we remind the student that the focus of experimental design in psychology is to use design and statistics as a *tool* to study important problems in psychology. Occasionally, experimental scientists become so preoccupied with design and statistical matters that they lose sight of the real purpose of experimental work: the advancement of knowledge through scientific inquiry. Experimental design, the application of statistics, and the study of important psychological problems are all a part of that purpose.

We will now turn to the five types of design paradigms using a single example:

Suppose that experimenters are interested in how the color of a wine influences how much a person enjoys it. They have developed a device that can change the color of wine without changing its taste. In the test, the natural color of one wine is dark ruby, which can be changed to a deep green. The dependent variable in this experiment is the rating of enjoyment on a 5-point scale; the independent variable is the color of the wine. Since *enjoyment* is not a precise term, it is essential that the experimenters operationally define the term to include the features that are important. For example, one might include taste, bouquet, and color as important features.

The box shows that even this simple experiment presents some challenging experimental design problems, including the temperature of the wine (should be held constant), the lighting (must not be too dark to see the wine nor too bright), the sequence (tasting one wine first may influence the evaluation of subsequent wines), the experience of the tasters (sophisticated wine buffs may evaluate differently than

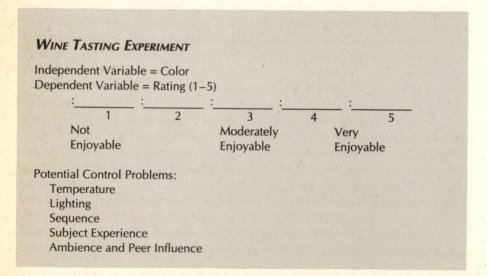

WINE TASTING EXPERIMENT

Independent Variable = Color
Dependent Variable = Rating (1–5)

:	:	:	:	:
1	2	3	4	5
Not Enjoyable		Moderately Enjoyable		Very Enjoyable

Potential Control Problems:
  Temperature
  Lighting
  Sequence
  Subject Experience
  Ambience and Peer Influence

"chug-a-luggers"), and the ambience (testing should be done individually rather than in groups where peer pressure may influence one's evaluation). The experiment may be conducted in several ways.

Independent Subject Design: Model 1

In one paradigm, called the **independent subject design** or **between-groups design** (Model 1), one group is given the artificially colored wine (the experimental group), while another group is given the naturally colored wine (the control group). In this design the subjects in one group are independent from the other group. The sample of subjects for this experiment is defined as 16 university students between the ages of 21 and 30. Although there are no rules governing the size of a sample, in general 8 subjects (or observations) per group for an experiment of this sort would be considered the minimum, and it would be prudent to have twice that number. From this sample, the subjects are randomly assigned to either the experimental or the control group: a **randomized subjects design.** For convenience, we have labeled the subjects S_1, S_2, S_3, . . . S_{16}. The arrangement of this procedure might look like the following:

In this experiment we have recorded some fictitious data, similar to the data that might be collected in an actual experiment. These data are further analyzed by means of a t-test in Appendix A to illustrate the computational procedures involved in that test. The inspection of raw data is sometimes fascinating in that it lets you

MODEL 1.   INDEPENDENT SUBJECT DESIGN

Experimental Group (artificially colored wine) DATA		Control Group (naturally colored wine) DATA	
S_1	1	S_9	5
S_2	2	S_{10}	4
S_3	2	S_{11}	3
S_4	1	S_{12}	5
S_5	2	S_{13}	5
S_6	1	S_{14}	4
S_7	1	S_{15}	4
S_8	4	S_{16}	5

Wine-tasting groups: Independent variable—color of wine
Dependent variable— evaluation of quality

know what subjects are actually doing, but it lacks mathematical precision. From the statistical analysis in Appendix A we see that the differences between the groups are statistically significant. While the preconceived notion that green wine is "yucky" is supported by the data, experimental psychologists view these matters with more cautious language. All one can state with scientific certainty is that the differences between the two groups are highly unlikely (i.e., less than 1 in 100) by chance alone. The next step is an inference, albeit a highly probable one, in which an experimenter might conclude that naturally colored ruby wine is preferred over the same wine that has been artificially colored to be green. The reasons for this preference may be cultural and further inferences are commonly made.

This paradigm is common in psychological experiments, and if the sample is large and representative enough, one might assume that the subject variables that might influence the results (e.g., having a large number of expert wine tasters or people from Ireland in one group) will be equally distributed between the two groups on the basis of random distribution.

Matched Subjects Design: Model 2

On the other hand, one may have reason to believe that a subject variable (such as wine-tasting experience) may be so critical to the results that the subjects should be matched on that variable. The assumption is that people who have had experience have developed a level of sensory discrimination that allows them to make critical judgments. Such a design is called a **matched subjects design** or **matched pair design** (Model 2) and might look like this:

MODEL 2. MATCHED SUBJECTS DESIGN

Experimental Group (artificially colored wine) DATA		Control Group (naturally colored wine) DATA	
S_{1a}	2	S_{9a}	4
S_{2b}	3	S_{10b}	5
S_{3c}	2	S_{11c}	5
S_{4d}	1	S_{12d}	5
S_{5e}	1	S_{13e}	4
S_{6f}	2	S_{14f}	4
S_{7g}	3	S_{15g}	5
S_{8h}	1	S_{16h}	4

Wine-tasting groups: Independent variable—color of wine
Dependent variable— evaluation of quality
Matching variable—wine-tasting experience

In this example we have labeled the subjects S_1, S_2, ... S_{16} but have added the subscript a, b, ... h to show that S_{1a} and S_{9a} are matched on some basis. In this case we have matched the subjects on wine-tasting experience so that each pair of subjects have about the same level of experience. Furthermore, in this example we may choose to select only people who are expert wine tasters, weeding out the chug-a-luggers. Other experiments may call for matching other attributes, such as sex, age, intelligence, running ability, or training. This design can be powerful provided that the matching variable is germane to the dependent variable. Also, the independent measure upon which matches are made must be valid. To match people on the basis of wine-tasting experience[10] or intelligence, for example, with an invalid test for these characteristics spoils the basic assumption of this method.

In Appendix A we have used the fictitious data to illustrate the computational procedures used in matched subject designs. This test is sometimes called the **correlated t-test** or **dependent t-test,** as the two groups are "co-related" or "dependent" in some important dimension. As the analysis in Appendix A indicates, the difference between the two groups is highly significant.

10. We can think of people who have much wine-drinking experience (i.e., they have consumed copious quantities of wine) but who are not wine connoisseurs!

Repeated Measure Design: Model 3

A third design, called a **repeated measure design** or a **within-subject design** (Model 3), is characterized by each subject being exposed to two or more experimental conditions. In our example, each subject would taste both kinds of wine and, as such, would serve as his or her own control. We can illustrate this procedure as follows:

MODEL 3. REPEATED MEASURE DESIGN

Experimental Condition (artificially colored wine) DATA		Control Condition (naturally colored wine) DATA	
S_1		S_1	
S_2		S_2	
S_3		S_3	
S_4		S_4	
S_5		S_5	
S_6		S_6	
S_7		S_7	
S_8		S_8	
S_9		S_9	
S_{10}		S_{10}	
S_{11}		S_{11}	
S_{12}		S_{12}	
S_{13}		S_{13}	
S_{14}		S_{14}	
S_{15}		S_{15}	
S_{16}		S_{16}	

Wine-tasting groups: Independent variable—color of wine
Dependent variable— evaluation of quality

As shown, S_1 tastes the experimental (colored) wine and the control (natural) wine. By allowing each subject to serve as his or her own control, this design makes it possible to gather more data because two measures rather than one are made for each subject. Reducing the number of subjects in an experiment may be practical with a limited subject sample.

Repeated Measure Design: Model 4

There are some problems that the previous design might create. One might be the sequence in which the substances are presented. It may be that when tasting wine the second taste may seem more enjoyable than the first, not because the taste is better but perhaps because the taster has a slightly rosier outlook on life in general. To safeguard against this, one could create a **balanced sequence** (Model 4). In this case, balancing the sequence would be achieved by having half the subjects taste the colored wine first, while the other half tasted the natural wine first. We can express this design as follows:

MODEL 4.   REPEATED MEASURE DESIGN

Control Condition (artificially colored wine) DATA		Experimental Condition (naturally colored wine) DATA	
S_1		S_1	
S_2		S_2	
S_3		S_3	
S_4		S_4	
S_5		S_5	
S_6		S_6	
S_7		S_7	
S_8		S_8	

Experimental Condition (artificially colored wine) DATA		Control Condition (naturally colored wine) DATA	
S_9		S_9	
S_{10}		S_{10}	
S_{11}		S_{11}	
S_{12}		S_{12}	
S_{13}		S_{13}	
S_{14}		S_{14}	
S_{15}		S_{15}	
S_{16}		S_{16}	

Wine-tasting groups: Independent variable—color of wine
Dependent variable— evaluation of quality

In this case, S_1 through S_8 would taste and evaluate the control wine first and then the experimental wine, while S_9 through S_{16} would reverse the sequence. Data could also be analyzed by statistical procedures that would identify the influence of sequencing. Further elaboration of the design is also possible. One could, for example, have a double or triple repeated design, but the considerations of these designs might be intoxicating.

Factorial Design: Model 5

In the previous models we have been concerned with the effect of a single independent variable on a dependent variable. Frequently, however, psychologists are interested in studying the effects of several independent variables on a dependent variable. Such factorial designs are very useful in experimental psychology, a fact which has been pointed out in Chapter 3. Although these designs are not specifically designed to control for subject variables, they are reiterated here to show how these variables may be used in a factorial design. In the present context, suppose that the experimenter chooses wine color and the type of grape used as the two independent variables. In this paradigm, the color of the wine would be either natural or burgundy, but in addition, three types of wines (derived from three different grapes) would be used. For the sake of illustration, a Pinot Noir, a Zinfandel, and a Chardonnay will be used.

A simple representation of this type of design can be shown in the following 2×3 matrix:

MODEL 5. FACTORIAL DESIGN

		Factor 1 (color)	
		level 1 (natural)	level 2 (burgundy)
Factor 2 (type of wine)	level 1 (Pinot Noir)	S_1, S_2, S_3, S_4	S_1, S_2, S_3, S_4
	level 2 (Zinfandel)	S_1, S_2, S_3, S_4	S_1, S_2, S_3, S_4
	level 3 (Chardonnay)	S_1, S_2, S_3, S_4	S_1, S_2, S_3, S_4

In this design, the same subjects are treated to all conditions, as in the repeated measure design. Thus, in Model 5 it is possible to integrate subject control techniques. But it should be pointed out that other subject control techniques, such as the independent subject design, may be used in a factorial experiment. This design could be illustrated as follows:

		Factor 1 (color)	
		level 1 (natural)	level 2 (burgundy)
Factor 2 (type of wine)	level 1 (Pinot Noir)	S_1, S_2, S_3, S_4	$S_{13}, S_{14}, S_{15}, S_{16}$
	level 2 (Zinfandel)	S_5, S_6, S_7, S_8	$S_{17}, S_{18}, S_{19}, S_{20}$
	level 3 (Chardonnay)	$S_9, S_{10}, S_{11}, S_{12}$	$S_{21}, S_{22}, S_{23}, S_{24}$

Use of Treatment and Control Groups

A control group is defined as a group of subjects similar to an experimental group that is exposed to all the conditions of an investigation except the experimental variable (independent variable). In some cases the control group and the experimental group should be drawn at random from the entire population in order to make generalizations about the results of an experiment.

Some of the techniques used to control extraneous variables have already been briefly examined. Let us consider in more detail the Paul (1966) experiment (see p. 24) on the treatment of speech phobia. The experimenter wanted to compare the effectiveness of two types of therapy. In addition, he had to consider what extraneous variables might vary as the level of a therapy varied. If an extraneous variable varied along with a therapy, then any improvement might be attributed to the extraneous variable rather than to the therapy itself. There are two well-known extraneous variables present when a subject receives some sort of therapy. First, it is well established that some people showing symptoms of a behavior problem will improve over time without receiving any specific treatment. This phenomenon is called **spontaneous remission**; there is a disappearance of symptoms that takes place spontaneously, that is, without any apparent treatment for the problem. In a therapy experiment the experimenter cannot be sure if the subject's improvement is due to the therapy itself or just to spontaneous remission.

Second, it is known that some people who think they are receiving treatment for their problems may show considerable improvement even when in fact they are not receiving treatment. This is called the **placebo effect.** The term comes from a Latin word meaning "to please," and was discovered by physicians who would give patients medically inert substances (e.g., sugar water) that resembled an active medication in order to please a patient rather than to provide physical benefit. Interestingly enough, they discovered that some patients improved upon receiving placebo medication, especially those whose illnesses seemed to be psychosomatic. Since this placebo effect also is assumed to occur in the treatment of psychological problems, any improvement in the therapy groups may have been due to this effect rather than to the therapy itself.

To separate the improvement due to spontaneous remission, to the placebo effect, and to the therapy itself, Paul (1966) used two control groups. Table 4.1 shows the four experimental groups as well as the variables influencing improvement in each of these groups.

As indicated in the table, the experimenter can now separate the effects of the various variables on the subjects' improvement. For example, we can subtract the amount of improvement shown in the placebo control group from that in each therapy group. This would give us an indication of the effectiveness of each therapy after we have eliminated the effect of the two extraneous variables.

Researchers will also frequently add treatment groups to an experimental design to ensure that the results are not caused by an extraneous variable. Suppose that in the Asch (1952) study on impression formation (see p. 21) Asch had used only a single treatment group, which received the positive adjectives first and the negative adjectives last. With this setup the subject's evaluation of the person would be generally positive, and this would support the hypothesis of a primary effect in

Table 4.1   VARIABLES INFLUENCING IMPROVEMENT IN THE FOUR EXPERIMENTAL GROUPS OF THE PAUL EXPERIMENT

EXPERIMENTAL GROUPS	VARIABLES PRESENT INFLUENCING IMPROVEMENT			PERCENT IMPROVEMENT
	THERAPY	PLACEBO	SPONTANEOUS REMISSION	
1. Behavior therapy	Yes	Yes	Yes	100
2. Insight therapy	Yes	Yes	Yes	60
3. Placebo	No	Yes	Yes	73
4. No treatment	No	No	Yes	32

impression formation. If only this single treatment group were used, several criticisms would arise. One would be that Asch's negative adjectives were not really very negative, and thus the person would have been evaluated positively regardless of the order. Another criticism would be that people tend to evaluate other people positively regardless of what information is given—that is, people look for good traits in others and tend to like others. If this hypothesis is valid, then Asch would have observed a positive evaluation regardless of the order or type of adjectives used. But the experimental design used by Asch eliminated these possible criticisms. One of his treatment groups received a positive-to-negative order of adjective presentation; the second treatment group received a negative-to-positive presentation. The subjects' evaluation of the person was positive in the first treatment and negative in the second treatment; therefore, the previously mentioned criticisms and explanations are not valid. Note also that they are demonstrated to be invalid because Asch used two treatment groups instead of one and obtained negative ratings from subjects.

It is difficult to formulate specific principles about the control of extraneous variables. An experimenter usually begins with some problem to be solved or some hypothesis to be tested. In designing an experiment the experimenter must keep in mind that any extraneous variables that could explain the results have to be eliminated. Certainly this would involve keeping all conditions except the independent variable constant, but it also may include the use of additional treatment or control groups. The best way for the student to learn what extraneous variables are most common in any specific research area is to read experiments in that area. In this manner the student can become aware of the designs used in that area and the extraneous variables that have to be controlled. Some experiments involving control problems follow. The purpose of presenting these examples is to illustrate some problems that have arisen in the past with the hope that they will aid the student in analyzing designs for control problems in the future.

Control Problem: Sleep Learning

➤ *CASE STUDY*

An experiment was conducted to determine if learning could take place during sleep. The material to be learned was the English equivalents of German words, and the subjects were 10 college students who reported that they had no knowledge of the German language. Each subject slept in a comfortable bed in a soundproof, air-conditioned laboratory room. Each subject retired about midnight, and at approximately 1:30 A.M. the experimenter entered the room and asked the subject if he or she was asleep. If there was no response, the experimenter turned on a recording that contained German words and their English equivalents: for example, "*ohne* means without." There were 60 different words on the recording, which was played continuously until 4:30 A.M. If the subject awoke during the night, he or she was to call out to the experimenter, and the recording would be stopped until the subject was asleep again. To test for learning, the 60 German words were played to the subjects in the morning, and after each word the subjects reported what they thought was the English equivalent of the word. The number of German words correctly identified was the dependent measure. The results indicated that the mean number of words correctly identified was nine (out of a possible 60), and the highest number correctly identified by any subject was 20. The experimenters interpreted these results as supporting the hypothesis that learning can occur during sleep.

The experiment has several important implications. At a theoretical level it suggests that during sleep the brain is actively processing information received by the senses. At a practical level it suggests that sleep learning may be an easy and effortless way to learn. It should be a boon to college students who, instead of staying up all night to cram for an exam, could simply turn on a tape recorder and go to sleep.

While these results are quite exciting, and one would like them to be valid, two major criticisms have been leveled against the experiment. The first was the failure to use a control group who had not been presented the learning material but who were given the recall test. While it is true that all subjects said that they had no knowledge of the German language, they may have been able to guess the meanings of some of the words. For example, the German word *Mann* means *man* in English. Furthermore, some German words are frequently used in English, particularly in old war movies—for example, *Schwein (pig), nein (no),* and *ja (yes).* Thus, the apparent effects of sleep learning may actually be due to the subjects' ability to guess some words and knowledge of other words; a control group would have checked for this.

A second criticism dealt with the experimenter's operational definition of sleep. Sleep was defined as what the subject did between 1:30 A.M. and 4:30 A.M. unless the subject reported being awake. But it is known that there are various levels of sleep, ranging from drowsiness to very deep sleep. It further is known that at the drowsiness level the subject has partial awareness of external stimuli, but at the level at which sleep technically begins there is little or no awareness of external stimuli. In this experiment there was no way of knowing what material was presented at what level of sleep. Therefore, it could be argued that any learning that occurred may have taken place at a drowsiness level rather than at a true sleep level.

➤ *CASE STUDY*

Simon and Emmons (1956) designed an experiment to correct for the previously mentioned problems. The materials to be learned consisted of 96 general information questions and their answers. They were presented in question form—for example, "In what kind of store did Ulysses S. Grant work before the war?" Then the answer was given: "Before the war, Ulysses S. Grant worked in a hardware store." Two groups of subjects were used: an experimental group, which was given the answers to the questions while sleeping, and a control group, which simply took the learning test without having the answers played. To begin the experiment both groups were given the questions and asked to guess the answers. Those questions that the subjects answered correctly were eliminated from the test. Next, the experimental group was presented with the questions and answers while asleep. During this period, recordings were made of their brain waves using an **electroencephalograph** (EEG). Because brain activity varies at different stages of sleep, EEG records enable experimenters to determine accurately the depth of sleep. As each answer was presented to the subject, the experimenter recorded the subject's level of sleep. Thus, the experimenter had a record of the level of sleep at which each answer was given.

In the morning the experimental group was tested on the material that was presented during the night. The multiple-choice test consisted of the question and five answers. The subject was to guess which of the answers was correct. A multiple-choice test was used since it probably is a better measure of sleep learning because the subject only has to recognize the correct answer rather than recall it. The control group also took this test.

After the test scores were received, the experimenters separated the questions for each experimental subject into categories determined by the level of sleep at which the answer was played. The experimenters used eight levels of sleep, which have been condensed into two categories in Table 4.2 where the percentage of correct answers in these categories is shown.

Table 4.2 PERCENTAGE OF ANSWERS CORRECT AT THREE SLEEP LEVELS

	LEVEL OF SLEEP		
	AWAKE	**DROWSY**	**ASLEEP**
Experimental group	92	65	23
Control group	24	23	23

The data indicate that considerable learning took place when the experimental subjects were awake, and moderate learning appeared at a drowsy level. But there was no apparent learning when the subjects were at a true sleep level. At this level the performance of the experimental group was the same as that of the control group, who had no learning experience. The 23% correct for the control group represents what could be expected if the subjects guessed.

In looking at the design of this experiment, it is important to note that the experimenters instituted several crucial control procedures to allow for a rather clear-cut test of sleep learning. First, to make sure that what appeared to be information learned during sleep was not actually information that was previously known, the experimenters gave all subjects a pretreatment test on the material and eliminated the answers subjects already knew. Second, since a certain proportion of the answers on the multiple-choice test could be gotten by guessing, the experimenters used a control group to find out what that percentage would be. Third, the experimenters identified different levels of sleep and noted what answers were presented at each level. This technique made it possible to separate material presented when the subject was awake, in a state of drowsiness, and in a state of true sleep. Using these control procedures, the results suggest that no learning takes place at a true sleep level.

Control Problem: Social Deprivation and Social Reinforcement

> ➤ *CASE STUDY*

It has been repeatedly demonstrated that for animals who have been deprived of food, the effectiveness of a food pellet as a reinforcer is considerably enhanced. An experiment was conducted to determine if the same results would occur with social deprivation and **social reinforcement** in children. Subjects were 6-year-olds in an elementary school.

The effectiveness of social reinforcement was measured by a marble game. The game consisted of a box with two holes in it in which the subject was to drop marbles one at a time into one of the two holes. For the first 4 minutes of the game the experimenter watched the subject play the game. For the next 10 minutes the experimenter verbally reinforced the subject every time he or she put a marble in the hole that was used least in the initial 4-minute period. The verbal reinforcement consisted of the experimenter saying "good" or "fine" every time the subject put a marble in that hole. The dependent variable was the amount of increase in placing marbles in the desired hole from the 4-minute to the 10-minute period.

To determine the effects of social deprivation the subjects were assigned randomly to one of three treatments. In the social deprivation treatment, the subjects were left alone in a room for 20 minutes prior to playing the game. In the nondeprivation treatment, the subjects started playing the game immediately. In the social satiation treatment, the subjects spent 20 minutes before the game talking with the experimenter while they were drawing and cutting out pictures.

The results indicated that the subjects in the deprivation treatment showed a greater increase in putting marbles in the desired hole than did either of the other two treatments. Furthermore, the increase was greater in the nondeprivation treatment than in the satiation treatment. The results were interpreted as supporting the hypothesis that the effectiveness of social reinforcement is influenced by conditions of social satiation or deprivation in a manner similar to that found for food or water deprivation.

The previous experiment has wide theoretical implications in that it suggests that social drives seem to be subject to the same laws that have been established using the primary appetitive drives, such as hunger. The experiment's authors also have developed a nice experimental situation to test their hypothesis in that the experimental manipulations seem fairly clear-cut and the dependent measure is easy to record without ambiguity. However, soon after the experiment was published, criticisms of the experiment began to appear, arguing that the results may have been caused by the failure to control extraneous variables.

One criticism of the experiment pointed out that when the experimenters were manipulating social deprivation, they also manipulated general sensory deprivation. For example, in the deprivation treatment the child was not only isolated from other people but also had no toys with which to play. In the satiation treatment the child not only interacted with the experimenter but also drew and cut out pictures. Thus, the experimenters manipulated an extraneous variable (general sensory deprivation) along with the social deprivation, and the results of the experiment might have been caused by this extraneous variable. Stevenson and Odom (1962) tested this alternative explanation of the results by comparing three groups of subjects. Before playing the marble game, one group of children was isolated and played with attractive toys for 15 minutes, one group was isolated for 15 minutes without toys, and a third group began playing the marble game immediately. The results indicated no difference in the task performance of the two isolation groups, but both groups had higher levels of performance (i.e., dropped more marbles in the desired hole) than the no-isolation group. Since there was no difference in the task performance of the two isolation groups (both of which were socially deprived, but only one of which was deprived of toys), then the higher performance must have been due to the social deprivation. This supports the original interpretation of the experiment, and the influence of the extraneous variable was apparently very minor.

A second group of researchers raised a different criticism of the experiment. The crucial point of this criticism was that being placed in a strange environment by a strange adult should arouse anxiety in 6-year-olds. The greatest anxiety should occur in the deprivation treatment, in which the subjects were left alone in a strange room for 20 minutes. The next highest level of anxiety should occur in the nondeprivation situation, in which the subjects were led directly to the game situation. The least anxiety should occur in the satiation treatment because after 20 minutes of friendly conversation with the experimenter the subject should be somewhat comfortable in the experimenter's presence. Since there is evidence demonstrating that heightened anxiety improves performance on some learning tasks (especially simple learning tasks), the results of the experiment could be explained by the difference in anxiety arousal in the three treatments, leaving no need to postulate some social drive.

To test this hypothesis, Walter and Parke (1964) used a 2 × 2 factorial design in which they used two levels of isolation (either leaving the child alone for 10 minutes or starting the game immediately) and two levels of anxiety arousal. In the low-anxiety condition the experimenter treated the subject in a pleasant and friendly manner, and in the high-anxiety treatment the experimenter treated the subject in a rather cold and abrupt manner. Using these treatments, (1) no statistically significant difference was found in performance between the two levels of isolation, which is evidence against the social drive interpretation; (2) subjects in the high-anxiety treatment performed better than subjects in the low-anxiety treatment, which supports the anxiety arousal interpretation; and (3) the interaction effect was not statistically significant. Thus, it appears that an extraneous variable (arousal level) may have determined the results of the original experiment, and the social deprivation interpretation may be invalid.

Control Problem: Perceptual Defense

The following experiment suggests that there is a process in the unconscious that determines whether a word is anxiety-provoking and that conscious recognition of such words is prevented or at least delayed. This is the notion of **perceptual defense,** and it has far-reaching implications concerning human behavior.

➤ *CASE STUDY*

It has long been suggested that the human body has certain mechanisms that protect it from anxiety-provoking stimuli. One such mechanism is called perceptual defense. An experiment was designed to test the perceptual defense hypothesis by presenting to subjects neutral and taboo words on a tachistoscope. The experimenter theorized that taboo words are anxiety-provoking, and while the subject may recognize them at an unconscious level, a perceptual defense mechanism would delay the subject's recognition of them at a conscious level. Based on this assumption, it was hypothesized that longer exposures would be necessary for the recognition of taboo words than for neutral words.

The subjects were eight male and eight female college students. Each subject was tested individually with both a male and a female experimenter present. Eleven neutral words (e.g., apple, trade) and seven taboo words (e.g., whore, bitch) were presented to each subject in a predetermined order. An ascending threshold method was used to determine the point at which the subject recognized the word. For each word, the shutter was set at a very fast exposure speed (0.01 seconds), and the exposure was gradually lengthened until the subject verbalized the word correctly. This process was repeated for each of the 18 words.

The mean threshold for the recognition of the neutral words was 0.053 seconds, and the mean threshold for the recognition of the taboo words was 0.098 seconds. The difference between the two means was statistically significant. Since it took longer exposures (higher thresholds) for the subjects to recognize the taboo words, the experimenter concluded that the perceptual defense hypothesis was supported.

It did not take long for other researchers to criticize the design. Howes and Solomon (1950) raised two methodological points. First, they suggested that the results may have been due to the subjects' reluctance to report a taboo word until they were absolutely positive of the identification of the word. The subjects might have been particularly reluctant to verbalize these words in front of an experimenter of the opposite sex. A second methodological point was that neutral words appear much more frequently in print than do taboo words, and therefore the subjects' quicker recognition of the neutral words may have occurred because they had seen these words more frequently. In a follow-up experiment Howes and Solomon (1951) demonstrated the validity of a word frequency hypothesis. They obtained a listing of the frequency at which some 30,000 words appear in print. They chose 60 words of varying frequencies (all nontaboo words) and determined the recognition thresholds of each word using a procedure similar to that used in the previous example. They found a high negative correlation (approximately -0.79) between the frequency at which the word appeared in print and its recognition threshold, that is, the more frequently the word appeared in print, the lower its recognition threshold. While this experiment demonstrates that word frequency is a plausible explanation for the results of the original experiment, it still can be argued that taboo words show a higher threshold even if frequency is controlled.

Postman, Bronson, and Gropper (1952) made a more direct test of the word frequency explanation by determining how frequently some taboo words appeared in print and matching them with neutral words that appeared in print at the same frequency. The list of words was presented to the subjects using a procedure similar to that of the previous example. The results indicated no support for the perceptual defense hypothesis. In fact, it was found that the recognition threshold for the taboo words was significantly lower than for the neutral words. This was probably due to an underestimation of the frequency of the taboo words. Although research in this area continues, it appears that the early experimental support for the perceptual defense phenomenon may have been due to the confounding variable of word frequency.

Control Problem: One-Trial Learning

A special type of control problem is found in cases where learning is based on a single trial (**one-trial learning**). This problem also can be found in similar cases. The following example illustrates the problem.

➤ *CASE STUDY*

When children are learning to read the alphabet, they are shown the letters A, B, C, and so on, while the teacher pronounces the letters. This is repeated until the child learns the association between the printed letter and the verbalized sound. There is a controversy among learning theorists as to whether this association is built up gradually (incremental process) or whether it occurs in an all-or-none fashion. The latter school of thought would say that if, after some learning trials, the child cannot verbalize a letter's name after being shown that letter, then no association between the letter and its name has taken place. The former school of thought would argue that some association has taken place, but that it is not yet of sufficient strength to allow the child to give the correct answer.

A rather ingenious experiment was performed to determine which of the previous theories was correct. The subjects' task was to learn eight nonsense syllable pairs. Each pair was presented to the subjects on a separate card. After a subject had seen all eight pairs, he or she then was shown the first nonsense syllable of each of the pairs and asked to give the second nonsense syllable. For example, the subject might have been shown POZ-LER. In the recall test the subject would be shown only the first syllable, POZ, and would have to supply the second syllable.

Two experimental groups were used. For the first experimental group, the experimenter replaced in each trial every nonsense syllable pair that had not been learned. For example, the subject was shown eight pairs and then given a learning test on these eight. Any pair that was not recalled was dropped, and a new pair was substituted into the list of eight. The eight cards (which then included only learned pairs and new pairs) were shown again to the subject, and another recall test was given. The experimenter again eliminated those pairs that were not recalled and substituted new pairs. This process continued until the subject could recall all eight pairs in a single trial. The second experimental group was treated like the first except that these subjects were shown the same eight cards on each trial. Thus, eight pairs were shown, a recall test was given, the same eight pairs were shown, another recall test was given, and so on. This process was repeated until the subject could recall all eight pairs in a single trial.

The experimenter then compared the number of trials it took to learn all eight pairs in each treatment group. The mean number of trials for perfect recall for both groups was exactly the same (8.1). Since all pairs learned in the first experimental group were learned in one trial (or else they were thrown out), and since there was no difference in the average number of trials necessary to learn the eight pairs, it was concluded that learning (i.e., associations) occurred in an all-or-none fashion. Stated a little differently, the experimenter argued that a gradual buildup of associations through repetition could have occurred in the second experimental group. If repetition is important in learning, then this second group should have learned the eight pairs faster than the first group, who learned eight pairs on a single trial without any repetition. Since there was no difference between the two groups, it was concluded that repetition is unnecessary for learning.

As might be expected, this experiment caused some excitement, particularly among those who support an incremental view of associative learning (since the experiment suggests that this view of learning is invalid). It was not long after the experiment was published that research criticizing it began to appear. One major criticism was that the more difficult pairs were probably dropped from the first treatment group, and thus the final list in this treatment consisted only of the easier pairs. No pairs were dropped from the second treatment group, and thus the final list learned by these subjects consisted of both easy and difficult pairs. The failure of the experiment to find quicker learning in the second treatment was due, therefore, to the fact that the list learned by this group was more difficult than the final list learned by the first treatment group.

In one test of the item-selection hypothesis, Underwood, Rehula, and Keppel (1962) repeated the previous procedure except that they added a control group that received lists composed of the pairs that subjects in the dropout condition received on their last learning trial. Subjects in this condition were given the same word pairs on each trial. The results indicated that subjects in this condition learned the pairs more rapidly than did subjects given a list containing a random sample of all of the pairs used. Thus, it appears that the subjects in the dropout condition of the original experiment were learning easier pairs, and this may explain the results.

DEFINITIONS

The following terms and concepts were used in this chapter. We have found it instructive for students to go through the material a second time to define each of the following:

balanced sequence	matched subjects design
between-groups design	one-trial learning
confounding variable	paradigm
contrasts	perceptual defense
correlated t-test	placebo effect
dependent t-test	random
electroencephalograph	randomized subjects design
extraneous variable	repeated measure design
holding conditions constant	social reinforcement
inferences	spontaneous remission
independent subject design	subject variable–subject variable
matched pair design	confound

EXERCISE

For each of the control problems described in this chapter, consider the experiment, the initial problem, and the corrected problem, and (1) identify the independent variable, (2) identify the dependent variable, and (3) explain how the design corrected for the initial control problem.

Chapter 5

Design Critiques I

Learn by doing.

Psychological research usually starts with a search of the literature to become familiar with previous studies. Then the researcher may find an issue that needs further investigation, which he or she can then formalize in terms of a hypothesis. The next stage may involve designing a valid, practical experiment. After all materials and subjects are lined up, the experimenter can collect and analyze data, and finally, conclusions can be made in the form of a discussion. As simple as the process may appear, there are pitfalls lurking at every stage.

In this section and in other sections to follow we will consider a special type of trap into which many seasoned as well as neophyte researchers fall. It is a trap basically caused by imperfect logic that has been manifest in either faulty design or faulty interpretation. Some students of experimental design identify these problems as requiring alternative explanations, as the results may be attributed to causes other than the one identified (see cartoon on p. 93). Experience leads us to believe that the more practice students have on these problems, the better they will be in discovering design and/or interpretative problems in their own research. In addition to reading and discovering the flaws in the problems presented here, it is useful for students to make up problems on their own.

On the following pages are a series of experiment briefs, each of which has one or more design problems. The problem occurs between what the experimenter did in the experiment and the conclusion that he or she arrived at on the basis of the results. We previously illustrated the basic principles of experimental design. Now our purpose is to show a series of fictitious studies and let students apply their

Suggest an Alternative Explanation

knowledge of experimental design in a critique of the studies. The briefs are a quick way to expose a variety of problems in a variety of research areas. No expertise or technical knowledge is needed in the research area being explored in the study; the problems can be recognized with a knowledge of the design principles previously discussed in the text.

In criticizing the design of the experiment briefs, students should only make use of the information given—inferences are not necessary. For example, if the experimenter uses a pencil-and-paper anxiety test, students should assume that the test is valid and reliable unless information to the contrary is given. There is at least one major defect in each brief, and students should concentrate their criticisms on this major problem. Students also should be specific as to the defect. For example, students should not just say that the experimenter should have used a control group but should point out exactly how this control group would be treated.

The following example illustrates how the briefs should be criticized:

➤ *EXAMPLE*

A certain investigator hypothesized that the hippocampus (a part of the brain) is related to complex thinking processes but not to simple thinking processes. He removed the hippocampus from a random sample of 20 rats. He had 10 of these rats learn a very simple maze and had the other 10 learn a very difficult and complex maze. The first group learned to run the maze without error within 10 tries (or trials). It took the second group at least 30 trials to run the maze without error. Based on these results, he concluded that his hypothesis had been confirmed—rats without a hippocampus have more trouble learning a complex task than they do a simple task.

In general, this experiment conforms to Model 1: the independent subject design. In that model subjects are randomly assigned to only one condition.

In criticizing this design, it appears reasonable to assume that any rat would take more trials to learn a complex maze than it would to learn a simple maze. Thus, the experimental results may have nothing to do with the removal of the hippocampus—rats with the hippocampus might show the same results. In other words, although two independent variables were intended (task difficulty and the presence or absence of the hippocampus), only one independent variable was varied. This criticism would suggest that in redesigning the experiment a 2×2 factorial design should be used: One factor should be the presence or absence of the hippocampus, and the second factor should be the complexity of the maze. The design is diagrammed here:

	SIMPLE MAZE	COMPLEX MAZE
Hippocampus intact	5 rats	5 rats
Hippocampus removed	5 rats	5 rats

This new design conforms to Model 5: factorial designs in which the effect of two independent variables is evaluated. This revised design would allow for a more reasonable test of the experimenter's hypothesis than did the original design. Although the new design might become more complex than the original design, it corrects the original defect. Before beginning the rest of the briefs, students may want to do the following exercise, which attempts to formulate questions that can be used to critique experiments.

EXERCISE

Formulate a series of questions that can be used to critique experiments. Each question should test the adequacy of one aspect of the experimental design. For example:

1. What is the independent variable? Are there (1) at least two levels of it or (2) an experimental group and a control group? If not, there is a design defect. In any experiment, one treatment has to be compared with another.
2. Is the independent variable a subject variable or a manipulated variable?

3. What is the dependent variable? How is it measured?
4. Assuming there are two levels of the independent variable, are all groups treated identically except for the experimental manipulation? If not, there is a confound in the experiment.
5. What type of design was used?

What other questions should be asked?

EXPERIMENT BRIEFS

1. An investigator attempted to ascertain the effects of hunger on aggression in cats. She took 10 cats, kept them in individual cages, and put them on a food deprivation schedule such that at the end of two weeks the cats weighed 80% of their normal body weight. She then put the cats in pairs for 15 minutes and watched to see if aggression or fighting would occur. In all cases, the cats showed the threat posture, and in most cases fighting occurred. The investigator concluded that hunger increases aggression in cats.

2. Psychologists working for food and beverage companies have always played a critical role in product development. In one experiment conducted by a consumer psychologist, subject preference for two types of cola was measured. The company had noticed that in one marketplace its brand of cola performed significantly worse than its leading competitor's cola. These data were particularly puzzling as on a nationwide basis its cola performed significantly better than its competitor's cola.

   The researchers were concerned that some local condition may have contributed to the rejection of their cola, so they set out to test this hypothesis. The experimental design was a repeated measure design in which each subject tasted two colas. One cola was marked Q (the competitor's brand) and the other was marked M (their brand). A random sample of citizens between the ages of 14 and 62 were asked to participate in the experiment. All subjects tasted brand Q and then brand M, and then gave their preference. Much to the surprise of the experimenters, the subjects reported an overwhelming preference for brand M. The authors concluded that the sample preferred their company's brand and that advertising must have contributed to the consumption of the competitor's brand in that area. Therefore, they suggested a multimillion-dollar advertising campaign to rectify the situation.

3. An experimenter wished to examine the effects of massed and distributed practice on the learning of nonsense syllables. He used three treatment groups and randomly assigned the subjects to one of the conditions. Group I practiced a list of 20 nonsense syllables for 30 minutes one day. Group II practiced the same list for 30 minutes per day for two successive days. Group III practiced the same list for 30 minutes per day for three succes-

sive days. Then the experimenter assessed each group's learning with a free-recall test. The mean recall of the 20 syllables for Group I was 5.2; for Group II, 10.0; and for Group III, 14.6. The means were significantly different from one another at the 0.01 level of significance, and the experimenter concluded that distributed practice is superior to massed practice.

4. A certain psychologist was looking for the cause of failure among college students. She took a group of former students who had flunked out and a group of students who had received good grades. She gave both groups a self-esteem test and found that the group that flunked out scored lower on the test than did the group that received good grades. She concluded that low self-esteem is one of the causes of college failure and suggested further that a person with low self-esteem probably expects to fail and exhibits defeatist behavior in college, which eventually leads to failure.

5. A psychologist designed a study to determine if people with high blood pressure could learn to control their blood pressure using biofeedback techniques. A device that records blood pressure was attached to each patient. Then the patient was given feedback as to his or her blood pressure level by a tone that decreased in loudness as the blood pressure decreased and increased in loudness as blood pressure increased. The patient was told to try to keep the tone as quiet as possible. Five patients with high blood pressure each received 10 half-hour sessions using biofeedback. All five lowered their blood pressure level considerably over the 10 sessions, and the researcher claimed success for this method.

6. An experiment was designed to test a hypothesis that stated that high-drive subjects would be able to learn a simple task much more quickly than would low-drive subjects. The hypothesis further stated that on a difficult task the opposite result would be found—low-drive subjects would learn the task more quickly. The experimenter's operational definition of *drive* was each subject's score on the Manifest Drive Scale. Twenty people who scored high on the scale (high-drive) and 20 people who scored low on the scale (low-drive) were given a difficult task to learn. The low-drive group learned the task more quickly than did the high-drive group, and the experimenter concluded that the hypothesis was correct.

7. An investigator set out to test the hypothesis that fear of punishment for poor performance has a detrimental rather than a facilitative effect on motor performance. As a measure of performance, the experimenter used a steadiness test in which a subject's task was to insert a stylus into a hole so that the stylus did not touch the sides of the hole. Each subject inserted the stylus into 15 different holes. The experimenter manipulated fear by threatening the subjects with electric shock if they performed poorly on the task. The experimenter strapped an electric shock apparatus to the leg of each subject before the subject performed the task but never shocked the subjects regardless of their performance. Subjects were randomly assigned to one of two conditions: One group was threatened with 50 volts of electricity—a mild-fear condition, and a second group was threatened with 100

volts of electricity—a high-fear condition. Contrary to the hypothesis, the high-fear subjects did not perform worse than the low-fear subjects—in fact, the means for both groups were approximately the same. Based on these results, the experimenter concluded that fear of punishment has little, if any, effect on motor performance.

8. An experimenter took 20 subjects who said they believed in astrology, gave them their horoscopes for the previous day, and asked them how accurate the horoscopes had been in predicting the previous day's occurrences. The subjects indicated their opinion on a 6-point scale that ranged from extremely accurate to extremely inaccurate. All 20 subjects reported their horoscopes as being accurate to some degree. The experimenter concluded that horoscopes are accurate.

9. A 2 × 3 factorial design was used to evaluate the effect of an experimental drug (Remoh) on the treatment of schizophrenia. Two patient classifications were used: (1) new admissions to a particular mental hospital and (2) patients who had been institutionalized for at least two years at that hospital. Both groups had not been hospitalized previously. Patients received one of three levels of Remoh—3 grams per day, 6 grams per day, or 9 grams per day. Subjects from each patient group were randomly assigned to one of the three dosage levels. There were 20 patients in each of the six groups. In addition to administering the drug, the experimenters also rated each patient each week on the presence or absence of schizophrenic symptoms. After two months, it was found that very few (10%) of the long-term patients in each group had improved, regardless of dosage level. But approximately 50% of the new patients had improved in each of the three dosage level groups. The researchers concluded that (1) Remoh is effective only for new arrivals and not for chronic cases and (2) a dosage of 3 grams per day is sufficient to maximize the effectiveness of the drug.

10. An interesting controversy in clinical psychology is whether psychosis is inherited or is caused by environmental experiences. One environmentalist hypothesized that children who live with psychotic parents would be prone to having the same problem later in life. To test this hypothesis, he took 1,000 psychotic adults and 1,000 normal adults and for each subject looked to see if either or both parents had been psychotic. Less than 1% of the normal adults had had psychotic parents, but over 30% of the psychotic adults' parents were psychotic. Based on this result, the experimenter concluded that psychosis is not inherited but that childhood experiences with a psychotic parent make a person especially prone to the disorder.

11. A group of investigators suspected that rats trained to run on a wheel against a drag would run significantly faster if fed a 20% sucrose solution along with their daily rations. The control was a group of rats who received only the daily rations. One hundred Mayflower rats arrived from Plymouth Rock Animal Breeders (know in the rat-running business as "designer rats") and were divided randomly into two groups. Fifty rats on

the normal rations ran on the wheel, followed by the 50 rats being fed the sucrose supplement (the order was determined by flipping a coin). The second group ran faster than the first, thus affirming the hypothesis that rats are energized when under a sugar high. The researchers generalized the results to grade school children who are fed a diet of junk food high in unprocessed sugar.

12. Recently, an association that represented police in New Jersey complained that the incidence of cancer was unusually high among police officers who used radar guns in tracking the speed of drivers. The association brought suit against the state, claiming that those officers who used these radar guns had an incidence of cancer 18% higher than a comparable group of government workers from the state park service who were randomly selected for comparison purposes. The comparison sample and the police officers were matched on educational level, age, gender, and years of service. The sample was large: 283 police officers and 231 park workers. Was the case justified?

13. A recent newspaper article announced, "Candy Cigarettes Influence Kids," and went on to report that children who buy candy cigarettes are much more likely to smoke later on than are children who don't buy them. The survey indicated that 7th-grade students who had purchased candy cigarettes at least twice were far more likely to have tried real cigarettes than were students who had not bought candy cigarettes. The results were so clear that school officials wanted to ban the sale of these candy cigarettes because, they contended, the candy cigarettes cause smoking in young children. The survey also indicated that in families where at least one parent smoked the children were much more likely to buy candy cigarettes and to have tried smoking on several occasions. Was the reaction by the school officials justified?

Chapter 6

Control of Subject Variables

The basic problem in the investigation of subject variables . . . is that whatever differences are observed in behavior may be caused by their confounded variables.
—Kontowitz, Roedigger, and Elmes (1988)

EQUALITY OF SUBJECTS IN TREATMENT GROUPS

Psychological experiments focus on the behavior of some specie of animal. Since psychological research uses subjects (human or otherwise), it is not surprising that psychologists have devoted considerable attention to controlling extraneous variables that are due to subjects' characteristics. In fact, a set of specific techniques has evolved that is applicable to a wide variety of research situations. This chapter will discuss the essential features of these techniques.

In research, the performance of one treatment group is often compared with that of another group. These groups consist of subjects who differ on a variety of traits that could influence the results. It is important that all treatment groups be approximately equal as to these various traits, so that the experimental results are attributable to the independent variable and not to the fact that subjects in one treatment were different in some trait (e.g., IQ) from subjects in another treatment.

Field studies provide the greatest possibility for results caused by subject differences rather than treatment differences. In any study in which subjects are studied in their natural groups or have volunteered specifically for one treatment or another, an important question is: How do the subjects in the various treatments differ? For example, consider a large manufacturing company that held leadership training courses for lower-level employees. These courses were run on a volunteer basis and took place at night on the employee's own time. In evaluating the effectiveness of this course 10 years later, it was found that those people who had taken the course had advanced further in the company than had those who had not taken the course. This was interpreted as supporting the effectiveness of the course, but an alternative explanation could be that the course may have attracted only those people who were highly motivated to advance in the company. Thus, the treatment group may have consisted of highly motivated people, and the control group may have consisted of unmotivated people. The course may have had little effect on advancement in the company; the results may have been due to differences in motivation in the two comparison groups.

This kind of problem can be avoided if the researcher assigns subjects to various treatment groups in a manner that ensures that the subjects are approximately equal in all relevant characteristics. There are three general techniques for accomplishing this. The first technique is the randomized subject design discussed in Chapter 4 where subjects are randomly assigned to the separate treatments. The random assignment allows the experimenter to be fairly certain that the subjects in all treatments are approximately equal as to the subject variables. This design incorporates the features of Model 1 introduced earlier. The second technique is the matched subject design (Model 2). In this design the experimenter scores each subject on some task or test and then assigns subjects to the various treatment groups so that the groups are equivalent with respect to these scores. The third technique is the within-subject design (Model 3). In this design each subject participates in all experimental treatments, which ensures that the treatment groups are equal as to the subject variables. A related technique is the repeated measure design with a balanced sequence (Model 4). This design also uses the same subjects in all treatments, but it also compensates for sequencing effects. In the first two techniques, randomized subject design and matched subject design, different subjects are assigned to the different experimental treatments. These designs are called **independent groups designs.** In the third technique, the within-subject design, the same subjects participate in all experimental treatments.

RANDOMIZED SUBJECT DESIGN: MODEL 1

The most common method of assigning subjects to treatments is random assignment. In its most rigid form this would mean that each subject has an equal opportunity (or probability) of being assigned to each treatment group. If an experiment consists of two treatment groups, a table of random numbers could be used to assign subjects to treatment conditions, or an experimenter could flip a coin for each subject. By using randomization procedures such as these, the experimenter could be fairly certain that, as a group, subjects in treatment A and subjects in treatment B would be approximately equal.

While this procedure is simple to administer and is consistent with the definition of random assignment, it also presents a serious problem. It is quite possible that the experimenter would end up with an unequal number of subjects in the two treatments. For example, it would be possible to end up with 15 subjects in treatment A and only five subjects in treatment B. This would be undesirable because the treatment mean based on the five in treatment B would probably be less stable than a mean based on more subjects. A second consideration is that some of the statistical analyses of results are simplified if an equal number (n) of subjects is used in each treatment group. Ideally, a procedure is needed that allows for some random assignment but also ensures an equal n in each of the treatment groups. Such procedures are commonly used, and although they do not conform with a rigid definition of random assignment, they are usually called random assignment or, perhaps more accurately, unbiased assignment.

In the Linder, Cooper, and Jones (1967) experiment (see p. 45), in which the subjects were given 50 cents or $2.50 for writing an essay, each subject was assigned to a treatment when he or she arrived at the lab booth. Experiments of this type usually take place over periods of weeks or months, and an unbiased selection procedure must take into account the possibility that subjects who report earlier for the experiment may have characteristics different from those of people who report later. For example, subjects who report later may have heard something about the experiment from subjects who reported earlier, or perhaps the more motivated subjects appear earlier. One procedure that takes this into account is **block randomization.** With this procedure an experiment is run in a series of blocks so that all treatments are represented within each block, but their order within each block is somewhat random. For example, in a two-treatment experiment the experimenter might flip a coin for the first subject who reports to the lab booth. If the coin lands on heads the subject is assigned to treatment A, and if the coin lands on tails the subject is assigned to treatment B. The next subject who reports to the lab booth is assigned to the other treatment. Thus, if heads comes up for the first subject, he or she is assigned to treatment A, and the second subject is assigned to treatment B. This is the first block. Then the coin is flipped for the third subject, and the fourth subject is assigned to the other treatment. This process is repeated until all the subjects have participated in the experiment. With this procedure the two treatments are equally distributed over the time period in which the experiment is conducted, and within each block there is unbiased assignment.

If more than two treatments are used in an experiment, block randomization can be used by putting slips of paper representing each treatment into a container and drawing one out as each subject reports for the experiment. For example, if six treatments are used, six slips of paper, each with a letter representing one of the six treatments, could be put into a container. As each subject reported for the experiment one slip of paper would be drawn out of the container, and the subject would be assigned to the treatment represented by the letter on that slip. The slip would not go back into the container, so the second subject would draw from the five remaining slips. This procedure would continue until all six slips had been drawn. At that point the slips would be put back into the container, and the procedure would begin all over again.

These procedures are especially applicable to experiments in which subjects report to the experiment over an extended period of time. But in another type of research, the subjects are available to the experimenter at the same time. For example, in the Paul (1966) experiment comparing the effectiveness of insight and behavior therapy in treating speech phobia, the experimenter had to assign 67 students who exhibited speech phobia to the four experimental conditions (see p. 24). And in the experiment on employment opportunity training discussed later in this chapter, there were 60 applicants for the course offered by the center, and the director somehow had to assign half of these applicants to the treatment group and half to the control group. There are several ways of accomplishing an unbiased assignment for this type of problem, but only a few of these will be discussed.

Suppose an experimenter has the names of 60 subjects and wants to assign 15 to each of four treatment groups. One procedure would be to use the slips of paper method. In this case 15 slips marked A would be placed in the container with 15 slips marked B, and so on. The names of the subjects would be listed on a sheet in alphabetical order, so as the experimenter went down the list, he or she would draw a slip of paper (without replacing it) for each name and assign that person to a treatment based on the letter on the slip. This procedure might be simplified by putting the names in random order (e.g., by putting each of the 60 names on a 3" × 5" index card and shuffling the cards) and then assigning the first name drawn to treatment A, the second name to treatment B, and so on.

In some experiments the experimenter might not have the subjects' names before the experiment begins, even though all subjects were reporting to the experiment at the same time. Suppose there are 60 subjects in a large room who must be assigned to four treatments. Probably the simplest way to assign groups would be to start at the front of the room and have the subjects count off by fours. All number ones go into one treatment group, all number twos go into a second group, and so on. This would seem to be an unbiased selection, provided that the subjects are not seated in some systematic manner. Frequently, subjects meet in a large room, and the various treatments are represented by the different test booklets passed out to the subjects. For example, Johnson and Scileppi (1969) wanted to study how someone's attitude would change when presented with plausible and implausible communications. These communications were in the form of written messages that appeared in test booklets along with attitude scales that were to be filled out by the subjects. Groups of 10 to 20 subjects were tested at the same time in classrooms. The procedure used in this experiment was to shuffle the test booklets before passing them out. In this way the experimenters were randomly assigning treatments to subjects rather than vice versa.

Similar procedures allow even rather complex assignment problems to be handled fairly easily. For example, a group problem-solving experiment was conducted in which subjects met in groups of three to solve a particular human relations problem. The independent variable was the type of instructions given to the group.

We shall call the two groups treatment A and treatment B. Thus, the problem was not only to randomly assign each three-person group to a treatments (either A or B) but also to randomly assign subjects into the three-person groups. To further complicate the assignment problem, one member of each three-person group was to be randomly assigned to be the leader of the group. Sixty subjects reported to a large classroom, and somehow the experimenter had to end up with 10 three-person groups in each of two treatments with a randomly assigned leader in each group. This assignment problem was easily solved by giving each subject a number from 1 to 60 upon entering the room. The experimenter had a deck of sixty 3" × 5" index cards, each with a number from 1 to 60 written on it. The experimenter then shuffled the cards and drew three from the top of the deck. The subject represented by the first number drawn was designated as the leader of the group, and the subjects represented by the other two numbers were designated as members of the group. The experimenter kept drawing blocks of three cards until all 20 groups were formed. The groups formed by the first 10 blocks of three cards were placed in treatment A, and the groups formed by the second 10 blocks of three cards were put in treatment B. Thus, a rather complicated assignment process was handled very quickly and in a reasonably unbiased manner.

Matched Subjects Design: Model 2

Another method of ensuring that all treatment groups are equal in terms of subject characteristics is called matching. With this method all subjects are measured on some test or task that is assumed to be highly related to the task used in the actual experiment. The subjects then are assigned to the various treatment groups on the basis of this pretest measure, so that the groups will be approximately equalized with respect to pretest scores. The experimenter can then be assured that all the treatment groups will be equal with respect to one subject characteristic that is believed to be highly related to performance on the actual experimental task.

Before discussing matching techniques note the assumptions made when using this procedure. First, researchers assume that they know what subject characteristic is highly related to performance on the experimental task. Second, they assume that they can get scores for each subject on this characteristic. There is always a danger that the first assumption is invalid and that the second condition cannot be met either because there is no good measure for the characteristic or because an experimenter cannot get the subjects' scores on that measure.

Pretest tasks or tests can usually be divided into two major types. The first are tasks or tests that are quite different from the experimental task but are assumed to be highly related to the task. For example, it might be assumed that intelligence is highly related to a learning task, so the experimenter might want to match subjects on intelligence. In this case the pretest could be an IQ test, which is quite different from the actual experimental task. The second type of pretest task or test is quite similar (or identical) to the experimental task.

> ➤ CASE STUDY

Lambert and Solomon (1952) trained 20 rats to run down a runway at the end of which was a goal box containing food. After 30 acquisition trials the 20 rats were divided into two groups. One group of 10 rats was blocked from going down the runway at a point near the goal box (treatment A), and the other group of 10 rats was blocked near the beginning of the runway (treatment B). The dependent variable was the number of trials it took each group of rats to extinguish the running re-sponse, which was considered extinguished when a rat did not leave the start box within a 3-minute period. During the last four acquisition trials, the researchers had recorded how long it took each rat to leave the start box. On the basis of these time scores the researchers assigned (matched) the 20 rats to the two treatment groups so that, as a group, the two groups were approximately equal in time taken to leave the start box. In this example the matching was done on the same task as was used in the experimental session.

Once experiments choose the matching variable and test the subjects, there are several ways they can then match them. One technique involves matched pairs. With this technique the experimenter takes two subjects with identical scores on the matching variable and assigns one to treatment A and the other to treatment B. Suppose a two-treatment learning experiment is to be conducted, and it is assumed that IQ is highly related to performance on the learning task. First, IQ scores would be obtained for a large number of subjects. The experimenter then would pick out two people with an IQ of 135 and put one in treatment A and one in treatment B. Then he or she would pick out two subjects with an IQ of 130 and put one in A and one in B. This process would be repeated until the experimenter had the desired number of subjects in each treatment group. Another example would be a drug therapy experiment on schizophrenics in which an experimenter might want to match on age. He or she could take two 50-year-old schizophrenics and put one in the drug therapy group and one in the placebo group. Next, he or she would take two 46-year-olds, and so on. In both examples there is a perfect matching of subjects in the two treatments. The same procedure could be used if there were more than two treatments; for example, the experimenter could pick three subjects with an IQ of 135 and assign one to each of the three treatment groups.

A difficulty with this technique is that it may be impossible to find subjects with matching scores. In the example it may be that only one subject with an IQ of 135 may be found. In such cases the researcher may choose to match on the basis of similar scores (e.g., a range of 130–140). But even this latter technique may prove to be difficult.

More frequently than not, precise matching may not be possible. Suppose that in a rat experiment similar to that of the Lambert and Solomon (1952) experiment the experimenter had eight rats and wanted to use a matched pair technique to assign the rats to two treatments (A and B) based on the length of time it took them to leave the start box on the last acquisition trial. The times of the eight rats (in seconds) were 20.5, 17.2, 10.7, 8.0, 7.2, 6.5, 4.3, and 3.2. It is obvious that precise matching is impossible since no two scores are alike, so the experimenter is forced to use an **ad lib matching** procedure in which he or she attempts to balance the scores. The following grouping of scores seems to be the best matching possible for the eight rats:

TREATMENT A	TREATMENT B
20.5	17.2
8.0	10.7
7.2	6.5
3.2	4.3
$\bar{X} = 9.72$	$\bar{X} = 9.67$

The means of both groups are approximately the same, and both groups contain high-, moderate-, and low-scoring rats. Still, this "solution" is not without its problems because data generated from such a technique are difficult to analyze with appropriate statistical techniques.

Another matching technique that may be used is a **random blocks technique.** Suppose that 80 schizophrenics were to be assigned to four treatment groups in a drug therapy experiment. The experimenter also wants to match them as to age, since he or she has some evidence that the older the patient, the less favorable will be the response to therapy. Using a random blocks technique the experimenter could rank the ages of the 80 patients and then take the four oldest patients and randomly assign one to each of the four treatments. Then the experimenter could take the four next-oldest patients and randomly assign each of them to one of the four treatments, and so on until all 80 subjects have been assigned. The experimenter is taking blocks of patients of approximately equal age and randomly assigning the patients within each block to a particular treatment.

Matching techniques can be very powerful in eliminating bias due to subject characteristics if (1) the experimenter knows what subject variables are highly related to the experimental task and (2) if he or she can get scores on these variables.

WITHIN-SUBJECT DESIGN: MODEL 3

Another way of ensuring that each treatment group is equal with regard to subject variables is to have all subjects participate in all treatments. In this setup it can be assumed that the treatment groups are equal as to subject variables. This type of design has been called a repeated measure design or a within-subject design. The latter term seems preferable and derives its name from the fact that comparisons between treatments are made within the same subject. This type of design was used in an experiment exploring the relationship between the frequency of a tone and the absolute loudness threshold. In this experiment each subject was tested at 10 frequency levels (25, 50, etc.). The mean absolute threshold for the four subjects at each frequency level formed the points of the curve. The within-subject design is most frequently used in this type of experiment—experiments in which subjects are required to make numerous judgments to different stimuli on different occasions. These different stimuli can be considered the independent variable.

REPEATED MEASURE DESIGN—SEQUENCE BALANCED: MODEL 4

The within-subject design ensures that subject characteristics are equal in all treatments, but this design brings up a problem with regard to the order in which treatments or stimuli will be presented to the subject. The results might change rather dramatically depending upon the order because the subject's judgment of a particular stimulus may be influenced by previous judgments. For example, a subject's judgment of the lightness or heaviness of an object is in part determined by the heaviness of previously judged objects. And in discrimination experiments subjects may become more accurate or proficient at a task as they make more judgments. This is usually called a **practice effect.** The opposite phenomenon, a **fatigue effect,** can occur if subjects become tired or bored and their proficiency decreases. To control for these effects several designs are available.

One method is the **Latin-square design.** Suppose 15 subjects are to judge three different stimuli (A, B, and C). Using a Latin-square design, different orders of stimuli presentation will be derived such that (1) each stimulus appears once for each group and (2) the stimuli appear in different sequences for each group. This is illustrated in Table 6.1.

Five subjects receive the stimuli in the order A, B, C; five receive them in the order B, C, A; and five receive them in the order C, A, B. Note that stimulus A is judged or responded to once in position 1, once in position 2, and once in position 3. The same is true for B and C. The means of the judgments of all 15 subjects to the three stimuli can then be compared. These means are independent of order effects because each stimulus appeared equally in each order position. Should the experimenter be interested in order effects, he or she can examine them by looking at the subjects' judgments of each stimulus as a function of the order.

Another procedure for controlling for order effects is a random blocks design similar to that described earlier in this chapter. Suppose that an experiment is conducted in which each subject is required to make 10 judgments on each of seven different stimuli. The seven stimuli would be treated as a block, and the order of presentation within the block would be determined by some randomization proce-

Table 6.1　THE LATIN-SQUARE DESIGN

	ORDER OF STIMULUS PRESENTATION		
	POSITION 1	POSITION 2	POSITION 3
1. Five subjects	A	B	C
2. Five subjects	B	C	A
3. Five subjects	C	A	B

dure. The first block would be presented to the subject, and after the subject responded to them the seven stimuli would be randomized again. This new order would constitute the second block. This procedure would be repeated eight more times to this subject until he or she had made 10 judgments on each of the seven stimuli. The same procedure would be repeated for each subject in the experiment.

The within-subject design often appears to be desirable because the same subjects are used in all treatments, which assures the researcher that treatments are equal as to subject variables. It also appears to be desirable because fewer subjects must be used. However, this design cannot be used in many experiments simply because the treatments themselves negate this possibility. If an experiment were performed to compare rats reared in isolation with rats reared in groups, it is obvious that the same subjects could not be used in both treatments. If a study were performed comparing high-IQ subjects with low-IQ subjects, or comparing one method of teaching French with a second method, it is apparent that independent groups of subjects would have to be used in each treatment. In some experimental problems it may be much easier to use independent groups rather than using the complicated procedure that may be needed for a within-subject design. The within-subject design seems to be most applicable to research problems in which subjects are required to make judgments on numerous different stimuli on different occasions, where these stimuli can be considered the independent variable. Another type of experiment that necessitates a within-subject design is one in which order effects are of major interest. Experimenters concerned with what happens when a subject switches from a high-reward situation to a low-reward situation, for example, would need the same subjects to participate in both conditions.

SUBJECT LOSS (ATTRITION)

We have discussed steps taken to ensure that subject characteristics in treatment groups are approximately equal. A related problem deals with **subject attrition,** or loss of subjects, in the various treatment groups of an experiment. This problem usually occurs in experiments in which subjects have to participate in more that one session, as subjects coming for the first session may not necessarily appear for the second session. Consider the following hypothetical example:

► *CASE STUDY*

An employment opportunities center offered a four-week course designed to teach young adults the techniques and procedures for finding jobs. Classes were held each day and covered such topics as where to go for a job, what type of job to look for, and how to fill out applications, as well as providing practice in taking psychological tests used for employee selection. Of the 60 young adults who signed up for the course, the director of the center randomly selected 30 to participate in the course and used the other 30 as a control group to test the effectiveness of the course. The people in the control group were simply told that there was no room in the course for them and that they would have to find jobs on their own. The dependent variable was the percentage of subjects in the two groups who had found jobs within a month after the course had ended. Of the 16 people (out of 30) who completed the course, 12 (75%) were employed within a month. Of the 30 control group subjects, 15 (50%) were employed within this period. The director of the center took these results as evidence of the effectiveness of his program.

Before criticizing this study, remember that research of this type is difficult to do well. However, it is important research. Training programs such as these should be evaluated as to their effectiveness—there are many types of programs, including therapy, counseling, remedial reading, leadership training, and so on, that are very popular, but so far there has been little research examining their effectiveness.

The main concern in the example is that the statistics are based on only those 16 subjects who completed the course out of the 30 who started. The real question is: Who were the subjects who dropped out? They could have been less motivated about getting a job, less intelligent and therefore unable to understand the materials in the course, or less emotionally stable and therefore unable to accept the routine of coming to the center every day. Thus, the poor employment prospects may have been weeded out of the treatment group. But it is difficult to drop out of a no-treatment control group, so all the poor employment prospects were still in that group. The head of the center might have been comparing a group of good employment prospects (the poor ones had dropped out) with a control group that consisted of both good and poor employment prospects. If this explanation is valid, then the effectiveness of the course is highly questionable.

In reality it is not known if this explanation is valid; however, it seems to be a plausible explanation for the results. Anytime subjects are lost from an experiment, particularly if the loss is greater in one treatment group than another, it is important to find out the characteristics of those subjects who dropped out. One possibility would be to look at the employment rate of those who dropped out. If it is only about 25 percent, then it would seem to support the explanation that the poor employment prospects were weeded out. If it is around 50 percent, then the course would seem to be fairly effective.

The primary danger with subject attrition is that a select subgroup—for example, low-IQ subjects—may drop out of one treatment group but remain in another treatment group. If this occurs, the treatment groups cannot be considered equal as to subject characteristics, and the results obtained may be due to subject differences rather than treatment differences. It is important to have as much relevant information as possible about subjects who drop out in order to ascertain if they are a select subgroup. If this information indicates that the dropouts are just a random sample of the original treatment group, the researcher may feel fairly confident that the results are not due to any particular subject differences. But even if the information indicates that the dropouts are a select subgroup, it may be possible to eliminate that subgroup in the other treatment groups and thus make the groups somewhat comparable.

Researchers can sometimes avoid problems of subject attrition by having subjects participate in only a single session. If subjects are needed for more than one session, a standard procedure is to inform them before the experiment begins that they will be needed for more than one session and ask that they participate only if they can attend a second session. Frequently, a phone call will bring in those who have forgotten. In experiments in which subjects are needed for repeated testing for perhaps 10 days or so (as in some perception experiments), it is customary to pay the subjects, but on the condition that they complete the required number of sessions. In animal experiments attrition usually is not a problem since the subjects reside in a laboratory colony and can be used at will. However, there is always a danger of loss of subjects due to sickness or death, particularly if deprivation or stress treatments are used.

> ## ➤ CASE STUDY
>
> Bayoff (1940) compared the performance of rats reared in isolation with rats reared in groups in a highly competitive stress situation. Ten of the isolation-reared rats died before the experiment was over, while only one of the group-reared rats died.

If possible, subject attrition problems should be avoided by using one of the techniques described previously. But if a study occurs over a long period of time and attrition problems seem inevitable, it is important to collect information on all subjects, including the dropouts. Ideally, this information should include (1) pretreatment information, which might include data on intelligence, motivation, adjustment, and so on; (2) data on the subjects' progress (for example, learning or performance data) up to the point at which they dropped out; and (3) posttreatment data on all subjects, such as employment rates in the previous example. With this information the results of an experiment may be more easily interpreted.

DEFINITIONS

The following terms and concepts were used in this chapter. We have found it instructive for students to go through the material a second time to define each of the following:

ad lib matching	Latin-square design
block randomization	practice effect
fatigue effect	random blocks technique
independent groups designs	subject attrition

Chapter 7

Design Critiques II

Practice makes perfect.

The following is another series of experiment briefs. There is at least one design problem in each brief, so students should critique and redesign each experiment as in Chapter 5. These briefs require little expert technical knowledge of the research area being covered.

EXPERIMENT BRIEFS

1. A statistics teacher wanted to compare two methods of teaching introductory statistics. One method relied heavily on teaching the theory behind statistics (the theory method). The other method consisted of teaching the students various statistical tests and explaining when to use each test (the cookbook method). The teacher found that a leading engineering school was using the theory method in all its introductory statistics classes and that a state teachers college was using the cookbook method in all its classes. At the end of each semester the teacher administered a standardized statistics test to both sets of classes. The results indicated that the classes that received the theory method performed far better than did the classes that received the cookbook method. The teacher concluded that the theory method was the superior method and that it should be adopted by statistics teachers.

2. In an effort to determine the effects of the drug chlorpromazine on the performance of schizophrenics, two clinical investigators randomly selected

20 acute schizophrenics from a mental hospital population. The patients were asked to order several stimuli along some dimension, such as ordering eight stimuli by weight. There were several tasks of this sort. The investigators used a within-subject design in which all subjects first performed the tasks after being injected with a saline solution (placebo) and then performed the tasks again several hours later after being injected with chlorpromazine. The results indicated that fewer errors were made in the chlorpromazine treatment, which suggested to the investigators that the drug facilitates more adequate cognitive functioning in this type of patient.

3. It was hypothesized that sensory deprivation inhibits the intellectual development of animals. To test this hypothesis an experimenter used two rats, each of which had just given birth to eight pups. One rat and her litter were placed in a large cage with ample space and objects to explore. The second rat's pups were separated from the mother, and each was placed in a separate cage. These cages were quite small, and the only objects they could see or hear were the four walls and the food dispenser. After five months, both groups were tested in a multiple-T maze using food as a reward. After 20 trials all of the nondeprived pups were running the maze without error, but the deprived pups were still making several errors. This latter group frequently froze and had to be prodded to move. The experimenter concluded that sensory deprivation inhibits intellectual development such that deprived rats did not have the intellectual ability to learn even a simple maze.

4. During World War II an investigator attempted to examine the hypothesis that punishment is more effective than reward for training people. The task she picked to test this hypothesis was the identification of enemy and friendly airplanes. She had subjects sit in front of what appeared to be a radar screen as the silhouettes of enemy and friendly airplanes were flashed on the screen in very short exposures (1 second). As each silhouette appeared, the subject had to respond by pressing one of two buttons—one button was marked "enemy" and the other was marked "friendly." Each subject participated in the experiment for two hours on five successive days. On the first day, after each stimulus the subjects were told if they had been right or wrong in their identification. Starting on the second day, the subjects were randomly assigned to one of two groups. In Group A the subjects were given 10 cents after every correct identification but were not punished for a wrong identification. In Group B each subject received an electric shock after every wrong identification but received nothing for a correct identification. This procedure continued for days three and four. The fifth day was considered the test day, and the subjects followed the same procedure except that no reward, punishment, or information was given to the subject. The number of correct identifications per 100 silhouettes was considered the test of the effectiveness of each training method. As expected, there was some loss of subjects over the five-day period; about 5% of the Group A subjects and about 35% of the Group B subjects had dropped out by the fifth day. The results indicated that on the 100

test trials given to each subject on the fifth day, the mean number of correct identifications for Group A was 80, and the mean for Group B was 92. The difference between the means was statistically significant. The experimenter concluded that the hypothesis had been confirmed and suggested that all training programs be based on punishment.

5. A YMCA official in a small town wanted some evidence to prove that his program was valuable in training future leaders. He went back to the group's membership records and got the names of those boys who were active members in his program 20 years earlier and also got the names of some boys who were not YMCA members. He compared the two groups as to present occupations, salaries, and so on and found that the YMCA group was doing much better. He concluded that this was due to the influence of his program.

6. A psychologist was interested in developing a test that would predict the success of prospective lawyers. She selected a random sample of lawyers listed in *Who's Who* under the assumption that they would be successful lawyers. She then sent the lawyers a mail questionnaire that contained several hundred questions. She then analyzed the questionnaires and compiled a profile of successful lawyers. The questionnaire was then given to a group of prospective law students, and those students whose scores differed significantly from the successful lawyer profile were advised not to pursue a law career.

7. A psychologist had a theory that as members of a group get to know each other better, the productivity of the group will increase up to a point and then will start to decrease slightly. The decrease ("the honeymoon is over" effect) would be at the point at which group members stopped acting in a highly cooperative manner and started jostling for power. To test this theory the psychologist formed groups of individuals who were strangers and had them perform a series of tasks. There were five tasks that each took 35 minutes to perform, and he gave the groups a 5-minute break between tasks. His results indicated that group productivity increased up to the fifth task, but for the fifth task there was a significant decrease in the group's performance. On the basis of this evidence he considered his theory supported.

8. A clinical researcher examined whether patient interviews or objective tests are better in the diagnosis of a patient's problems and determining how long a patient would remain in the hospital. The experiment took place at a large mental hospital. In one group, 10 clinical psychology students each interviewed six new patients for one to two hours. In the other group, 60 patients were given a battery of standardized psychological tests, and the test results were interpreted by three clinical psychologists who had several years of experience in interpreting tests for the hospital. Each psychologist interpreted the test results for 20 patients. Both groups were asked to list each patient's major problems and to assign the patient to a diagnostic category (e.g., process schizophrenic, reactive schizophrenic). They were also

asked to predict how long the patient would be in the hospital before improving enough to be released. Results indicated that the interviews were 67% accurate in predicting diagnostic categories and 22% accurate in predicting length of stay. The tests group was 83% accurate in predicting diagnostic categories and 65% accurate in predicting length of stay. The experimenter concluded that interviews are of questionable value in both diagnosis and prediction of outcome and should be discontinued.

9. An experimenter wanted to test the hypothesis that males are more creative than females. She also hypothesized that the male superiority in creativity would be heightened under conditions involving ego. The design used was a 2 × 2 factorial design in which one variable was sex and the other variable was high and low ego involvement. She manipulated ego involvement by telling half the subjects that the task was a measure of intelligence and that their scores would be posted on a bulletin board (high ego involvement). She told the other half of the subjects that she wanted to test the reliability of a task she was developing and that they should not put their names on the answer sheets (low ego involvement). Her test of creativity was an "unusual uses" test in which a person is given the name of an object (e.g., hammer) and has to write as many different unusual uses for that object as he or she can in 5 minutes. Twenty-five males and 25 females participated in each of the two ego-involvement conditions. The males were members of a senior ROTC class, and the females came from sorority pledge classes. Two objects were used: (1) an army compass and (2) a monkey wrench. Subjects were given 5 minutes for each object. The results indicated that the mean number of unusual uses for the two objects for males was 4.1 under low ego involvement and 7.6 under high ego involvement. The means for the females were 3.2 under low ego involvement and 2.4 under high ego involvement. Since both the main effects and the interaction effect were statistically significant, the experimenter concluded that her hypotheses were supported.

10. A researcher was asked to conduct a quick survey in three large cities to find out what political issues or problems were important to the voters. The survey results were to be used by a political candidate in developing his campaign. The researcher randomly selected names from the phone books and called as many as she could reach between 9 A.M. and 5 P.M. on Monday and Tuesday. The results were collated and presented to the candidate with the statement that they were a valid representation of the attitudes of the voters in the three cities.

11. A psychologist noticed that following the nuclear disaster at Chernobyl in the Ukraine, there was a palpable increase in the incidence of extramarital sex among those who had been exposed to the radioactive fallout. Some people he interviewed told of orgies and other sometimes bizarre forms of sexual activities following the accident. The psychologist concluded that in radioactive fallout a secret type of love potion might be found. He applied for a grant to test his hypothesis with rats.

12. An experimental psychologist working with a basketball coach decided to test a theory of motor learning knowledge and visual feedback on basketball players and college students. In the experiment subjects threw darts at a dart board. In one condition, as soon as the subject released the dart the lights were turned off so the subject could not see the consequence of the throw. In the other condition, the lights remained on. Then all the subjects were given a long questionnaire that probed their knowledge of kinesthetic feedback mechanisms. The athletes did better than the students on the dart throwing task in both the feedback and nonfeedback conditions, but they did worse on the written task. All groups did better in the feedback condition than the nonfeedback condition, and the interactions were not significant. The experimenters concluded that motor skill does not improve with greater knowledge of kinesthetic theory but does improve through knowledge of results. Is the conclusion flawed? Why or why not?

EXERCISE

In the exercise in Chapter 5 students were asked to formulate a series of questions that could be used to critique experiments. Students should add questions to that list based on the material discussed in Chapter 6.

Chapter 8

Ethics of Experimental Research

Honesty is the best policy.

Throughout the history of American psychology, the issue of research ethics has been a subject of concern and debate. On one hand, a psychologist is a scientist dedicated to the discovery of laws of behavior. The purpose of experimental research in psychology is to enhance our knowledge of the psychological characteristics of the human species. In order to do this, psychologists often use human and animal subjects in their experiments. Some would argue that in certain instances the development of valid laws of psychology requires that a subject be deceived or in some way physically harmed. This point of view is based on the principle that the pursuit of knowledge must continue unabated. After all, it is possible that the ultimate truths revealed in psychological research may be of great benefit to humankind. But a psychologist is also bound by a code of ethics in which the psychological and physical safety of subjects is rigidly safeguarded.

In many psychological experiments it is possible to advance our understanding of psychological characteristics while clearly protecting the psychological and physical well-being of the subject. While an experimenter should still be sensitive to the ethical problems that may arise in experiments, those concerns are not as serious as in many other experiments in which the subject may be harmed in some way.

At times, experimenters have used questionable means to obtain results. Such abuses of human and animal subjects have caused great concern among psychologists and some members of the public. (See Larson, 1982, for a recent trial of an

115

animal researcher.) Because of these concerns the Committee on Scientific and Professional Ethics has been formed by the American Psychological Association. Through many years of work, this organization has developed a series of ethical principles.

Each of the principles that concerns psychologists is presented in this chapter. Following the principles, we describe case studies in which the principles are illustrated. Some of these cases (in our opinion) represent clear violations of the principles, while others do not. These cases are presented for analysis and possible class discussion. In the original document, published by the American Psychological Association, further examples and comments are provided, which should be consulted when ethical questions are raised.

ETHICAL PRINCIPLES OF PSYCHOLOGISTS AND CODE OF CONDUCT[11]

In this section the parts of the *Ethical Principles of Psychologists* that deal with research are paraphrased.

PREAMBLE
Psychologists work to develop a valid and reliable body of scientific knowledge based on research. They may apply that knowledge to human behavior in a variety of contexts. In doing so, they perform many roles, such as researcher, educator, diagnostician, therapist. . . . Psychologists respect the central importance of freedom of inquiry and expression in research, teaching, and publication. . . .

PLANNING RESEARCH
(a) Psychologists design, conduct, and report research in accordance with recognized standards of scientific competence and ethical research.

(b) Psychologists plan their research so as to minimize the possibility that results will be misleading.

(c) In planning research, psychologists consider its ethical acceptability under the Ethics Code. If an ethical issue is unclear, psychologists seek to resolve the issue through consultation with institutional review boards, animal care and use committees, peer consultations, or other proper mechanisms.

(d) Psychologists take reasonable steps to implement appropriate protections for the rights and welfare of human participants, other persons affected by the research, and the welfare of animal subjects.

11. This version of the *Ethical Principles of Psychologists and Code of Conduct* (formerly titled *Ethical Standards of Psychologists*) was adopted by the American Psychological Association in August 1992. The revised version contains both substantive and grammatical changes from the *Ethical Standards of Psychologists* previously adopted in 1981. Inquiries concerning the *Ethical Principles of Psychologists* should be addressed to the Office of Ethics, American Psychological Association, 750 First Street, NE, Washington, DC 20002. These revised ethical principles apply to psychologists, students of psychology, and others who do work of a psychological nature under the supervision of a psychologist. They are also intended for the guidance of nonmembers of the Association who are engaged in psychological research or practice. While the full code is not published here, it is available from the APA and appeared in the December 1992 issue of *American Psychologist*.

RESPONSIBILITY

(a) Psychologists conduct research competently and with due concern for the dignity and welfare of the participants.

(b) Psychologists are responsible for the ethical conduct of research conducted by them or by others under their supervision or control.

(c) Researchers and assistants are permitted to perform only those tasks for which they are appropriately trained and prepared.

(d) As part of the process of development and implementation of research projects, psychologists consult those with expertise concerning any special population under investigation or most likely to be affected.

COMPLIANCE WITH LAW AND STANDARDS

Psychologists plan and conduct research in a manner consistent with federal and state law and regulations, as well as professional standards governing the conduct of research, and particularly those standards governing research with human participants and animal subjects.

INSTITUTIONAL APPROVAL

Psychologists obtain from host institutions or organizations appropriate approval prior to conducting research, and they provide accurate information about their research proposals. They conduct the research in accordance with the approved research protocol.

RESEARCH RESPONSIBILITIES

Prior to conducting research (except research involving only anonymous surveys, naturalistic observations, or similar research), psychologists enter into an agreement with participants that clarifies the nature of the research and the responsibilities of each party.

INFORMED CONSENT TO RESEARCH

(a) Psychologists use language that is reasonably understandable to research participants in obtaining their appropriate informed consent (except as provided in Standard 6.12, Dispensing With Informed Consent). Such informed consent is appropriately documented.

(b) Using language that is reasonably understandable to participants, psychologists inform participants of the nature of the research; they inform participants that they are free to participate or to decline to participate or to withdraw from the research; they explain the foreseeable consequences of declining or withdrawing; they inform participants of significant factors that may be expected to influence their willingness to participate (such as risks, discomfort, adverse effects, or limitations on confidentiality, except as provided in Standard 6.15, Deception in Research); and they explain other aspects about which the prospective participants inquire.

(c) When psychologists conduct research with individuals such as students or subordinates, psychologists take special care to protect the prospective participants from adverse consequences of declining or withdrawing from participation.

(d) When research participation is a course requirement or opportunity for extra credit, the prospective participant is given the choice of equitable alternative activities.

(e) For persons who are legally incapable of giving informed consent, psychologists nevertheless (1) provide an appropriate explanation, (2) obtain the participant's assent, and (3) obtain appropriate permission from a legally authorized person, if such substitute consent is permitted by law.

DISPENSING WITH INFORMED CONSENT

Before determining that planned research (such as research involving only anonymous questionnaires, naturalistic observations, or certain kinds of archival research) does not require the informed consent of research participants, psychologists consider applicable regulations and institutional review board requirements, and they consult with colleagues as appropriate.

INFORMED CONSENT IN RESEARCH FILMING OR RECORDING

Psychologists obtain informed consent from research participants prior to filming or recording them in any form, unless the research involves simply naturalistic observations in public places and it is not anticipated that the recording will be used in a manner that could cause personal identification or harm. ·

OFFERING INDUCEMENTS FOR RESEARCH PARTICIPANTS

(a) In offering professional services as an inducement to obtain research participants, psychologists make clear the nature of the services, as well as the risks, obligations, and limitations. (See also Standard 1.18, Barter [With Patients or Clients].)

(b) Psychologists do not offer excessive or inappropriate financial or other inducements to obtain research participants, particularly when it might tend to coerce participation.

DECEPTION IN RESEARCH

(a) Psychologists do not conduct a study involving deception unless they have determined that the use of deceptive techniques is justified by the study's prospective scientific, educational, or applied value and that equally effective alternative procedures that do not use deception are not feasible.

(b) Psychologists never deceive research participants about significant aspects that would affect their willingness to participate, such as physical risks, discomfort, or unpleasant emotional experiences.

(c) Any other deception that is an integral feature of the design and conduct of an experiment must be explained to participants as early as is feasible, preferably at the conclusion of their participation, but no later than at the conclusion of the research. (See also Standard 6.18, Providing Participants With Information About the Study.)

SHARING AND UTILIZING DATA

Psychologists inform research participants of their anticipated sharing or further use of personally identifiable research data and of the possibility of unanticipated future uses.

MINIMIZING INVASIVENESS

In conducting research, psychologists interfere with the participants or milieu from which data are collected only in a manner that is warranted by an appropriate research design and that is consistent with psychologists' roles as scientific investigators.

PROVIDING PARTICIPANTS WITH INFORMATION ABOUT THE STUDY

(a) Psychologists provide a prompt opportunity for participants to obtain appropriate information about the nature, results, and conclusions of the research, and psychologists attempt to correct any misconceptions that participants may have.

(b) If scientific or humane values justify delaying or withholding this information, psychologists take reasonable measures to reduce the risk of harm.

HONORING COMMITMENTS

Psychologists take reasonable measures to honor all commitments they have made to research participants.

CARE AND USE OF ANIMALS IN RESEARCH

(a) Psychologists who conduct research involving animals treat them humanely.

(b) Psychologists acquire, care for, use, and dispose of animals in compliance with current federal, state, and local laws and regulations, and with professional standards.

(c) Psychologists trained in research methods and experienced in the care of laboratory animals supervise all procedures involving animals and are responsible for ensuring appropriate consideration of their comfort, health, and humane treatment.

(d) Psychologists ensure that all individuals using animals under their supervision have received instruction in research methods and in the care, maintenance, and handling of the species being used, to the extent appropriate to their role.

(e) Responsibilities and activities of individuals assisting in a research project are consistent with their respective competencies.

(f) Psychologists make reasonable efforts to minimize the discomfort, infection, illness, and pain of animal subjects.

(g) A procedure subjecting animals to pain, stress, or privation is used only when an alternative procedure is unavailable and the goal is justified by its prospective scientific, educational, or applied value.

(h) Surgical procedures are performed under appropriate anesthesia; techniques to avoid infection and minimize pain are followed during and after surgery.

(i) When it is appropriate that the animal's life be terminated, it is done rapidly, with an effort to minimize pain, and in accordance with accepted procedures.

REPORTING OF RESULTS

(a) Psychologists do not fabricate data or falsify results in their publications.

(b) If psychologists discover significant errors in their published data, they take reasonable steps to correct such errors in a correction, retraction, erratum, or other appropriate publication means.

PLAGIARISM

Psychologists do not present substantial portions or elements of another's work or data as their own, even if the other work or data source is cited occasionally.

PUBLICATION CREDIT

(a) Psychologists take responsibility and credit, including authorship credit, only for work they have actually performed or to which they have contributed.

(b) Principal authorship and other publication credits accurately reflect the relative scientific or professional contributions of the individuals involved, regardless of their relative status. Mere possession of an institutional position, such as Department Chair, does not justify authorship credit. Minor contributions to the research or to the writing for publications are appropriately acknowledged, such as in footnotes or in an introductory statement.

(c) A student is usually listed as principal author on any multiple-authored article that is substantially based on the student's dissertation or thesis.

DUPLICATE PUBLICATION OF DATA

Psychologists do not publish, as original data, data that have been previously published. This does not preclude republishing data when they are accompanied by proper acknowledgment.

SHARING DATA

After research results are published, psychologists do not withhold the data on which their conclusions are based from other competent professionals who seek to verify the substantive claims through reanalysis and who intend to use such data only for that purpose, provided that the confidentiality of the participants can be protected and unless legal rights concerning proprietary data preclude their release.

PROFESSIONAL REVIEWERS

Psychologists who review material submitted for publication, grant, or other research proposal review respect the confidentiality of and the proprietary rights in such information of those who submitted it.

RESEARCH WITH HUMANS

The use of human subjects in a psychological experiment poses special problems. A psychologist is both a scientist and a member of society. Sometimes, in the zealous pursuit of scientific truths, experimenters may become so wrapped up in their research that they overlook some ethical considerations regarding human participants. This is a grievous mistake and will ultimately reflect poorly on experimenters and psychological research in general.

In the early days of psychological research, few ethical guidelines were available save the researcher's personal ethical code and the laws of society. In fact, some research conducted during that period would not be allowed by current standards, and as a consequence, some prospective participants today are wary of volunteering for experiments. But some researchers have argued that the standards that have since evolved are too restrictive and forbid the collection of important data. Once again, new ethical standards will likely evolve.

This section presents the principles that govern research using human participants. We have also included an example of a consent form and one case study. We suggest that each point be discussed in class and that students be encouraged to write their own cases exemplifying both ethical and unethical research for each principle. Also, if contemplating research with human subjects, the complete *Ethical Principles of Psychologists* should be read thoroughly.

RESEARCH WITH HUMAN PARTICIPANTS[12]

The decision to undertake research rests upon a considered judgment by the individual psychologist about how best to contribute to psychological science and human welfare. Having made the decision to conduct research, the psychologist considers alternative

---

12. From *Ethical Principles in the Conduct of Research with Human Participants* (1982). Washington, DC: American Psychological Association.

directions in which research energies and resources might be invested. On the basis of this consideration, the psychologist carries out the investigation with respect and concern for the dignity and welfare of the people who participate and with cognizance of federal and state regulations and professional standards governing the conduct of research with human participants.

A. In planning a study, the investigator has the responsibility to make a careful evaluation of its ethical acceptability. To the extent that the weighing of scientific and human values suggests a compromise of any principle, the investigator incurs a correspondingly serious obligation to seek ethical advice and to observe stringent safeguards to protect the rights of human participants.

B. Considering whether a participant in a planned study will be a "subject at risk" or a "subject at minimal risk," according to recognized standards, is of primary ethical concern to the investigator.

C. The investigator always retains the responsibility for ensuring ethical practice in research. The investigator is also responsible for the ethical treatment of research participants by collaborators, assistants, students, and employees, all of whom, however, incur similar obligations.

D. Except in minimal-risk research, the investigator establishes a clear and fair agreement with research participants, prior to their participation, that clarifies the obligations and responsibilities of each. The investigator has the obligation to honor all promises and commitments included in that agreement. The investigator informs the participants of all aspects of the research that might reasonably be expected to influence willingness to participate and explains all other aspects of the research about which the participants inquire. Failure to make full disclosure prior to obtaining informed consent requires additional safeguards to protect the welfare and dignity of the research participants. Research with children or with participants who have impairments that would limit understanding and/or communication requires special safeguarding procedures.

Consent Form

The *Ethical Principles of Psychologists* has several basic components, including: (1) that the agreement be clear and explicit, (2) that the terms be fair and not exploitive, and (3) that the investigator honor the agreement. An example of an informed consent form follows:

```
        Consent Form for ''Color Research''

My name is Bill Simon. I am a student working on an advanced
degree in experimental psychology.

You have been asked to participate in a psychological experi-
ment, which we call ''Color Research.'' The purpose of the
experiment is to measure your reaction time when matching
colors with color names, colors, or associates of colors. No
deception is involved in this experiment. You will be asked
to look at colors on a TV screen and press a key, which will
```

measure your reaction time. The entire experiment will take 20 minutes or less. The task is similar to looking at a TV, and we foresee no risks or discomforts. You will receive class credit for your participation in the experiment.

The results of this experiment may be presented at professional meetings or published in the scientific literature. Your name will not be used in the reporting of results. Only group data will be used; however, your scores and name will be coded for a possible follow-up study or reanalysis of the data. All personal data will be kept confidential.

If you wish to withdraw from the experiment you may do so at any time without penalty.

Following the experiment I will discuss the results of the experiment with you.

If you have any questions please feel free to ask me or the supervisor of this research, Dr. Elizabeth Kane, Department of Psychology, University of Western California, phone 882-5968. Thank you.

I, _____ _____, understand that
    (First name)   (Last name) Please print.
my participation in this experiment is voluntary and that I may refuse to participate or withdraw from the experiment at any time without penalty.

_____   _____
Signature Participant   Date

_____   _____
Signature Experimenter   Date

E. Methodological requirements of a study may make the use of concealment or deception necessary. Before conducting such a study, the investigator has a special responsibility to (1) determine whether the use of such techniques is justified by the study's prospective scientific, educational, or applied value; (2) determine whether alternative procedures are available that do not use concealment or deception; and (3) ensure that the participants are provided with sufficient explanation as soon as possible.

F. The investigator respects the individual's freedom to decline to participate in or to withdraw from the research at any time. The obligation to protect this freedom requires careful thought and consideration when the investigator is in a position of authority or influence over the participant. Such positions of authority include, but are not limited to, situations in which research participation is required as part of employment or in which the participant is a student, client, or employee of the investigator.

G. The investigator protects the participant from physical and mental discomfort, harm, and danger that may arise from research procedures. If risks of such consequences exist, the investigator informs the participant of that fact. Research procedures likely to cause serious or lasting harm to a participant are not used unless the failure to use these procedures might expose the participant to risk of greater harm or unless the research has great potential benefit and fully informed and voluntary consent is obtained from each participant. The participant should be informed of procedures for contacting the investigator within a reasonable time period following participation should stress, potential harm, or related questions or concerns arise.

> ## ➤ Case Study

In an experiment reported by Smith, Tyrell, Coyle, and William (1987) in the *British Journal of Psychology*, the effects of experimentally induced colds and influenza on human performance was measured to determine whether minor illnesses "alter the efficiency of human performance."

The experiment involved recruiting volunteers who stayed at the Common Cold Unit for 10 days. They were housed in groups of two or three and isolated from outside contacts. Following a three-day quarantine period the subjects were inoculated with nose drops containing either the virus or a placebo. An incubation period of 48–72 hours followed. Then each participant was assessed by a clinician who evaluated the severity of the illness. Objective measures included temperature, number of paper tissues used, and the quantity of nasal secretion.

Then two performance tasks were done. In one the subjects were to detect and respond quickly to target items that appeared at irregular intervals (a detection task). The second task tested hand-eye coordination.

The results indicated that influenza impaired performance on the detection task but not in the hand-eye coordination task. Colds generally had the reverse effect.

The procedures used in this experiment were approved by the local ethical committee, and the informed consent of the volunteers was obtained. All subjects were screened to exclude pregnant women and people who took sleeping pills, tranquilizers, or antidepressant medicines. The subjects also took a medical examination, including a chest X-ray, and anyone who failed the examination was excluded. The subjects were not paid but received food, accommodation, traveling expenses, and pocket money. Other clinical trials were also conducted.

Students should discuss the ethics of this experiment. Did the experimenters follow acceptable (APA) standards? Were the subjects coerced? Was the risk to benefit ratio worthwhile? (Keep in mind that other tests were done.) Were there alternative means available to collect the data? Were the subjects treated in a way consistent with the APA principles? Ask students if they would volunteer for this experiment. Would they collect the data for this experiment? Why or why not? Discuss this case (or the original article) in class.

H. After the data are collected, the investigator provides the participant with information about the nature of the study and attempts to remove any misconceptions that may have arisen. Where scientific or humane values justify delaying or withholding this information, the investigator incurs a special responsibility to monitor the research and to ensure that there are no damaging consequences for the participant.

I. Where research procedures result in undesirable consequences for the individual participant, the investigator has the responsibility to detect and remove or correct these consequences, including long-term effects.

J. Information obtained about a research participant during the course of an investigation is confidential unless otherwise agreed upon in advance. When the possibility exists that others may obtain access to such information, this possibility, together with the plans for protecting confidentiality, is explained to the participant as part of the procedure for obtaining informed consent.

CASE STUDIES

In this section we present several research projects, some of which are questionable and others that seem to conform to the principles discussed previously. Try to find the flaws and acceptable standards in these projects.

Case Study 1

An experimenter was interested in the personality traits of subjects who had scored well in a test of ESP. The gist of the experiment was to have subjects (receivers) report their impressions of an ESP card that was being viewed by another person (sender). The cards were randomly ordered, and the sender was completely concealed from the receivers. Several receivers scored well, but their personality measures did not differ from the rest of the group. Nevertheless, the experimenter thought he had identified a group of subjects that was unusually sensitive to receiving ESP signals, so he conducted four additional experiments with these subjects. On the first three experiments the subjects scored significantly better than would be expected by chance alone, but on the fourth experiment the group's performance was no better than would be expected by chance. In reviewing the results, the experimenter decided to report only the first three experiments, attributing the results on the fourth experiment to sender and/or receiver fatigue.

This case raises both procedural and ethical problems.

Case Study 2

While working at a major eastern university, an experimental psychologist trained in research design and physiological psychology was approached by a large company that produced a health food cereal. The company asked the psychologist to design an experiment that would demonstrate the effectiveness of a cereal in reducing common ailments (e.g., colds) and absenteeism. The company agreed to pay the researcher a large sum of money if she would design the experiment and lend her name to the conclusion in a subsequent advertising campaign. The psychologist agreed to do so but stipulated that she would have final approval of the advertising copy. She would not be directly involved in the collection of data but was assured that it would be done according to her exact specifications. Although she had little knowledge or training in nutrition, she felt that her background in experimental design and physiological psychology was sufficient to design a valid experiment.

Should the psychologist have accepted this offer? Why or why not?

Case Study 3

In an investigation of higher moral standards an experimental psychologist was interested in the strength of subjects' moral convictions. An experiment was designed in which subjects were told that a child was in critical need of a drug that could be derived from a fungus found in a particular limestone cave. The fungus grew in abundance in this cave, and only a small portion was needed for treatment. However, the owner of the cave refused to allow anyone to use the fungus. The experimenter found that a large percentage of subjects reported that they would trespass to obtain samples of the fungus. In a second part of the experiment, the researcher asked some of the subjects to obtain the fungus illegally. In justifying the procedure, he argued that a higher moral principle was served and that the results would have a major impact on the knowledge of civil disobedience.

Did the researcher's plan conform to the code? Could this research be modified to achieve the psychologist's aim and yet conform to the ethical principles?

Case Study 4

A research psychologist, working on problems of human memory, developed a superior mnemonic technique. He decided to test the technique by becoming a contestant on a television quiz show. After several successful appearances, he decided to write a book on the technique. Because he was well known after his appearance on the television quiz show, he agreed to have his picture on the cover and in the advertising for the book, along with the statement, "You can learn the extraordinary memory technique of Dr. Joe Brown, too!"

What specific ethical issues are raised by this example?

Case Study 5

At a meeting of lawyers, a social psychologist was asked to present the results of her recent research on the decision-making process of juries. In one of her studies, she interviewed each member of a jury involved in a celebrated murder trial. In the study, the identity of each member of the jury was carefully concealed, but she did discuss the deliberative processes of subgroups. For example, the jury had among its members seven women, two African-Americans, one foreign-born Italian-American, an architect, and a truck driver, so the researcher referred to the voting and deliberative patterns of these groups. When questioned about the ethical propriety of revealing the findings, she said the names of the jurors had not been used and the jurors were now public figures whose opinions were no longer private.

What are your views?

Case Study 6

A doctoral candidate at a large midwestern university was completing his dissertation on the relationship between mothers' religious attitudes and the bed-wetting tendencies of their children. His sample consisted of 48 white mothers between the ages of 20 and 28 who were members of a specific religious group. Their children were healthy. He had nearly completed his study when four subjects dropped out of the experiment. Since the dissertation was due soon, he decided to recruit subjects from his friends and their wives. He was careful to make sure that all new subjects were identical on the designated attributes.

The dissertation was successful and he was awarded a doctoral degree. He is now a valued member of a department of psychology at a large midwestern university.

Did the candidate act unethically?

Case Study 7

"One of the best graduate seminars I took was in industrial psychology from good ol' Professor B. J. Smith," a colleague told another professor. "Each member of the seminar was given a very specific hypothetical problem concerning the design of work spaces and its effect on productivity. We did an extensive review of the literature and designed an experiment. We even anticipated the results and analyzed and discussed them. The professor gave us the problems and we did all the work, but it was an excellent learning experience."

Several years later the same two colleagues saw each other at a psychology meeting. "Do you remember the story I told you about good ol' Professor Smith?" one began.

"Yes, I do . . . " the other said. "Something about a good seminar you had with him."

"Well, the old fraud," the first said, seething, "he took our research ideas, put them into practice, and published the results in the latest issue of the *Journal of Important Industrial Research*. See, here is a copy."

The design in the article was similar to the one submitted by the seminar group, and the article even cited the very sources in the introductory material that were contained in the original paper.

Did the professor act unethically?

Case Study 8

A social psychology experiment wanted to use a test instrument called "Study of Basic Attitudes" to see if attitudes are related to scholastic achievement. A group of subjects reported to a psychology investigator from a small liberal arts college to complete the attitude scale. But when the subjects arrived the principal investigator was absent. Because the test was simple to administer—in effect it was self-administering—the experimenter decided to have the secretary administer it. The secretary had no formal background in psychometrics but was briefed in the proper procedures for the administration of the test. Following the collection of data, the psychologist discussed the test and its results with the subjects.

Is the procedure questionable from an ethical standpoint?

Case Study 9

The identification of criminal offenders by eyewitnesses is considered an important social and psychological issue. To study it, a researcher decided to stage a crime in the presence of eyewitnesses and then ask them for a description of the perpetrator. The experiment was conducted in a fast-food outlet, and all employees carefully rehearsed the staged crime. The crime was committed by an actor who entered the store, displayed an unloaded handgun, and demanded all the money from the cash register. He told the employees not to call the police and, in making his getaway, shouted, "The first one out the door is going to get blown away." Immediately after the thief left, the researcher and his associates entered the store with a questionnaire, which they distributed to the patrons. The questions dealt with the physical appearance of the thief, whether or not the person had a weapon, and what he or she said. Then each patron was presented with a series of photographs and asked to identify the thief.

Each patron was thoroughly debriefed after the questionnaire was completed, and the important social and psychological issues were discussed. An opportunity was provided for further debriefing and counseling, but no subject indicated a need for further intervention.

Comment on the experiment from an ethical standpoint.

Case Study 10

In a study of the effects that vitamin A has on rats' maze-learning ability in a semidarkened environment, a researcher had reason to believe that enhanced performance would occur under minimum dosages but that at higher dosages performance would decrease. The experimenter selected four levels of vitamin A ingestion. The highest level had been shown by previous research to be toxic to rats, but the researcher argued that to demonstrate the hypothesized results such levels were necessary.

Previous research had also suggested that higher levels of vitamin A interfere with maze performance, but the hypothesis had not been tested empirically. Thus, the results would reveal something new and would have important scientific implications.

The rats were well cared for except for the high level of vitamin A ingestion in one group.

Upon collecting the minimum amount of data necessary for analysis, the experiment was terminated and the rats were rapidly and painlessly killed.

Comment on this experiment.

As the above guidelines show, it is impossible to cover all the ethical questions that might arise in the course of experimental work in psychology, just as civil laws cannot cover all contingencies of human conduct. But the guidelines we have presented do provide a general structure that can be applied to a wide range of specific situations. These principles are subject to a degree of interpretation and, as such, may be interpreted differently by different investigators. When planning research that might involve ethical questions, the researcher (especially the neophyte researcher) should seek the opinions of other scholars before interpreting ethical principles. Sometimes there is a fine line between ethical and unethical experimentation, and the advice of others may help determine whether a piece of experimental work is advisable.

Chapter 9

The Psychological Literature: Reading for Understanding and as a Source for Research Ideas

Science is organized knowledge.

—Herbert Spencer

The scientific literature in psychology has expanded greatly during the last few years. In fact, the number of articles, books, and technical reports exceeds the capacity of any researcher to examine all the sources in his or her own specialty, let alone in all of psychology. Because so much literature is available and because the human mind is limited in its ability to collect and store information, becoming an expert in a particular field in psychology might seem impossible. (See Solso, 1987a.) But while the task may at first seem difficult, given the guidelines in this chapter it is not impossible.

As we will show, the psychology literature is organized—and because electronic storage of information in computers is becoming more prevalent it is becoming much more thoroughly organized. Given several principles that show how this information is organized, students will find it relatively easy to focus their reading of the literature, which is an important step toward the development of ideas in psychology.

IDEAS, HUNCHES, AND THE PSYCHOLOGICAL LITERATURE

Earlier in this book we discussed the source of knowledge. Frankly, finding a truly original idea may appear to be an overwhelming task, especially for new scientists. Unfortunately, no single surefire pathway will inevitably lead to the development of a new idea. But we do have some suggestions that, when combined with an inquiring mind, will enhance the likelihood of an experimental hunch materializing. Those suggestions are as follows:

1. Read the psychological literature selectively and critically.
2. Ask "what if" questions while reading.
3. Test ideas by discussing them with a friend or colleague.
4. Gather some relevant data about the idea.

First, consider the matter of selective reading. Suppose someone is searching for information about experimental psychology. That person could indiscriminately read all the literature available, but that would be impossible, even for the most voracious reader. A far more efficient procedure would be to select a topic of interest—be it perceptual motor processes, thumb sucking by children, language development in chimpanzees, learning mathematics, the use of personality tests in industry, or music and pain (see box)—and then find relevant literature on that topic. Fortunately, the current literature in psychology (and other branches of knowledge) is highly organized. All anyone needs to know is the *system* of organization.

➤ ELECTRONIC LIBRARIES

In the past, getting a comprehensive review of the literature in any field required going to a library and paging through indexes, reference books, and thousands of journals. It may have even required going to several libraries to find references that were not in one library. This process used to take several days or even several weeks to complete. But with the advent of the personal computer and computerized data bases many literature searches can now be completed in under an hour.

A computerized data base can be defined as an organized collection of facts in computer-readable form. Imagine the complete *Psychological Abstracts* (1966–present) in a computer data base. Using a computer or a terminal, and a modem anyone can call the electronic library (usually a local phone number). After logging on, uncomplicated commands can be used to access information (see sample search that follows). With these commands the data base's contents can be searched at a rate of thousands of items per second. Most data base services also provide on-line ordering procedures for references found during a computer search.

There are several electronic data base vendors available to the consumer (e.g., Dialog, The Source, Bibliographic Retrieval Service, CompuServe, etc.). These on-line services offer many different data bases to choose from and valuable on-line help systems. When choosing a data base service, several different aspects should be considered: when the data base can be accessed (day or evening), what data bases are needed (e.g., medical, engineering, psychological, etc.), and how charges are determined (e.g., per reference, computer time, evening and day rates, etc.).

The following is an example of a search using Dialog's nighttime service.

```
                    Knowledge Index:


? begin psycl
2/5/88 18:14:20 EST
Now in PSYCHOLOGY (PSYC) Section
  PsycINFO (PSYCI) Database
  (Copyright 1982 American Psychological Association)


? find music and pain

        3106 MUSIC
        5067 PAIN
         S1          21 MUSIC AND PAIN


? display s/1/1/1-21


   1/L/1
   74-16157
   Music therapy in pain management.
   Bailey, Lucanne M.
   Memorial Sloan-Kettering Cancer Ctr, New York, NY
   Journal of Pain & Symptom Management, 1986 Win Vol 1
   (1) 25-28 ISBN: 08853924
   Journal Announcement: 7406
   Language: ENGLISH Document Type: JOURNAL ARTICLE
   Suggests that music therapy can be used to treat pain
   and suffering by promoting relaxation, alteration in
   mood, a sense of control, and self-expression. Music
   therapy techniques are individually devised, with
   consideration given to the patient's physical,
   emotional, and psychological needs; coping abilities;
   and prior musical experiences. Active involvement is
   encouraged to facilitate cognitive and emotional
   expression, as well as involvement in pain management.
   (13 ref) (PsycINFO Database Copyright 1987 American
   Psychological Assn. all rights reserved)
```

```
Descriptors: PAIN (36150); MUSIC THERAPY (32670);
RELAXATION (43697); EMOTIONAL STATES (16900)
Identifiers: music therapy, relaxation & mood
alteration & control & self expression, pain patients
Section Headings: 3300 (TREATMENT AND PREVENTION)

1/L/2
```

PsycLIT, *Psychological Abstracts,* and Other Indexes

Research information in psychology is available through two sources. One involves library searches, which we will discuss later. The other is abstracted information that is accessible through computer-based information services (see box). These services provide a wide range of information quickly and conveniently for a fee usually based on on-line computer time. Anyone who owns a computer and modem can log on to these systems, and sometimes departments of psychology or libraries will either provide computer time or perform electronic searches for students. These computer-based indexes most likely are the way information will be accessed in the future, and the student of experimental psychology should explore the prodigious capacity of these systems.

PsycLIT.   PsycLIT is a data base provided by the APA on a "platinum disk" that is available in most research libraries. The format is CD-ROM. The disk covers literature from nearly 50 countries and includes citations and summaries of journal articles published since 1974 and of book chapters and books published since 1987. An example of *PsycLIT*'s format is shown in Figure 9.1.

PsycLIT is a user friendly disk that puts the researcher in contact with articles, chapters, and books on specific topics such as schizophrenia, weight loss, neurotransmitters, dogmatism, memory for music, and nearly any other topic of interest in psychology. Once the researcher locates the abstracts through PsycLIT, he or she can then access the full articles from the library's holdings. PsycLIT offers the most comprehensive collection of research abstracts available, and all researchers—from the beginner to the expert—should explore the vast references available in this easy to use and time-saving computer disk.

Library Searches

The second source of research information is library-based journals, books, and indexes. Books are excellent sources for finding an overview of a selected topic. If someone is interested in, say, thumb sucking in children, that person might consult any of a number of developmental psychology or child clinical psychology books. But if that person is interested in basic research in that field, he or she would have to consult the current experimental literature in the field.

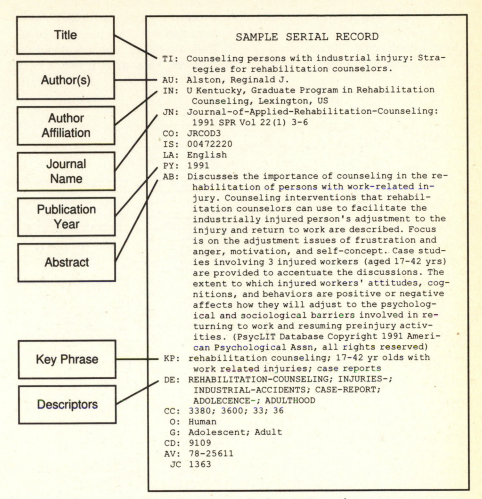

SAMPLE SERIAL RECORD

TI: Counseling persons with industrial injury: Stra-
    tegies for rehabilitation counselors.
AU: Alston, Reginald J.
IN: U Kentucky, Graduate Program in Rehabilitation
    Counseling, Lexington, US
JN: Journal-of-Applied-Rehabilitation-Counseling:
    1991 SPR Vol 22(1) 3-6
CO: JRCOD3
IS: 00472220
LA: English
PY: 1991
AB: Discusses the importance of counseling in the re-
    habilitation of persons with work-related in-
    jury. Counseling interventions that rehabil-
    itation counselors can use to facilitate the
    industrially injured person's adjustment to the
    injury and return to work are described. Focus
    is on the adjustment issues of frustration and
    anger, motivation, and self-concept. Case stud-
    ies involving 3 injured workers (aged 17-42 yrs)
    are provided to accentuate the discussions. The
    extent to which injured workers' attitudes, cog-
    nitions, and behaviors are positive or negative
    affects how they will adjust to the psycholog-
    ical and sociological barriers involved in re-
    turning to work and resuming preinjury activ-
    ities. (PsycLIT Database Copyright 1991 Ameri-
    can Psychological Assn, all rights reserved)
KP: rehabilitation counseling; 17-42 yr olds with
    work related injuries; case reports
DE: REHABILITATION-COUNSELING; INJURIES-;
    INDUSTRIAL-ACCIDENTS; CASE-REPORT;
    ADOLECENCE-; ADULTHOOD
CC: 3380; 3600; 33; 36
 O: Human
 G: Adolescent; Adult
CD: 9109
AV: 78-25611
 JC 1363

Labels at left: Title, Author(s), Author Affiliation, Journal Name, Publication Year, Abstract, Key Phrase, Descriptors

Figure 9.1   An example of a PsycLIT record.

Information can also be found in indexes. A primary index of the psychological literature—both domestic and foreign—is *Psychological Abstracts,* published by the APA each month, with an index volume published every six months and cumulative indexes published every year. Using the abstracts is simple, and the more practice someone has with them, the more efficient will be that person's search.

Suppose someone is interested in a perceptual motor process such as hand-eye coordination. In the subject index of the *Psychological Abstracts* that person would find an entry called "Perceptual Motor Processes" (see Figure 9.2). Following the entry, other related entries would be suggested (e.g., "See also Perceptual Motor Coordination, Rotary Pursuit, . . ."). Then a series of numbers would be presented that would correspond to article abstracts. An abstract is only a bare

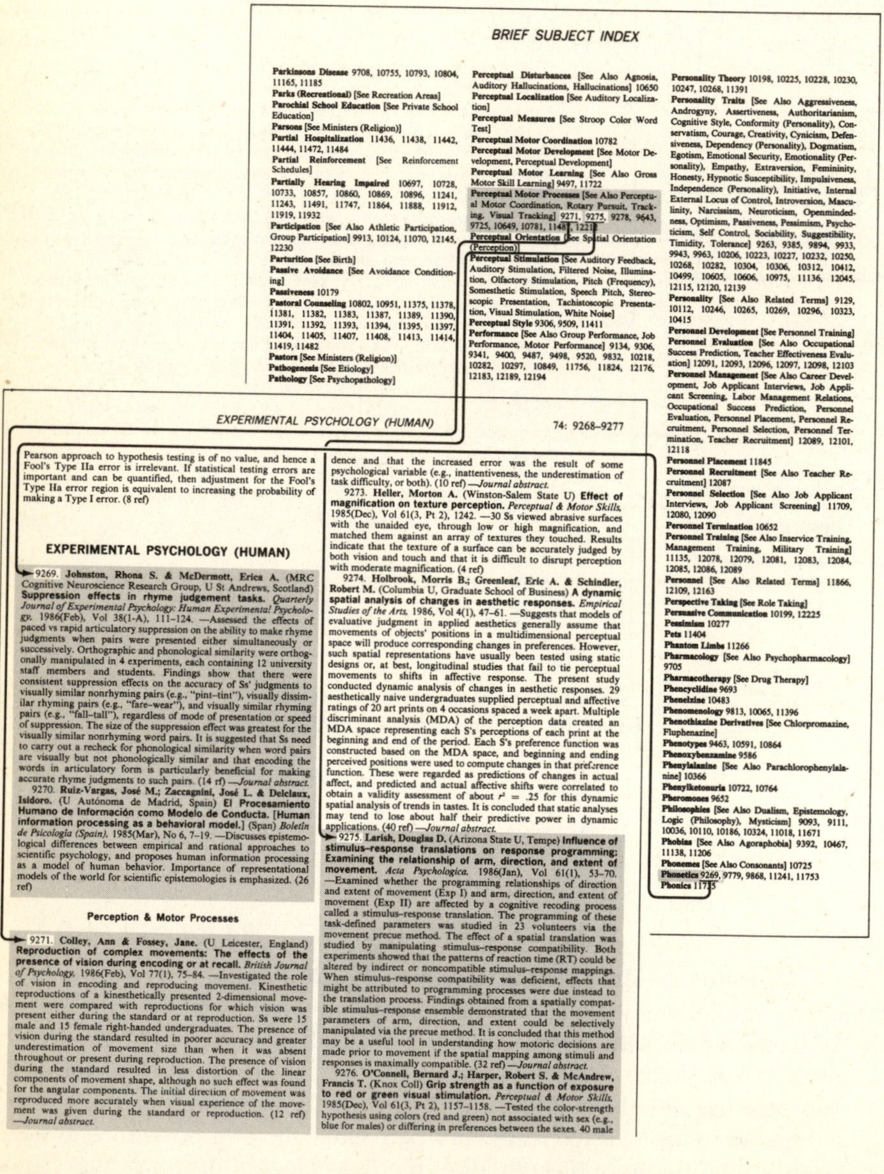

BRIEF SUBJECT INDEX

Parkinsons Disease 9708, 10755, 10793, 10804, 11165, 11185
Parks (Recreational) [See Recreation Areas]
Parochial School Education [See Private School Education]
Parsons [See Ministers (Religion)]
Partial Hospitalization 11436, 11438, 11442, 11444, 11472, 11484
Partial Reinforcement [See Reinforcement Schedules]
Partially Hearing Impaired 10697, 10728, 10733, 10857, 10860, 10869, 10896, 11241, 11243, 11491, 11747, 11864, 11888, 11912, 11919, 11932
Participation [See Also Athletic Participation, Group Participation] 9913, 10124, 11070, 12145, 12230
Parturition [See Birth]
Passive Avoidance [See Avoidance Conditioning]
Passiveness 10179
Pastoral Counseling 10802, 10951, 11375, 11378, 11381, 11382, 11383, 11387, 11389, 11390, 11391, 11392, 11393, 11394, 11395, 11397, 11404, 11405, 11407, 11408, 11413, 11414, 11419, 11482
Pastors [See Ministers (Religion)]
Pathogenesis [See Etiology]
Pathology [See Psychopathology]
Perceptual Disturbances [See Also Agnosia, Auditory Hallucinations, Hallucinations] 10650
Perceptual Localization [See Auditory Localization]
Perceptual Measures [See Stroop Color Word Test]
Perceptual Motor Coordination 10782
Perceptual Motor Development [See Motor Development, Perceptual Development]
Perceptual Motor Learning [See Also Gross Motor Skill Learning] 9497, 11722
Perceptual Motor Processes [See Also Perceptual Motor Coordination, Rotary Pursuit, Tracking, Visual Tracking] 9271, 9275, 9278, 9643, 9725, 10649, 10781, 11481, 12212
Perceptual Orientation [See Spatial Orientation (Perception)]
Perceptual Stimulation [See Auditory Feedback, Auditory Stimulation, Filtered Noise, Illumination, Olfactory Stimulation, Pitch (Frequency), Somesthetic Stimulation, Speech Pitch, Stereoscopic Presentation, Tachistoscopic Presentation, Visual Stimulation, White Noise]
Perceptual Style 9306, 9509, 11411
Performance [See Also Group Performance, Job Performance, Motor Performance] 9134, 9306, 9341, 9400, 9487, 9498, 9520, 9832, 10218, 10282, 10297, 10849, 11756, 11824, 12176, 12183, 12189, 12194
Personality Theory 10198, 10225, 10228, 10230, 10247, 10268, 11391
Personality Traits [See Also Aggressiveness, Androgyny, Assertiveness, Authoritarianism, Cognitive Style, Conformity (Personality), Conservatism, Courage, Creativity, Cynicism, Defensiveness, Dependency (Personality), Dogmatism, Egotism, Emotional Security, Emotionality (Personality), Empathy, Extraversion, Femininity, Honesty, Hypnotic Susceptibility, Impulsiveness, Independence (Personality), Initiative, Internal External Locus of Control, Introversion, Masculinity, Narcissism, Neuroticism, Openmindedness, Optimism, Passiveness, Pessimism, Psychoticism, Self Control, Sociability, Suggestibility, Timidity, Tolerance] 9263, 9385, 9894, 9933, 9943, 9963, 10206, 10223, 10227, 10232, 10250, 10268, 10282, 10304, 10306, 10312, 10412, 10499, 10605, 10606, 10975, 11136, 12045, 12115, 12120, 12139
Personality [See Also Related Terms] 9129, 10112, 10246, 10265, 10269, 10296, 10323, 10415
Personnel Development [See Personnel Training]
Personnel Evaluation [See Also Occupational Success Prediction, Teacher Effectiveness Evaluation] 12091, 12093, 12096, 12097, 12098, 12103
Personnel Management [See Also Career Development, Job Applicant Interviews, Job Applicant Screening, Labor Management Relations, Occupational Success Prediction, Personnel Evaluation, Personnel Placement, Personnel Recruitment, Personnel Selection, Personnel Termination, Teacher Recruitment] 12089, 12101, 12118
Personnel Placement 11845
Personnel Recruitment [See Also Teacher Recruitment] 12087
Personnel Selection [See Also Job Applicant Interviews, Job Applicant Screening] 11709, 12080, 12090
Personnel Termination 10652
Personnel Training [See Also Inservice Training, Management Training, Military Training] 11135, 12078, 12079, 12081, 12083, 12084, 12085, 12086, 12089
Personnel [See Also Related Terms] 11866, 12109, 12163
Perspective Taking [See Role Taking]
Persuasive Communication 10199, 12225
Pessimism 10277
Pets 11404
Phantom Limbs 11266
Pharmacology [See Also Psychopharmacology] 9705
Pharmacotherapy [See Drug Therapy]
Phencyclidine 9693
Phenelzine 10483
Phenomenology 9813, 10065, 11396
Phenothiazine Derivatives [See Chlorpromazine, Fluphenazine]
Phenotypes 9463, 10591, 10864
Phenoxybenzamine 9586
Phenylalanine [See Also Parachlorophenylalanine] 10366
Phenylketonuria 10722, 10764
Pheromones 9652
Philosophies [See Also Dualism, Epistemology, Logic (Philosophy), Mysticism] 9093, 9111, 10036, 10110, 10186, 10324, 11018, 11671
Phobias [See Also Agoraphobia] 9392, 10467, 11138, 11206
Phonemes [See Also Consonants] 10725
Phonetics 9269, 9779, 9868, 11241, 11753
Phonics 11735

EXPERIMENTAL PSYCHOLOGY (HUMAN) 74: 9268–9277

Pearson approach to hypothesis testing is of no value, and hence a Fool's Type IIa error is irrelevant. If statistical testing errors are important and can be quantified, then adjustment for the Fool's Type IIa error region is equivalent to increasing the probability of making a Type I error. (8 ref)

EXPERIMENTAL PSYCHOLOGY (HUMAN)

9269. **Johnston, Rhona S. & McDermott, Erica A.** (MRC Cognitive Neuroscience Research Group, U St Andrews, Scotland) **Suppression effects in rhyme judgement tasks.** *Quarterly Journal of Experimental Psychology: Human Experimental Psychology.* 1986(Feb), Vol 38(1-A), 111–124. —Assessed the effects of paced vs rapid articulatory suppression on the ability to make rhyme judgments when pairs were presented either simultaneously or successively. Orthographic and phonological similarity were orthogonally manipulated in 4 experiments, each containing 12 university staff members and students. Findings show that there were consistent suppression effects on the accuracy of Ss' judgments to visually similar nonrhyming pairs (e.g., "pint–tint"), visually dissimilar rhyming pairs (e.g., "fare–wear"), and visually similar rhyming pairs (e.g., "fall–tall"), regardless of mode of presentation or speed of suppression. The size of the suppression effect was greatest for the visually similar nonrhyming word pairs. It is suggested that Ss need to carry out a recheck for phonological similarity when word pairs are visually but not phonologically similar and that encoding the words in articulatory form is particularly beneficial for making accurate rhyme judgments to such pairs. (14 rf) —*Journal abstract.*

9270. **Ruiz-Vargas, José M.; Zaccagnini, José L. & Delclaux, Isidoro.** (U Autónoma de Madrid, Spain) **El Procesamiento Humano de Información como Modelo de Conducta. [Human information processing as a behavioral model.]** (Span) *Boletín de Psicología (Spain).* 1985(Mar), No 6, 7–19. —Discusses epistemological differences between empirical and rational approaches to scientific psychology, and proposes human information processing as a model of human behavior. Importance of representational models of the world for scientific epistemologies is emphasized. (26 ref)

Perception & Motor Processes

9271. **Colley, Ann & Fossey, Jane.** (U Leicester, England) **Reproduction of complex movements: The effects of the presence of vision during encoding or at recall.** *British Journal of Psychology.* 1986(Feb), Vol 77(1), 75–84. —Investigated the role of vision in encoding and reproducing movement. Kinesthetic reproductions of a kinesthetically presented 2-dimensional movement were compared with reproductions for which vision was present either during the standard or at reproduction. Ss were 15 male and 15 female right-handed undergraduates. The presence of vision during the standard resulted in poorer accuracy and greater underestimation of movement size than when it was absent throughout or present during reproduction. The presence of vision during the standard resulted in less distortion of the linear components of movement shape, although no such effect was found for the angular components. The initial direction of movement was reproduced more accurately when visual experience of the movement was given during the standard or reproduction. (12 ref) —*Journal abstract.*

dence and that the increased error was the result of some psychological variable (e.g., inattentiveness, the underestimation of task difficulty, or both). (10 ref) —*Journal abstract.*

9273. **Heller, Morton A.** (Winston-Salem State U) **Effect of magnification on texture perception.** *Perceptual & Motor Skills,* 1985(Dec), Vol 61(3, Pt 2), 1242. —30 Ss viewed abrasive surfaces with the unaided eye, through low or high magnification, and matched them against an array of textures they touched. Results indicate that the texture of a surface can be accurately judged by both vision and touch and that it is difficult to disrupt perception with moderate magnification. (4 ref)

9274. **Holbrook, Morris B.; Greenleaf, Eric A. & Schindler, Robert M.** (Columbia U, Graduate School of Business) **A dynamic spatial analysis of changes in aesthetic responses.** *Empirical Studies of the Arts,* 1986, Vol 4(1), 47–61. —Suggests that models of evaluative judgment in applied aesthetics generally assume that movements of objects' positions in a multidimensional perceptual space will produce corresponding changes in preferences. However, such spatial representations have usually been tested using static designs or, at best, longitudinal studies that fail to tie perceptual movements to shifts in affective response. The present study conducted dynamic analysis of changes in aesthetic responses. 29 aesthetically naive undergraduates supplied perceptual and affective ratings of 20 art prints on 4 occasions spaced a week apart. Multiple discriminant analysis (MDA) of the perception data created an MDA space representing each S's perceptions of each print at the beginning and end of the period. Each S's preference function was constructed based on the MDA space, and beginning and ending perceived positions were used to compute changes in that preference function. These were regarded as predictions of changes in actual affect, and predicted and actual affective shifts were correlated to obtain a validity assessment of about $r^2 = .25$ for this dynamic spatial analysis of trends in tastes. It is concluded that static analyses may tend to lose about half their predictive power in dynamic applications. (40 ref) —*Journal abstract.*

9275. **Larish, Douglas D.** (Arizona State U, Tempe) **Influence of stimulus–response translations on response programming: Examining the relationship of arm, direction, and extent of movement.** *Acta Psychologica,* 1986(Jan), Vol 61(1), 53–70. —Examined whether the programming relationships of direction and extent of movement (Exp I) and arm, direction, and extent of movement (Exp II) are affected by a cognitive recoding process called a stimulus–response translation. The programming of these task-defined parameters was studied in 23 volunteers via the movement precue method. The effect of a spatial translation was studied by manipulating stimulus–response compatibility. Both experiments showed that the patterns of reaction time (RT) could be altered by indirect or noncompatible stimulus–response mappings. When stimulus–response compatibility was deficient, effects that might be attributed to programming processes were due instead to the translation process. Findings obtained from a spatially compatible stimulus–response ensemble demonstrated that the movement parameters of arm, direction, and extent could be selectively manipulated via the precue method. It is concluded that this method may be a useful tool in understanding how motoric decisions are made prior to movement if the spatial mapping among stimuli and responses is maximally compatible. (32 ref) —*Journal abstract.*

9276. **O'Connell, Bernard J.; Harper, Robert S. & McAndrew, Francis T.** (Knox Coll) **Grip strength as a function of exposure to red or green visual stimulation.** *Perceptual & Motor Skills,* 1985(Dec), Vol 61(3, Pt 2), 1157–1158. —Tested the color-strength hypothesis using colors (red and green) not associated with sex (e.g., blue for males) or differing in preferences between the sexes. 40 male

summary of an article, but it should provide enough information for someone to decide if an article would be relevant or not. If an abstract is relevant, the person could find the original article and read the details of the experiment. The abstracts provide a complete description of the journal, year, volume, and page numbers.

Social Science Citation Index

The *Social Science Citation Index* is another very useful library tool, and yet it is frequently unknown to many students. Every year it codes over 70,000 articles published by more than 2,000 journals throughout the world. Such a comprehensive index would be impossible without computers. The *SSCI* has four separate indexes: a subject index, an author index, a corporate index, and a citation index. Depending on the nature of someone's research, each of these indexes could be useful. The subject index is an organized list of the major words from the titles of articles. For example, if an article is titled "Reaction Time for Children in a Concept Formation Task," it might be indexed under "Reaction Time," "Children," and "Concept Formation." The author index lists the name of the author, the title of the article, the journal in which the article appeared, the volume and page numbers of the article, the year the article appeared, the number of references, the author's address, and a list of the sources cited in the article. An article might appear as follows:

```
KRUSHINSKI, K.
      REACTION TIME FOR CHILDREN IN A CONCEPT FORMATION
      TASK
        J CHILD PSY        15  6  89    12R  N1
        PAX MEDICAL CENTER, LINDSEY AVE., HERSHEY, PA.
      LAIRD, R.    85   J EXP PSY    01   20   12
      (etc.)
```

Students can also use the *Citation Index*. This index lists the citations of an article. The assumption is that the theme of a published article will be carried through in other researchers' work. As a result, after becoming more familiar with specific areas of psychology, students will begin to recognize seminal papers in those fields because contemporary work in the various areas of psychology frequently cites these original works. Thus, if someone were interested in iconic memory and knew the original work was done by George Sperling, that person could find out which contemporary researchers had cited that article in the *Citation Index*. A page of the *Citation Index* is reproduced in Figure 9.3. This page lists citations concerning the topic of mental rotation. Roger Shepard and Jacqueline Metzler published an important article on that topic in *Science* in 1971. The citation of that article is shown in the index. Following the date, journal, volume, and page of that article are the name, journal, volume, page, and year of sources in which the article was cited.

Several other indexes from areas related to psychology (e.g., biosciences, medicine, education, and general literature) can be found in most college libraries. In fact, all major abstracts are now available on computer data bases and can be accessed on-line for a fee. One particularly useful index for biomedical and clinical

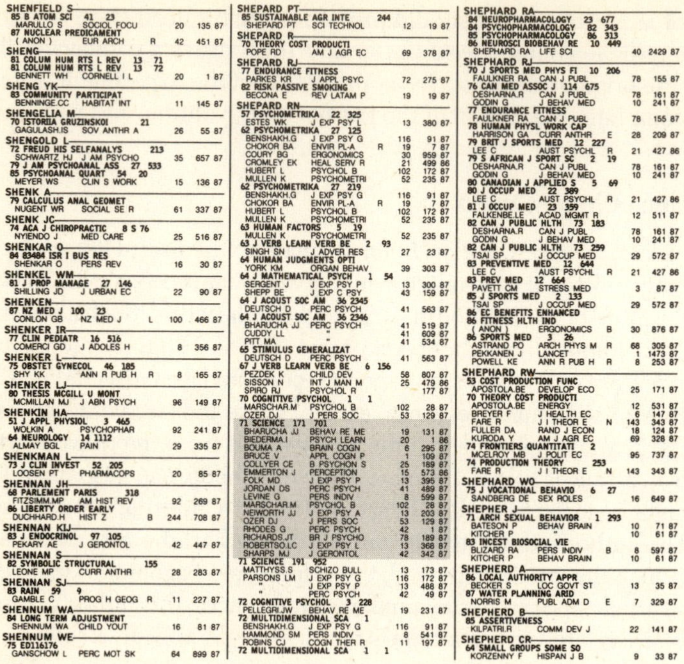

SHENFIELD S					
85 B ATOM SCI 41 23					
MARULLO S	SOCIOL FOCU		20	135	87
87 NUCLEAR PREDICAMENT					
(ANON)	EUR ARCH	R	42	451	87
SHENG					
81 COLUM HUM RTS L REV 13 71					
81 COLUM HUM RTS L REV 13 72					
BENNETT WH	CORNELL I L		20	1	87
SHENG YK					
83 COMMUNITY PARTICIPAT					
BENNINGE.CC	HABITAT INT		11	145	87
SHENGELIA M					
70 ISTORIIA GRUZINSKOI 21					
GAGULASH.IS	SOV ANTHR A		26	55	87
SHENGOLD L					
72 FREUD HIS SELFANALYS 213					
SCHWARTZ HJ	J AM PSYCHO		35	657	87
79 J AM PSYCHOANAL ASS 27 533					
85 PSYCHOANAL QUART 54 20					
MEYER WS	CLIN S WORK		15	136	87
SHENK A					
79 CALCULUS ANAL GEOMET					
NUGENT WR	SOCIAL SE R		61	337	87
SHENK JC					
74 ACA J CHIROPRACTIC 8 S 76					
NYIENDO J	MED CARE		25	516	87
SHENKAR O					
84 83484 ISR I BUS RES					
SHENKAR O	PERS REV		16	30	87
SHENKEL WM					
81 J PROP MANAGE 27 146					
SHILLING JD	J URBAN EC		22	90	87
SHENKEN					
87 NZ MED J 100 23					
CONLON GB	NZ MED J	L	100	466	87
SHENKER IR					
77 CLIN PEDIATR 16 516					
COMERCI GD	J ADOLES H		8	356	87
SHENKER L					
75 OBSTET GYNECOL 46 185					
SHY KK	ANN R PUB H	R	8	165	87
SHENKER LJ					
80 THESIS MCGILL U MONT					
MCMILLAN MJ	J ABN PSYCH		96	149	87
SHENKIN HA					
51 J APPL PHYSIOL 3 465					
WOLKIN A	PSYCHOPHAR		92	241	87
64 NEUROLOGY 14 1112					
ALMAY BGL	PAIN		29	335	87
SHENKMAN L					
73 J CLIN INVEST 52 205					
LOOSEN PT	PHARMACOPS		20	85	87
SHENNAN JH					
68 PARLEMENT PARIS 318					
FITZSIMM.MP	AM HIST REV		92	269	87
86 LIBERTY ORDER EARLY					
DUCHHARD.H	HIST Z	B	244	708	87
SHENNAN KIJ					
83 J ENDOCRINOL 97 105					
PEKARY AE	J GERONTOL		42	447	87
SHENNAN S					
82 SYMBOLIC STRUCTURAL 155					
LEONE MP	CURR ANTHR		28	283	87
SHENNAN SJ					
83 RAIN 59 9					
GAMBLE C	PROG H GEOG	R	11	227	87
SHENNUM WA					
84 LONG TERM ADJUSTMENT					
SHENNUM WA	CHILD YOUT		16	81	87
SHENNUM WE					
75 ED116176					
GANSCHOW L	PERC MOT SK		64	899	87
SHEPARD PT					
85 SUSTAINABLE AGR INTE 244					
SHEPARD PT	SCI TECHNOL		12	19	87
SHEPARD R					
70 THEORY COST PRODUCTI					
POPE RD	AM J AGR EC		69	378	87
SHEPARD RJ					
77 ENDURANCE FITNESS					
PARKES KR	J APPL PSYC		72	275	87
82 RISK PASSIVE SMOKING					
BECONA E	REV LATAM P		19	19	87
SHEPARD RN					
57 PSYCHOMETRIKA 22 325					
ESTES WK	J EXP PSY L		13	380	87
62 PSYCHOMETRIKA 27 125					
BENSHAKH.G	J EXP PSY G		116	91	87
CHOKOR BA	ENVIR PL-A	R	19	7	87
COURY BG	ERGONOMICS		30	959	87
CROMLEY EK	HEAL SERV R		21	499	86
HUBERT L	PSYCHOL B		102	172	87
MULLEN K	PSYCHOMETRI		52	235	87
62 PSYCHOMETRIKA 27 219					
BENSHAKH.G	J EXP PSY G		116	91	87
CHOKOR BA	ENVIR PL-A	R	19	7	87
HUBERT L	PSYCHOL B		102	172	87
MULLEN K	PSYCHOMETRI		52	235	87
63 HUMAN FACTORS 5 19					
MULLEN K	PSYCHOMETRI		52	235	87
63 J VERB LEARN VERB BE 2 93					
SINGH SN	J ADVER RES		27	23	87
64 HUMAN JUDGMENTS OPTI					
YORK KM	ORGAN BEHAV		39	303	87
64 J MATHEMATICAL PSYCH 1 54					
SERGENT J	J EXP PSY P		13	300	87
SHEPP BE	J EXP C PSY		43	159	87
64 J ACOUST SOC AM 36 2345					
DEUTSCH D	PERC PSYCH		41	563	87
64 J ACOUST SOC AM 36 2346					
BHARUCHA JJ	PERC PSYCH		41	519	87
CUDDY LL	"		41	609	87
PITT MA	"		41	534	87
65 STIMULUS GENERALIZAT					
DEUTSCH D	PERC PSYCH		41	563	87
67 J VERB LEARN VERB BE 6 156					
PEZDEK K	CHILD DEV		58	807	87
SISSON N	INT J MAN M		25	479	86
SPIRO RJ	PSYCHOL R			177	87
70 COGNITIVE PSYCHOL 1 1					
MARSCHAR.M	PSYCHOL B		102	28	87
OZER DJ	J PERS SOC		53	129	87
71 SCIENCE 171 701					
BHARUCHA JJ	BEHAV RE ME		19	131	87
BIEDERMA.I	PSYCH LEARN		20	1	86
BOUMA A	BRAIN COGN		6	295	87
BRUCE V	APPL COGN P		1	109	87
COLLYER CE	B PSYCHON S		25	189	87
EMMERTON J	PERCEPTION		15	573	86
FOLK MD	J EXP PSY P		13	395	87
JORDAN DS	PERC PSYCH		41	489	87
LEVINE G	PERS INDIV		8	599	87
MARSCHAR.M	PSYCHOL B		102	28	87
NEIWORTH JJ	J EXP PSY A		13	203	87
OZER DJ	J PERS SOC		53	129	87
RHODES G	PERC PSYCH		42	1	87
RICHARDS.JT	BR J PSYCHO		78	189	87
ROBERTSO.LC	J EXP PSY L		13	368	87
SHARPS MJ	J GERONTOL		42	342	87
71 SCIENCE 191 952					
MATTHYSS.S	SCHIZO BULL		13	173	87
PARSONS LM	J EXP PSY G		116	172	87
"	J EXP PSY P		13	488	87
"	PERC PSYCH		42	49	87
72 COGNITIVE PSYCHOL 3 228					
PELLEGRI.JW	BEHAV RE ME		19	231	87
72 MULTIDIMENSIONAL SCA 1					
BENSHAKH.G	J EXP PSY G		116	91	87
HAMMOND SM	PERS INDIV		8	541	87
ROBINS CJ	COGN THER R		11	197	87
72 MULTIDIMENSIONAL SCA 1 1					
SHEPHARD RA					
84 NEUROPHARMACOLOGY 23 677					
84 PSYCHOPHARMACOLOGY 82 343					
85 PSYCHOPHARMACOLOGY 86 313					
86 NEUROSCI BIOBEHAV RE 10 449					
SHEPHARD RA	LIFE SCI		40	2429	87
SHEPHARD RJ					
70 J SPORTS MED PHYS FI 10 206					
FAULKNER RA	CAN J PUBL		78	155	87
76 CAN MED ASSOC J 114 675					
DESHARNA.R	CAN J PUBL		78	161	87
GODIN G	J BEHAV MED		10	241	87
77 ENDURANCE FITNESS					
FAULKNER RA	CAN J PUBL		78	155	87
78 HUMAN PHYSL WORK CAP					
HARRISON GA	CURR ANTHR	E	28	209	87
79 BRIT J SPORTS MED 12 227					
LEE C	AUST PSYCHL	R	21	427	86
79 S AFRICAN J SPORT SC 2 19					
DESHARNA.R	CAN J PUBL		78	161	87
GODIN G	J BEHAV MED		10	241	87
80 CANADIAN J APPLIED S 5 69					
80 J OCCUP MED 22 389					
LEE C	AUST PSYCHL	R	21	427	86
81 J OCCUP MED 23 359					
FALKENBE.LE	ACAD MGMT R		12	511	87
82 CAN J PUBLIC HLTH 73 183					
DESHARNA.R	CAN J PUBL		78	161	87
GODIN G	J BEHAV MED		10	241	87
82 CAN J PUBLIC HLTH 73 259					
TSAI SP	J OCCUP MED		29	572	87
83 PREVENTIVE MED 12 644					
LEE C	AUST PSYCHL	R	21	427	86
83 PREV MED 12 664					
PAVETT CM	STRESS MED		3	87	87
85 J SPORTS MED 2 133					
TSAI SP	J OCCUP MED		29	572	87
86 EC BENEFITS ENHANCED					
86 FITNESS HLTH IND					
(ANON)	ERGONOMICS	B	30	876	87
86 SPORTS MED 3 26					
ASTRAND PO	ARCH PHYS M	R	68	305	87
PEKKANEN J	LANCET		1	1473	87
POWELL KE	ANN R PUB H	R	8	253	87
SHEPHARD RW					
53 COST PRODUCTION FUNC					
APOSTOLA.BE	DEVELOP ECO		25	171	87
70 THEORY COST PRODUCTI					
APOSTOLA.BE	ENERGY		12	531	87
BREYER F	J HEALTH EC		6	147	87
FARE R	J I THEOR E	N	143	343	87
FULLER DA	RAND J ECON		18	124	87
KURODA Y	AM J AGR EC		69	328	87
74 FRONTIERS QUANTITATI 2					
MCELROY MB	J POLIT EC		95	737	87
74 PRODUCTION THEORY 253					
FARE R	J I THEOR E	N	143	343	87
SHEPHARD WO					
75 J VOCATIONAL BEHAVIO 6 27					
SANDBERG DE	SEX ROLES		16	649	87
SHEPHER J					
71 ARCH SEXUAL BEHAVIOR 1 293					
BATESON P	BEHAV BRAIN		10	71	87
KITCHER P	"		10	61	87
83 INCEST BIOSOCIAL VIE					
BLIZARD RA	PERS INDIV	B	8	597	87
KITCHER P	BEHAV BRAIN		10	61	87
SHEPHERD A					
86 LOCAL AUTHORITY APPR					
BECKER S	LOC GOVT ST		13	35	87
87 WATER PLANNING ARID					
NORRIS M	PUBL ADM D	E	7	329	87
SHEPHERD B					
85 ASSERTIVENESS					
KILPATRI.R	COMM DEV J		22	141	87
SHEPHERD CR					
64 SMALL GROUPS SOME SO					
KORZENNY F	HISPAN J B		9	33	87

Figure 9.3 Unnumbered page of *Citation Index*. Reprinted by permission.

psychology research is *Index Medicus*. It does not contain abstracts, but it does contain titles and notations of articles.

Reading and Understanding Psychological Articles

After finding several articles pertinent to an area of interest, how does someone go about digesting the material therein? Most psychological articles are technical reports littered with arcane jargon, and it takes practice to understand them. To better understand the peculiar language and unfamiliar structure of psychology, we suggest two methods:

1. Read with an inquiring mind.
   - The first stage is to skim an article to gain an overview of the problem, the design, and the conclusions.
   - Read the article for details.
   - Review the article and identify the most important points.
   - Ask questions when reading. For example, in the first part of an article someone might ask, What is the relationship between past results and the current experiment? Someone could also question the methods, the results, and the conclusions that the investigator reaches.

2. Become familiar with the structure of articles in psychology—the structure is amazingly consistent.
   - First is the title and affiliation of the author.
   - Second is a brief summary or abstract.
   - Third is the body of the experiment, which includes a review of previous studies.
   - Fourth is a description of the experiment and the results.
   - Fifth is a discussion of the findings with the references.

Students should look at an article and try to identify these components.

A Form for Recording Articles

Comprehending even complex experiments in psychology is a teachable skill. In addition to reading with an inquiring mind and learning the structure of articles, psychology students need to organize their notes and thoughts about an article. The form that follows shows one way to organize notes.

After reading an article students should fill out a copy of the form, identifying the topic, the author(s), the title, the publication from which the article was taken, the problem being investigated and how the researcher(s) did the task, the results and interpretation, and any criticisms the student has, including additional research necessary to answer some of the questions raised.

We have found that many students keep these brief reviews and refer to them later in their careers. Even professional psychologists have found this form useful. Students can copy this form on a personal computer, photocopy the form, or develop a form of their own. Those who consistently record their summaries of articles will have a sizable collection of reviews in a short time, and this may prove useful later when developing research ideas. The summaries also offer a strong reminder of previous readings.

REVIEW OF ARTICLE FORM

TOPIC _____   DATE _____

AUTHOR(S) _____

BRIEF TITLE _____

SOURCE _____

PROBLEM _____

_____

_____

HOW INVESTIGATED _____

_____

_____

_____

RESULTS _____

_____

_____

_____

_____

INTERPRETATION _____

_____

_____

CRITICISM _____

_____

_____

ADDITIONAL RESEARCH TO DO _____

_____

_____

_____

Chapter 10

Conducting Research and Writing a Research Paper

Hofstadter's Law: It always takes longer than you expect, even when you take into account Hofstadter's Law.
                                        —*Douglas R. Hofstadter*

This chapter is about doing research and writing a report about that research. In one sense, these are the most complicated steps in the experimental process, but if an experiment has been designed carefully, these steps follow logically. Figure 10.1 shows the essential stages. It should be noted, however, that these steps are only guidelines. Many times in an actual research program the researcher must backtrack, start over, reconsider hypotheses, read new literature, go back to the drawing board, redesign the experiment, take a long and thoughtful walk, seek help, abandon pet ideas, think new thoughts, consider the practical aspects of the proposal, and always apply logic, creative thought, and hopefully wisdom throughout every part of the process. First, consider the matter of doing research.

DOING RESEARCH

Steps Involved

Before beginning an experiment, a researcher should be reasonably familiar with the previous findings in the field. It may be that someone else has done the

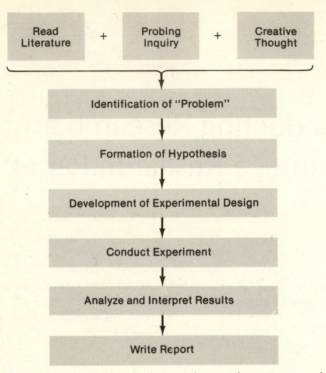

Figure 10.1   Steps involved in planning, doing, and reporting research.

experiment. But, as was mentioned in the previous chapter, the vast amount of literature in psychology makes it possible to know what has been done and, more importantly, what needs to be done. The information contained in previous studies may also suggest a new **problem**—a term that means a question proposed for solution. Practically, a problem suggests that either the current state of knowledge is lacking in some critical dimension or a gap in knowledge exists. A **hypothesis,** or testable proposition, can be constructed from a problem. After a problem is identified and a hypothesis is developed, a researcher can carefully plan an experiment that will test the hypothesis and solve the problem. In this stage, proper attention should be given to design and control (see previous chapters). The final step is writing a research paper.

As straightforward as this scheme may sound—from literature, to needs, to problem, to hypothesis, to plan, to experiment, to report—we remind students of Murphy's Law[13]: "Anything that can go wrong will go wrong," and its corollary:

13. See also Hofstadter's Law: beginning of the chapter.

"Murphy was an optimist." Be prepared to have some disappointments; all researchers fail. But researchers can minimize the likelihood of a total fiasco through careful planning.

Plan Research Projects That Are Realistic

A key ingredient in planning an experiment that can be completed is the development of a manageable project. If someone is interested in human brain research but has neither the skill nor the equipment necessary, or if someone is interested in the mating behavior of whales in their natural habitat but lives in Nebraska, it is likely that the results would only be publishable in the *Journal of Frustrating and Quixotic Ventures.* It is not that bizarre thoughts are prohibited from the experimental laboratory or that one should not think expansively. The point is that people who plan grandiose research programs must be prepared to endure the hardships involved in carrying them out, be it by studying brain surgery or moving to California.

Numerous credible research projects may involve fancy equipment that is unavailable in many laboratories. But if the research is important, it may be worthwhile to seek funds through grants or an institution's equipment moneys. But sometimes this process can take longer than doing the research. An alternative would be to go to a laboratory (such as a medical school) or setting (such as the ocean) where the project could be done.

Another practical consideration is the availability of subjects. Nebraska doesn't have any whales. Likewise, there aren't many left-handed piccolo players in Minot, North Dakota, or Gaboon vipers in Amherst, Massachusetts. In some areas, it may even be difficult for researchers to find usable rats, college-aged subjects, schizophrenics, bed wetters, or third graders. Also, students should be mindful of the ethical considerations involved in working with subjects (see Chapter 8).

Another consideration is whether the proposed research can be done within the time available. In our experience, this component more than any other has caused students (and their professors) the most grief. Experimental projects usually take longer to complete than first anticipated, which may be due to equipment failure, the necessity to replan an experiment, the failure of subjects to show up for an experiment, natural disasters, unreliable coworkers, new discoveries in the field, power failures, spring break, and the like. Good research requires time and dedication, and we know of no prominent researcher who has not devoted a major portion of his or her lifetime to science.

Do a Pilot Experiment

Even if a research project is methodically planned to the most minute detail, it is still important to run test trials in order to become thoroughly acquainted with the technique and procedure. It also may be necessary to debug the methodology and/or equipment.

Planning a Research Program

Throughout the years that we have done research with students and collaborators, we have found it useful to follow a prescribed form in planning and thinking about experimental research. One such form follows. It includes the name of the research, the experimenter(s), the topic, the problem, and the significance. Then a brief space is provided for previous research, theory, design, and results, and a section is included for a brief description of the experimental paradigm and the analysis of data. Then space follows for a list of the materials needed, including subjects, and finally, space is provided for an anticipated timetable and a list of the responsibilities of the researchers. (A brief note: In order to avoid hassles with collaborators it is important to spell out in detail the responsibilities of each member of a research team. Then the collaborators can discuss who will be listed as an author, the order of authorship,[14] who will be mentioned in footnotes, and who will receive no acknowledgment.)

This form is an outline that can be expanded. It also can be stored in computer memory, which will increase the flexibility of its use.

In practice, we keep track of our research proposals by using different-colored sheets of paper. When contemplating a project, we begin by filling out a blue proposal form. If the project develops to the point where it appears that we will actually do the experiment, we then use the green form (we "go green"). We also keep a running file of reprints, diagrams, the list of equipment, and data collected.

Writing a Research Paper

The final stage of experimentation is the preparation of a research paper—after all, results that are not reported to the scientific community are of little use. After a researcher has formulated a problem (usually after reading the literature in the field), developed and conducted an experiment, and analyzed the data, then he or she faces the problem of how to best communicate the findings to others. The investigator is encouraged to prepare his or her report in a style consistent with the current editorial policies of contemporary journals in psychology. Nearly all psychological journals use the style found in the APA's *Publication Manual* (1983), so the serious researcher should obtain a copy of this guide. The style of the journal articles discussed in this section is consistent with this manual.

Good scientific writing takes practice. It is unlikely that anyone's first attempt at writing a scientific paper will be acceptable, so we encourage students to rewrite and revise their material frequently. Even experienced writers must constantly re-

---

14. If two or more authors contribute equally to a project, order of authorship is sometimes determined by chance, as by flipping a coin. Sometimes contributors to an ongoing program will alternate the order of names on papers. The matter should be one not of ego but of attributing credit (or blame) for the research in direct proportion to the actual contribution made by each member of the team.

RESEARCH PROPOSAL

NAME OF RESEARCH_____   EXPERIMENTER(S) _____
                                                  _____
                                                  _____

                                    DATE_____

TOPIC _____
_____

PROBLEM _____
_____

SIGNIFICANCE _____
_____

PREVIOUS RESEARCH
    THEORY _____
_____

    DESIGN _____
_____

    RESULTS _____
_____

EXPERIMENTAL PARADIGM _____
_____

    ANALYSIS OF DATA _____
_____

MATERIALS _____
_____
_____

TIMETABLE AND RESPONSIBILITIES
PREPARATION OF MATERIALS

	PREPARATION OF MATERIALS_____ BY _____
PHASE 1	LITERATURE SEARCH_____ BY _____
	PILOT (DRY RUN)_____ BY _____
PHASE 2	RUN EXPERIMENT_____ BY _____
	ANALYSIS OF DATA_____ BY _____
PHASE 3	WRITE PAPER_____ BY _____
	SUBMIT FOR PUBLICATION_____ BY _____

© *R. Solso*

write and revise their material. Some guidelines follow that should help in preparing a scientific paper, but we still emphasize the importance of practice.

Writing Style

Scientific writing must communicate ideas clearly. To do this, the APA *Publication Manual* makes a few suggestions.

> To achieve clarity, good writing must be precise in its words, free of ambiguity, orderly in its presentation of ideas, economical in expression, smooth in flow, and considerate of its readers. A successful writer invites readers to read, encourages them to continue, and makes their task agreeable by leading them from thought to thought in a manner that evolves from clear thinking and logical development. (1974, p. 25)

A component of good scientific writing is the selection of words that convey a precise meaning. Ambiguous words should be avoided. In casual conversation it may be acceptable to use such expressions as "for the most part," "very few," "I would estimate," "an intelligence test was used," or "animals were deprived." But in scientific writing the author should use words and phrases that can be operationally defined. For example, "very few" leaves the reader mystified as to how many, and "an intelligence test was used" does not specify which intelligence test was used.

In addition, ideas should be developed logically. This is particularly important when introducing the research and discussing the results, but it is also important when describing the scientific method. To illustrate this, consider the review of literature section of a research paper, which is normally the first major section. The purpose of this section is to introduce the reader to previous research on the subject so that the research question being addressed is in a meaningful context. It is impossible and distracting to recount all related research, so the writer must select the most pertinent studies and present them in such a way that the reader can understand the development of previous experiments and/or theories. (Part Two of this book contains many excellent examples of how other authors have logically developed reviews of literature.)

Careful attention should also be given to grammar and paragraph development, and awkward word sequences should be avoided. Sentences should be technically correct, but just as important, they should state clearly the intended meaning.

A common error is to incorporate too many thoughts in a single paragraph. A paragraph should have a controlling idea that is supported by every sentence in the paragraph. Then the paragraphs should be organized around a major principle. If the topic is "the auditory feedback in bats in a dark room" or "children's dreams and bed-wetting" or "the influence of green and blue packaging on the sales of laundry detergent," all components of the paragraphs should revolve around that theme.

As the previous discussion suggests, the author has great freedom in choosing a writing *style* but has much less freedom in the overall *structure* of the report. The structure that has evolved in American psychology follows a tightly prescribed model, which follows. (This model is discussed further in Part Two.)

Title, author(s), and affiliation
Abstract
Review of literature, problem, and hypothesis to be tested
Method
    Subjects
    Apparatus or materials
    Procedure
Results
Discussion (including a summary)
References

Consider each of these topics:

1. **Title, author(s), and affiliation.** The title should be a clean, brief statement of the subject under inquiry. Authors frequently suggest a title such as "The Effects of ____ on ____," which gives the reader a clue as to the independent and dependent variables. The title should be followed by the author's name and affiliation.

2. **Abstract.** Here the author should summarize the article. Although the abstract usually contains less than 150 words, the researcher should try to identify the problem, method, and results. The abstract should interest and inform the reader in a form that makes the article easier to understand.

3. **Review of literature, problem, and hypothesis to be tested.** Articles typically begin with a brief historical review of major issues and previous findings in the area. The results of other researchers should be presented in a sequence so that the current experiment follows logically from a series of experimental developments. The statement of the problem and the hypothesis are frequently contained in a single paragraph, if not in a single sentence. An "if ____ then ____" structure works well for presenting a hypothesis. The independent and dependent variables frequently are identified in this section.

4. **Method.** The method section should describe the experimental design in sufficient detail to allow replication of the experiment. There are usually three parts to this description: (1) the subjects, (2) the apparatus or materials, and (3) the procedure. All variables that may affect the results should be identified here, along with a description of how they will be controlled. If a procedure has been described previously by another researcher, then it is permissible simply to cite the original source.

5. **Results.** In this section, all relevant data generated by the experiment and the analysis of these data should be reported. All relevant data should be presented in the text, but they may also be presented in tabular or graphic form. Because of space limitations, the material should be presented in the most succinct form. Since it is usually impractical to present the data for every subject, an experimenter should employ those statistical techniques in summarizing the results that most clearly convey the findings.

6. **Discussion.** The purpose of this section is to interpret the results of the experiment. This should include pointing out any reservations about the results, noting similarities or differences between the results and previous findings, suggesting future research, and noting the implications of the results to theory and/or practice in the specific research area.
7. **References.** All material cited in the article should be noted in this section. The general format is as follows: author's name, article title, journal, publication year, volume, and pages. (For specific examples, see the APA *Publication Manual* or the examples at the end of the articles in Part Two.)

REPORTING DATA: QUANTIFICATION OF OBSERVATIONS

Experimental results can be reported verbally ("The racially prejudiced subjects saw more instances of aggression in the environment"), statistically ("The probability of obtaining these data by chance is less than 1 in 100, or $p <$ .01), or graphically. In this book only the barest of statistical analyses are discussed,[15] not because they are unimportant—they are—but because in this book we emphasize research design in psychology. Other books deal with statistics in great detail.

Quantification of Observations

In almost every psychological experiment researchers collect numerical data, which can be conceptualized in two categories: descriptive data and statistical data. Conveniently, these two types of data correspond to the two major types of experimental investigation: observational studies and experimental studies.

Observational studies, as discussed earlier, include naturalistic observations, **statistical studies,** and correlational studies. An example of naturalistic observation might be the recording of the frequency of high-pitched sounds made by a male blue-winged warbler before and after the mating season. Parameter estimates are frequently numerical descriptions of a sample of subjects, such as voters, from which generalizations are drawn. Correlational studies are designed to determine the relationship between two or more variables, such as grades in college and grades in high school, blood pressure and income, creativity and IQ, and so on. In general, the purpose of observational studies is to give an accurate description of a naturally occurring phenomenon. Results of these studies are frequently reported in terms of **descriptive statistics,** which are mathematical techniques used to characterize subjects that are observed. Experimenters may, for example, choose to characterize

15. In Appendix A there is a brief description of the mathematical procedures used in some of the statistical tests covered in this book.

the intelligence of the sophomore class at the University of Michigan. Those data (IQ test results, for instance) could be summarized in terms of a mean score and a deviation score or in the form of a graph. (In psychological journal articles both statistical data and graphic data are reported in order to accurately present the results of a study.)

Experimental studies involve the manipulation of one or more variables whose effect is then measured statistically. In both experimental and observational studies, though, an additional type of statistical application might be applied. These are called **inferential statistics,** which are mathematical techniques in which a sample of information is gathered and generalizations are made based on that sample. One could, for example, study the effect of an ad campaign for a strawberry-flavored breakfast cereal on people in northeastern Ohio with the aim of generalizing the results to a larger population, such as the entire country.

Statistics.   The testing of hypotheses with statistical tests is an important part of modern experimental psychology. A wide range of these problems has already arisen in previous sections and will appear again in the second part of this text. Determining which statistical test to apply to a certain type of data is a major theme in courses in statistics, and with fundamental knowledge of the tests and their assumptions, anyone can properly apply the correct test to the correct situation. The task is made even easier today with the many computer packages that will calculate a wide variety of statistical tests in seconds. These packages allow an experimental psychologists to calculate complex equations, such as those for F, t, and χ^2, and other, sometimes exotic, statistics, without knowing the mathematical steps involved. Data in, data out!

But an element of controversy surrounds the application of statistics to psychological studies—not in terms of their use (there is general agreement that statistics are indispensable in much of experimental psychology) but in terms of the level of sophistication required to apply statistics appropriately. Some argue that only after one takes several statistics courses can he or she be competent enough to use statistical tests judiciously. But others contend that only the bare rudiments of statistical theory are necessary before one can aptly apply statistical tests with the help of the many computer programs available today. While we believe that a basic course in statistics is vital for the experimental psychologist, we also endorse the use of the many computer packages, which circumvent many arduous computational procedures. But basic knowledge of statistics is still critical not only because it allows the experimenter to select the appropriate statistical test, but also because it allows the investigator to interpret data precisely. Such knowledge also helps when interpreting the results of others.

Bar Graphs.   Data can also be represented in a bar graph or histogram. The histogram is a particularly vivid graphic form that allows the reader to form a quick impression of the results of an experiment.

➤ CASE STUDY

An example demonstrating how a histogram can effectively portray information is shown in an experiment by Czeisler, Moore, Ede, and Coleman (1982) in which shift workers in a minerals and chemical corporation were given work schedules that either corresponded to their natural circadian schedules (day/night biological schedules) or did not. Worker preference was then measured for two schedules: one in harmony with the circadian schedule and one not in harmony with the schedule. In this example the dependent variable was worker satisfaction, which was measured on questionnaires that were distributed three months after the introduction of the independent variable—out of harmony with the natural circadian schedule (Schedule A) or in harmony with the natural circadian schedule (Schedule B). As Figure 10.2 clearly shows, the workers preferred the harmonious schedule.

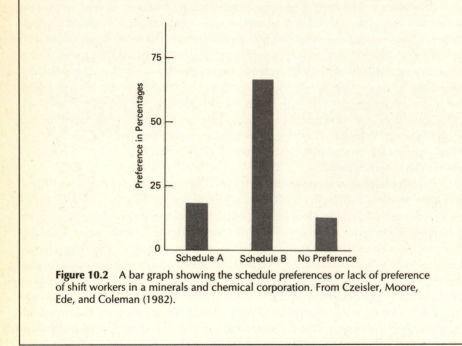

Figure 10.2   A bar graph showing the schedule preferences or lack of preference of shift workers in a minerals and chemical corporation. From Czeisler, Moore, Ede, and Coleman (1982).

Typically, the manipulated element (cause) is depicted on the horizontal plane, while the consequence (effect) is shown on the vertical axis.

In this example the data along the ordinate, that is, the horizontal axis (the vertical axis is called the abscissa), are discrete—meaning the categories represented on the horizontal plane are distinct and noncontinuous. In many psychology experiments the variables may be more or less continuous. These variables could include time, IQ, running speed, delay of reinforcer, or amount or frequency of drug administration.

An example of a histogram that uses continuous data can be seen in the dramatic portrayal of the incidence of cancers among men who stop using cigarettes. The relationship between the time that elapses after one quits smoking and the likelihood of developing cancer is clearly shown in the three histograms in Figure 10.4 on page 151. These figures also add present smokers and nonsmokers on the extreme ends of the ordinate, although strictly speaking, these two classes are non-continuous and out of place with the continuous data (years of smoking cessation). But their inclusion as two types of base data (or, in this case, a type of control) gives a fuller picture of the effect of stopping smoking on the risk of cancer. Note that the study shows that the risk of lung, oral, and esophagus cancer is actually greater for those who had quit smoking 1–3 years before the study than for present smokers.

Figures and Functions.   A figure can also illustrate the relationship between two or more variables. For example, a curve can be drawn any time an experimenter has values for a dependent variable and for several levels of some other variable (usually the independent variable). Plotting curves allows the researcher to see the relationships between two variables. For example, the shapes of learning curves can be compared to determine more precisely the effects of certain types of reinforcement on conditioning.

At a more sophisticated level, figures can be used to derive a mathematical equation that describes the relationship between two variables. This mathematical equation, called a **function,** is simply a shorthand method of describing the relationship, so that once the function is derived, all one has to do is substitute the value for one variable to find the value of the second variable.

Two examples demonstrating functional relationships follow. These are classic studies, one concerning audition and the other concerning memory.

▶ *Case Study*

Several audition experiments have been conducted to determine the relationship between the frequency (pitch) of a tone (measured in cycles per second) and a subject's sensitivity to the intensity of that tone. Hirsch (1966) investigated this question by presenting a pure tone to a subject through earphones. The tone was presented at a clearly audible level, and then the intensity (measured in decibels) was gradually decreased until the subject signaled that he or she could no longer hear the tone. The decibel level at which this signal was given was called the descending threshold. The procedure was then reversed, with the same tone being presented at an inaudible intensity that was gradually increased until the subject signaled that he or she could hear the tone. This decibel level was called the ascending threshold. This procedure was repeated several times for each subject using the same tone.

In this experiment, the dependent variable is the absolute threshold, which is the decibel level above which the subject will hear the tone and below which he or she will not hear it. The absolute threshold for a given tone is determined by averaging the threshold values obtained in the descending and ascending trials. For example, if a tone of 100 cycles per second is presented to a subject, the values of the descending threshold for three trials might be 40, 45, and 42 decibels, and the values of the ascending threshold might be 48, 49, and 52 decibels. To get the absolute threshold we would add these six values (276) and divide by the number of values (6), which would yield an absolute threshold of 46 decibels for that tone for that subject.

After this had been determined, the procedure could be repeated using a different frequency. Once the absolute threshold for several tones of varying frequencies had been ascertained, the experimenter could plot a curve. Figure 10.3 has been constructed from data presented by Stevens and Davis (1938). In this study, four subjects were tested at 10 frequency levels.

The points on the curve represent the mean absolute threshold for the four subjects at each level. For example, at the frequency of 50 cycles per second, the absolute threshold values for the four subjects were 59, 60, 63, and 66 decibels. To get the mean, these four values were added (248), and the sum was divided by the number of values (4), which yielded a mean of 62 decibels. This was done for each of the levels tested. The points were then connected to form a curve.

The results indicate that the subjects were rather insensitive to low-frequency tones. Sensitivity increased as frequency increased up to a point, with the subjects being most sensitive to tones that ranged between 1,000 and 3,000 cycles per second. Sensitivity decreased outside that range.

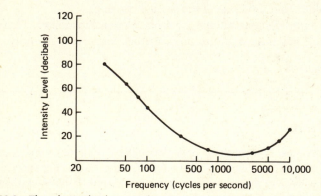

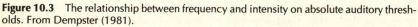

Figure 10.3   The relationship between frequency and intensity on absolute auditory thresholds. From Dempster (1981).

The independent variable in the previous example is frequency, measured in cycles per second, and the dependent variable is the absolute threshold, measured

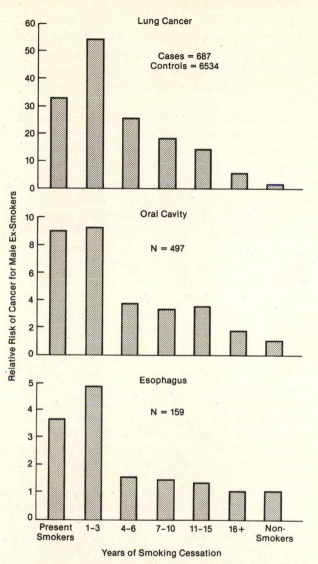

Figure 10.4   Male ex-smokers' relative risk of three types of cancer, by years since quitting. From the American Health Foundation's control study of 3,716 cancer cases and 18,000 controls (Wynder & Stellman, 1977). Copyright 1977 by Cancer Research, Inc. Reprinted by permission.

on a decibel scale. When plotting a curve it is customary to place the dependent variable on the vertical axis (ordinate) and the independent variable on the horizontal axis (abscissa), with increasing values as one moves away from the point at which the two axes intersect. This procedure was followed in plotting the curve in the previous example. The resulting curve clearly illustrates the relationship between the two variables.

In the previous example, there were 10 levels of the independent variable (frequency), and each subject was tested at all 10 levels. The curve in Figure 10.4 combines the performance of the subjects at each of these 10 levels. Another type of curve frequently encountered in psychological research is one in which each subject is tested at only one level of the independent variable. However, the subject is tested at this level over a series of trials or time periods.

The second example of a functional relationship comes from developmental psychology and memory research. The question posed asks, Does memory for digits increase as a function of age? The results of an experiment by Dempster (1981) indicate that as age increases so too does the ability to remember digits. The results are shown in Figure 10.5.

➤ CASE STUDY

It is hard to think of human behavior without considering memory, and recently memory has been a major focus of cognitive psychologists. Research in any field often begins with the study of simple elements and then advances to more complex elements. In memory research, psychologists have for many decades examined a person's memory span for immediate events. Studies of this sort have been integrated into larger theories of memory, especially the theory of short-term memory—a hypothetical memory system that stores a limited amount of information for about 12 seconds.

A task that is commonly used in this area of memory research is memory for digits—the digit span test. This test is so common that it is part of several popular individual intelligence tests.

Digit span tests require subjects to repeat a series of numbers that have been presented. If the experimenter reads 4-8-3-6-9, the subject's task is to repeat the numbers 4-8-3-6-9.

An intriguing question in memory research is whether memory span for digits increases as a function of age, and if so, how that relationship might appear.

In a study by Dempster (1981), children of various ages were given a digit span test. The results are shown in Figure 10.5 and indicate that digit span increases as a function of age. Why this occurs is an even more fascinating puzzle. The answer may be found in the ability of older subjects to group information into more meaningful chunks. An older subject may chunk a series into two well-known mathematical progressions (4-8 and 3-6-9), which would reduce the information load substantially. Everyone uses these types of memory tricks, even if it's done subconsciously.

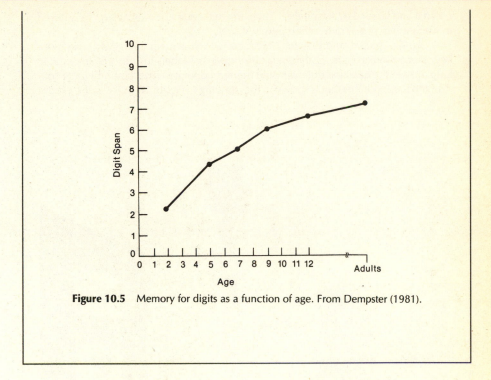

Figure 10.5 Memory for digits as a function of age. From Dempster (1981).

In addition to demonstrating a functional relationship, this example also raises the issue of stimulus control. If the purpose of the experiment is to study pure memory (i.e., memory that is not based on the chunking of meaningful material), then the experimenter should try to avoid the use of any series that could be meaningfully organized. This can be very difficult.

Illustrations. In addition to the use of statistics, a researcher can use numerous visual forms to present information. In some instances, these graphic forms can be superior to a verbal presentation. Consider the photographs and drawing used in the following example.

▶ *CASE STUDY*

Humans have an unusual ability to remember photographs, and some psychologists have argued that such an ability may be related to the verbal labeling of visual stimuli. This argument suggests, for example, that someone is able to remember a picture of a flashlight because when that person saw a flashlight he or she recognized it and labeled it "flashlight." Thus, the real code (the way an article is stored in memory) is verbal rather than purely visual.

Sands, Lincoln, and Wright (1982) studied visual memory in the rhesus monkey. Since monkeys are thought not to have verbal labels for objects, they could help answer the question of how visual memory operates. The experimenters constructed the apparatus that is shown in the drawing in Figure 10.6. Next, the experi-

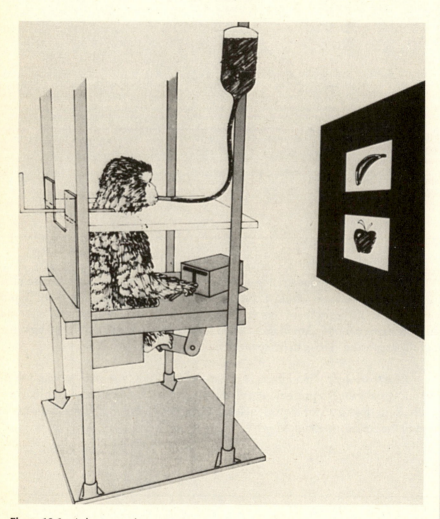

Figure 10.6   A rhesus monkey operating an apparatus with a lever in a test of visual memory conducted by Sands, Lincoln, and Wright (1982).

menters presented their subjects with two photographs at a time. The photographs were presented one above the other (a banana and an apple are illustrated). The monkey was trained to slide a lever to the right or left to indicate whether the objects were the same or different. Reinforcement was delivered by liquid in the bottle. Some of the faces that were used are shown in Figure 10.7.

Figure 10.7   Some of the faces shown to the rhesus monkey in Sands, Lincoln, and Wright's study of visual memory (1982).

In this example, the use of a drawing and several photographs helps the reader grasp the technique used while greatly facilitating the reproduction of the experiment.

The researchers found that a visual stimulus, such as a face, could be differentiated by monkeys, which suggested that visual memory in primates can be represented by a visual code.

Next we turn to a more detailed example of how experimental results can be reported in manuscript form.

EXAMPLE OF A MANUSCRIPT

This section contains an actual research article that has been changed slightly from the published version. The article is presented in the form of a manuscript prepared for submission to a professional journal, and it provides a model of good experimental design and demonstrates proper use of APA style. As students read the manuscript, they should pay close attention to (1) the title, authors, and affiliation, (2) the abstract, (3) the review of the literature, the problem, and the hypothesis, (4) the method (including subjects, apparatus or materials, and procedure), (5) the results (presented both in statistical and graphic forms), (6) the discussion, and (7) the references. We have labeled these sections in the article, but on a submitted manuscript they would not be labeled.

Several features of the paper, written by Loftus and Palmer (1974), are particularly noteworthy. One is the abstract. The authors have done three essential things: They have told the reader something about the methodology, they have reported the basic results, and they have interpreted the results within a theoretical framework.

Also noteworthy is the review of literature, which is preceded by a series of questions designed to stimulate the reader's interest in the topic. This is not required, but in this experiment it serves an important function. The authors then present several findings of other researchers, and then describe the problem. Then they briefly describe the method, which is followed by an informal hypothesis. In describing the method, the authors give explicit information about what was done to whom. Then the results are presented in verbal, statistical, graphic, and tabular form. Of particular interest in this manuscript is the discussion. The authors briefly review the findings and then give two interpretations of the data. Then they discuss the findings and interpretation within the context of previous findings and theories. Their discussion is to the point and does not go beyond the results. The final section contains the references; students can use this section as a model when listing references in their own writings.

This experiment clearly portrays many of the essential features of design and style. When students write their own experiments either for class work or for submission to a professional journal, they may wish to refer to this manuscript as a model.

SUBMISSION OF MANUSCRIPT

After a manuscript has been written, the author submits it to a professional journal for review and possible publication. Selecting the appropriate journal for a manuscript can be difficult, especially for the beginning researcher. In general, researchers prefer to publish their work in accepted, well-known, respected journals rather than "schlock" periodicals. Of course, the difference between good and bad publications is a matter of interpretation. Journals devoted to empirical studies that are published by the American Psychological Association (e.g., *Journal of Ex-*

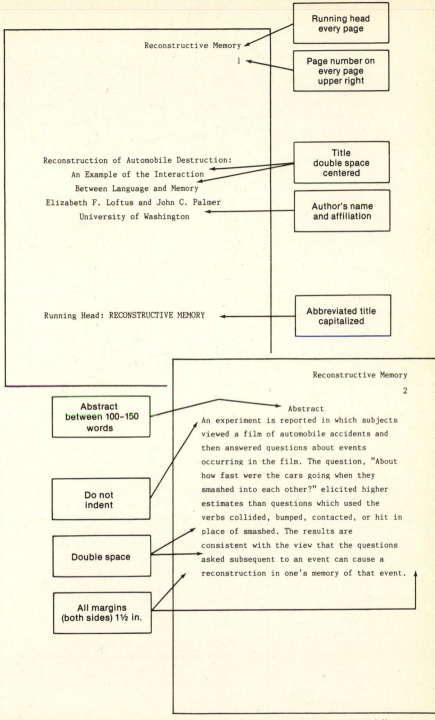

Figure 10.8 A model of experimental design and APA style *(continues on following 5 pages)*.

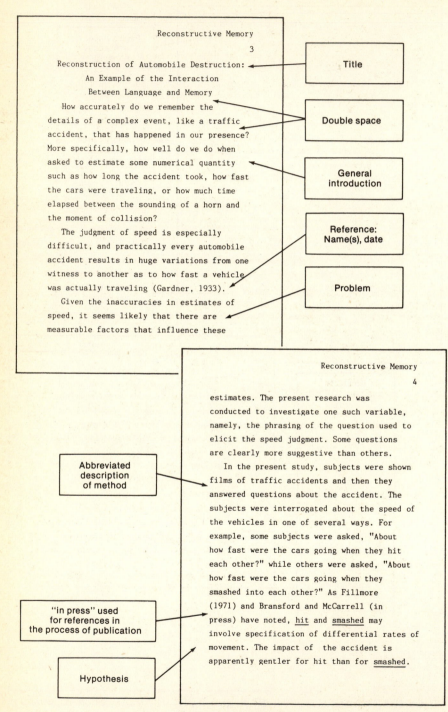

Reconstructive Memory

3

Reconstruction of Automobile Destruction:
An Example of the Interaction
Between Language and Memory

How accurately do we remember the
details of a complex event, like a traffic
accident, that has happened in our presence?
More specifically, how well do we do when
asked to estimate some numerical quantity
such as how long the accident took, how fast
the cars were traveling, or how much time
elapsed between the sounding of a horn and
the moment of collision?

The judgment of speed is especially
difficult, and practically every automobile
accident results in huge variations from one
witness to another as to how fast a vehicle
was actually traveling (Gardner, 1933).

Given the inaccuracies in estimates of
speed, it seems likely that there are
measurable factors that influence these

Title

Double space

**General
introduction**

**Reference:
Name(s), date**

Problem

Reconstructive Memory

4

estimates. The present research was
conducted to investigate one such variable,
namely, the phrasing of the question used to
elicit the speed judgment. Some questions
are clearly more suggestive than others.

In the present study, subjects were shown
films of traffic accidents and then they
answered questions about the accident. The
subjects were interrogated about the speed of
the vehicles in one of several ways. For
example, some subjects were asked, "About
how fast were the cars going when they hit
each other?" while others were asked, "About
how fast were the cars going when they
smashed into each other?" As Fillmore
(1971) and Bransford and McCarrell (in
press) have noted, hit and smashed may
involve specification of differential rates of
movement. The impact of the accident is
apparently gentler for hit than for smashed.

**Abbreviated
description
of method**

**"in press" used
for references in
the process of publication**

Hypothesis

Figure 10.8   *(continued)*

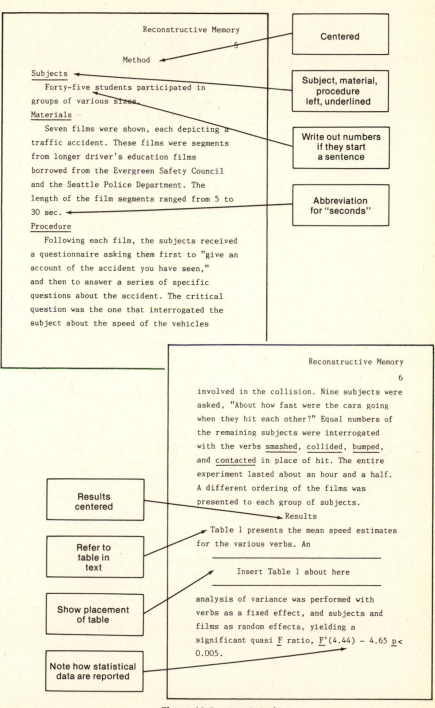

Reconstructive Memory

5

Method

Centered

Subjects

Forty-five students participated in groups of various sizes.

Materials

Seven films were shown, each depicting a traffic accident. These films were segments from longer driver's education films borrowed from the Evergreen Safety Council and the Seattle Police Department. The length of the film segments ranged from 5 to 30 sec.

Procedure

Following each film, the subjects received a questionnaire asking them first to "give an account of the accident you have seen," and then to answer a series of specific questions about the accident. The critical question was the one that interrogated the subject about the speed of the vehicles

Subject, material, procedure left, underlined

Write out numbers if they start a sentence

Abbreviation for "seconds"

Reconstructive Memory

6

involved in the collision. Nine subjects were asked, "About how fast were the cars going when they hit each other?" Equal numbers of the remaining subjects were interrogated with the verbs smashed, collided, bumped, and contacted in place of hit. The entire experiment lasted about an hour and a half. A different ordering of the films was presented to each group of subjects.

Results

Table 1 presents the mean speed estimates for the various verbs. An

Insert Table 1 about here

analysis of variance was performed with verbs as a fixed effect, and subjects and films as random effects, yielding a significant quasi F ratio, $F'(4,44) - 4.65$ $p < 0.005$.

Results centered

Refer to table in text

Show placement of table

Note how statistical data are reported

Figure 10.8 *(continued)*

Reconstructive Memory

7

The speed estimates for the various verbs are shown in Figure 1. Four of the seven films

---

Insert Figure 1 about here

> Show placement of figure

were staged crashes; the original purpose of these films was to illustrate what can happen to human beings when cars collide at various speeds. One collision took place at 20 mph, one at 30, and two at 40. The mean estimates of speed for these four films were: 37.7, 36.2, 39.7, and 36.1, respectively.

Discussion

> Centered

The results of this experiment indicate that the form of a question (in this case changes in a single word) can markedly and systematically affect a witness's answer to that question. The actual speed of vehicles controlled little variance in subject

Reconstructive Memory

8

reporting, while the phrasing of the question controlled considerable variance.

> Interpretation of data

Two interpretations of this finding are possible. First, it is possible that the differential speed estimates result merely from response-bias factors. A subject is uncertain whether to say 30 mph or 40 mph, for example, and the verb smashed biases his response toward the higher estimate. A second interpretation is that the question form causes a change in the subject's memory representation of the accident. The verb smashed may change a subject's memory such that he "sees" the accident as being more severe than it actually was. If this is the case, we might expect subjects to "remember" other details that did not actually occur but are commensurate with an accident occurring at higher speeds.

Figure 10.8   *(continued)*

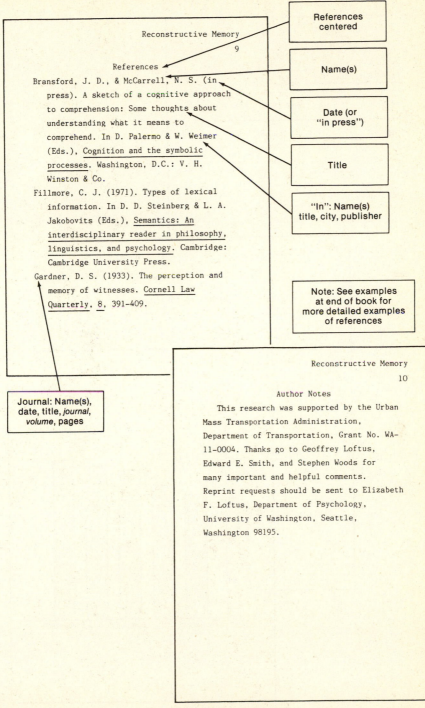

Reconstructive Memory

9

References centered

References

Name(s)

Bransford, J. D., & McCarrell, N. S. (in press). A sketch of a cognitive approach to comprehension: Some thoughts about understanding what it means to comprehend. In D. Palermo & W. Weimer (Eds.), Cognition and the symbolic processes. Washington, D.C.: V. H. Winston & Co.

Date (or "in press")

Title

Fillmore, C. J. (1971). Types of lexical information. In D. D. Steinberg & L. A. Jakobovits (Eds.), Semantics: An interdisciplinary reader in philosophy, linguistics, and psychology. Cambridge: Cambridge University Press.

"In": Name(s) title, city, publisher

Gardner, D. S. (1933). The perception and memory of witnesses. Cornell Law Quarterly, 8, 391–409.

Note: See examples at end of book for more detailed examples of references

Journal: Name(s), date, title, *journal*, *volume*, pages

Reconstructive Memory

10

Author Notes

This research was supported by the Urban Mass Transportation Administration, Department of Transportation, Grant No. WA–11–0004. Thanks go to Geoffrey Loftus, Edward E. Smith, and Stephen Woods for many important and helpful comments. Reprint requests should be sent to Elizabeth F. Loftus, Department of Psychology, University of Washington, Seattle, Washington 98195.

Figure 10.8   *(continued)*

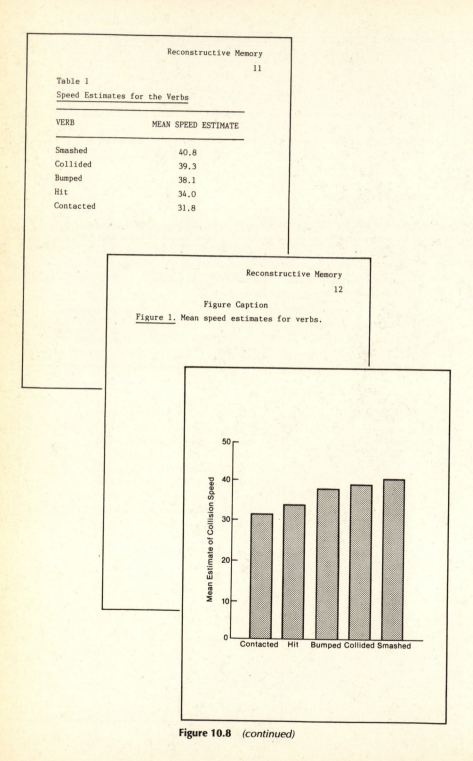

Reconstructive Memory

11

Table 1

Speed Estimates for the Verbs

VERB	MEAN SPEED ESTIMATE
Smashed	40.8
Collided	39.3
Bumped	38.1
Hit	34.0
Contacted	31.8

Reconstructive Memory

12

Figure Caption

Figure 1. Mean speed estimates for verbs.

Figure 10.8   *(continued)*

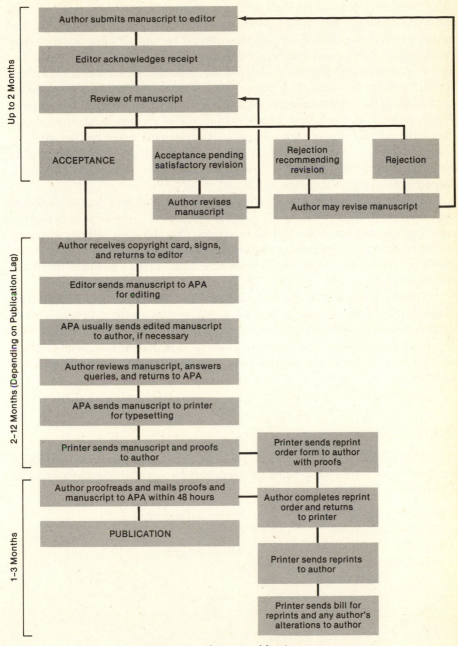

Up to 2 Months

Author submits manuscript to editor

Editor acknowledges receipt

Review of manuscript

ACCEPTANCE

Acceptance pending satisfactory revision

Rejection recommending revision

Rejection

Author revises manuscript

Author may revise manuscript

2–12 Months (Depending on Publication Lag)

Author receives copyright card, signs, and returns to editor

Editor sends manuscript to APA for editing

APA usually sends edited manuscript to author, if necessary

Author reviews manuscript, answers queries, and returns to APA

APA sends manuscript to printer for typesetting

Printer sends manuscript and proofs to author

Printer sends reprint order form to author with proofs

1–3 Months

Author proofreads and mails proofs and manuscript to APA within 48 hours

Author completes reprint order and returns to printer

PUBLICATION

Printer sends reprints to author

Printer sends bill for reprints and any author's alterations to author

Figure 10.9   The APA publication process.

perimental Psychology, which has four separate editions, *Child Development, Journal of Personality & Social Psychology, Developmental Psychology, Journal of Comparative Psychology,* and others) are unusually well regarded and widely read. The American Psychological Society publishes several high-quality journals including *Psychological Science* and *Current Directions in Psychological Science.* Also, numerous groups publish periodicals that include important research articles, including the Psychonomic Society (e.g., *Memory & Cognition, Perception & Psychophysics, Bulletin of the Psychonomic Society*) and the British Psychological Society (e.g., *British Journal of Psychology, Quarterly Journal of Experimental Psychology*). Many other journals also publish works (e.g., *American Journal of Psychology, Cognitive Psychology, Neuropsychology, Brain and Cognition, Cortex, Journal of Speech & Hearing Research, Science, Journal of Experimental Analysis of Behavior, Journal of Memory & Language,* and so on).

This list is not inclusive. Many other high-level publications deal with selected topics in experimental psychology. It is the responsibility of researchers to find the most appropriate journals for their manuscripts. One clue is to find out where researchers working on similar problems are publishing their work. This can be determined by reading and reviewing the literature. Many times a common theme (or several themes) pervade a journal.

Once an article goes to a journal, the review process may take up to six months, although many editors attempt to reach a decision on a manuscript within two months. An outline of the editorial process used by the APA is shown in Figure 10.9.

Part Two of this book deals with actual experiments that have been carefully selected from the psychological literature to demonstrate a variety of topics and experimental techniques.

DEFINITIONS

The following terms and concepts were used in this chapter. We have found it instructive for students to go through the material a second time to define each of the following:

descriptive statistics	inferential statistics
experimental studies	problem
function	statistical studies
hypothesis	

Analysis of Experiments

In Part Two several research articles are reviewed. These articles were carefully selected to represent a variety of topics and design procedures.

After conducting an experiment, the psychologist's next obligation is to report its results to the scientific community. In practice, this means preparing a research paper for a journal. Thus, the ability to write such papers and read them with understanding is one of the most important demands facing the professional psychologist.

Because journal space is limited and there are so many other demands on a scientist's time, clarity and conciseness are crucial in the presentation of research results. At the same time, these articles are aimed at other scientists, particularly those working in the same problem area, making a knowledge of that area's research history and its methodologies a prerequisite. This can make journal articles difficult going, both for students and for psychologists who have different research specialties.

In Part Two the articles are presented in a form that has been edited slightly from the original. Most articles are accompanied by an analysis that attempts to explain exactly why the researcher did what he or she did. By repeated exposure to and careful analysis of such papers, the student can begin to feel comfortable reading research reports and may develop an understanding of the principles of experimental design in psychology. Four articles are not accompanied by an analysis; they are to be analyzed by you, the student.

In the articles that we have analyzed, the analysis corresponding to the appropriate section of the article appears below it. Many of the experiments in this section were selected because they provide an example of a "Special Issue." These issues are set apart from the main text in our analysis of the article. Each of the issues listed below is addressed in at least one article.

There is no right way to read an article, but some suggestions follow that might help you to better understand research reports.

SPECIAL ISSUES	ARTICLES
1. control problems	"Cola Tasting"—Chapter 11
2. field-based experiments	"Distance and Rank"—Chapter 12
3. unobtrusive measures	"Distance and Rank"—Chapter 12
4. poster presentations	"Fanning Old Flames"—Chapter 13
	"Teeth Grinding and Arousal Tendencies"—Chapter 14
5. selection of subjects	"Hormones and Toy Preferences"—Chapter 16
6. research with animals	"Maternal Behavior"—Chapter 17
7. small n experiments	"Creative Porpoise"—Chapter 19
8. attribution theory	"Humor"—Chapter 20
10. single-subject design in clinical studies	"Therapy for Anger"—Chapter 21
11. multiple experiments	"Birdsong Learning"—Chapter 23
12. subject variables	"Note Taking"—Chapter 24
13. clinical research	"Weight Loss"—Chapter 25

First, read the title and try to establish the general category within psychology that the study is investigating. Consider an article such as "Two-Phase Model for Human Classical Conditioning" (Prokasy & Harsanyi, *Journal of Experimental Psychology,* 1968, *78,* 359–368). Most people have had some experience with classical conditioning or at least know something about the famous Pavlovian studies with dogs. So from the title you could infer that these authors have apparently studied learning by using a conditioning method, and that they have developed some model to describe their results. In addition to simply reading the title, it may help to try to raise concrete questions about the problem and procedures. In this example, you might ask, How did the authors condition the humans—with a bell and meat powder? What are two possible phases of classical conditioning?

It may also help to know that most journal articles contain an abstract at the beginning of the paper where the major findings and the method used are briefly described. A careful reading of this, coupled with an inquiring mind, will facilitate an understanding of the paper.

Then, a quick scan of the entire paper is suggested. You should place greater emphasis on the review of literature, the hypothesis, and the discussion than on specific details of the methods and results sections. These steps should give you a general impression of the author's intent. You may also want to return to the abstract to reestablish a point of view, but the critical aspect of scientific reading is a probing inquiry. When reviewing and reading, *question!*

Finally, read the article in its entirety, sorting out data from discussion. During this reading, you should examine how the author controlled for variables and how he or she specifically treated the data.

In reading an article, it may also help to ask some of the following questions, plus many more that would be determined, in part, by the responses to the questions.

1. What is this research all about?
2. What is the general problem?
3. What are the results of others?
4. What is the hypothesis?
5. What materials does the author use?
6. How does the author operationally define his or her variables?
7. What controls does he or she use?
8. How does he or she analyze the data?
9. How does he or she interpret the data?
10. Whom does he or she cite as relevant investigators in the area?
11. Can this study be improved?
12. What additional work needs to be done?
13. Can the data be interpreted in another way?
14. What have I learned from this article?
15. What new research can now be developed?

Some articles are so complex or use so much jargon that they may be difficult to understand. This problem is common among new students (as well as for some experienced students). If this happens, repeating the process suggested above should help. Sometimes students also find it helpful to review an article for a friend and allow him or her to ask questions.

For convenience' sake the table that follows summarizes some of the relevant characteristics of the articles in Part Two. We suggest that you look at the table prior to reading each article for an overview of the article.

ABBREVIATED TITLE	PSYCHOLOGICAL SUBJECT MATTER	SOURCE OF HYPOTHESIS	SETTING	TREATMENT OF SUBJECTS	STATISTICAL/DESCRIPTIVE ANALYSIS	SUBJECTS	INDEPENDENT VARIABLE	DEPENDENT VARIABLE
Cola Tasting Chapter 11	Perception	Applied research	Laboratory	Paired comparisons within subjects	χ^2, table, percentages	Humans (college students)	Different colas	Identification of cola
Distance and Rank Chapter 12	Personal space	Hypothesis	Field research	Designated groups (rank)	F, linear trend rank order correlation tables, χ^2	Humans (military personnel)	Rank	Distance when initiating a conversation
Flaming Old Flames Chapter 13	Physiological/Emotions	Hypothesis	Laboratory	Assignment via questionnaire	F, graph	Human (college students)	Thought suppression	Physiological responses (SCL)
Teeth Grinding Chapter 14 (to be analyzed by the student)								
Picture Memory Chapter 15	Cognitive memory	Hypothesis testing/theory	Laboratory	Random group (2 groups)	Table, percentages, t, z	Humans (college students)	Thematic clue	Memory for pictures
Toy Preferences Chapter 16	Child/physiological	Hypothesis	Hospital	Designated subjects	t, graph	Children	Hormone level	Toy preference
Maternal Behavior Chapter 17	Comparative motivation	Hypothesis testing/theory	Laboratory	Random group (4 groups)	F, Mann-Whitney, Duncan's range test, graphic	Rats	Injection of blood	Maternal behavior

ABBREVIATED TITLE	PSYCHO-LOGICAL SUBJECT MATTER	SOURCE OF HYPOTHESIS	SETTING	TREATMENT OF SUBJECTS	STATIS-TICAL/ DESCRIPTIVE ANALYSIS	SUBJECTS	INDEPEN-DENT VARIABLE	DEPENDENT VARIABLE
Seven Dwarfs Chapter 18 (to be analyzed by the student)								
Creative Porpoise Chapter 19	Creative learning	Hypothesis testing/ theory (demonstration)	Laboratory/ seminatural setting	Single subject	Graphic, correlation	Porpoises	Reinforcement	Novel behavior
Humor Chapter 20	Social	Hypothesis testing	Laboratory	Random group	F, table	Humans (college students)	Type of joke	Attitude attribution
Therapy for Anger Chapter 21	Clinical	Applied clinical	Clinical	Single subject	Figures, correlation	Hospitalized patients	Therapy	Anger control/ reduction
Prosocial Behavior Chapter 22 (to be analyzed by the student)								
Birdsong Learning Chapter 23	Intersensory processing	Hypothesis	Laboratory/ 2 experiments	Assigned groups	Table, F, t, percentages	Humans (college students)	Presentation means	Name birdsongs
Note Taking Chapter 24	Educational Psychology	Hypothesis testing/ applied	Natural setting	Random group (5 groups)	F, table, correlation, t	Humans (college students)	Note taking and note reviewing	Recall of lecture
Weight Loss Chapter 25	Clinical	Hypothesis testing/ applied clinical	Natural setting/ minor laboratory intervention	Random group (5 groups)	F, Newman-Keuls, percentages, averages	Human adults	Type of "therapy"	Weight loss
Parents' Views Chapter 26 (to be analyzed by the student)								

Chapter 11

Cola Tasting

INTRODUCTION

In a recent test of cola beverages conducted for a leading business newspaper, 70% of the tasters were mistaken as to which of four cola beverages they were tasting. Yet, many people swear allegiance to a particular soft drink.

The first experiment in Part Two deals with the identification of cola beverages and presents some interesting design problems. As you read this experiment ask yourself what variables, both experimental and cultural, may influence a person's preference for one cola drink over another.

The purpose of including this experiment is (1) to introduce you to some of the issues involved when controlling variables in a psychological study and (2) to heighten your awareness of how complex the relationships are between a real world stimulus (cola drink) and the psychological evaluation of the stimulus.

SPECIAL ISSUES

Control Problems

We previously suggested that there are two types of control. The first is when the experimenter makes something occur and the second is when the experimenter prevents something from occurring (extraneous variables). We have discussed the first type, so let us now look at the extraneous variables that were controlled in this experiment.

Relations between pairs. One cola may be easy to identify when presented with a second cola but difficult to distinguish from a third. For example, Coke may be easy to identify when it is paired with Royal Crown but difficult to identify when it is paired with Pepsi. To avoid such problems, each cola was paired with the other colas an equal number of times. Another problem is that within each pair presentation, the first cola may be easier to identify than the second, or vice versa. To eliminate this problem each cola could have been presented an equal number of times in the first and second position. While Thumin did not systematically control for this, it could easily have been incorporated into the design to strengthen the procedure.

Order effects. We already know that when subjects make a series of judgments, order effects can appear. Subjects may become more sensitive with practice, or they may become fatigued so that their judgments become poorer. In taste experiments it is very probable that the subject's judgments will become poorer since the previous cola may not have been completely washed out of his or her mouth after each trial, and the residue may confuse later judgments. To control for this Thumin presented the stimulus pairs in random order. This procedure is effective for controlling order effects.

Stimulus accumulation. Related to the preceding point, the subject washed out his or her mouth with water after each pair. This control procedure attempted to eliminate the confusion of taste from one pair to the next. It is reasonable to assume that a subject could not judge a given pair effectively if the taste of the previous pair was still in his or her mouth. Furthermore, clean cups were used for each presentation to eliminate the possibility of an accumulation of cola in the cups.

Visual cues. Thumin was testing to see whether or not the subjects could identify the colas by taste. To do this he had to eliminate any other cues that might help the subject identify the cola. One such set of cues are visual cues; for example, Coke may look different from Pepsi. To eliminate visual cues, experimenters should blindfold subjects in experiments of this type to eliminate any possibility of visual cues.

Temperature. Most people drink colas that are cooled to a temperature of 5°C (46°F). Their experience with tasting cola is limited to this temperature; that is, the taste of Coke is really the taste of Coke at 5°C. To allow for this, Thumin presented all colas at a constant temperature of 5°C.

Guessing. In many judgment experiments, a problem arises when the subjects guess. Some subjects will guess when they do not know the correct identification, and some subjects will simply say that they do not know. To avoid these problems Thumin told all subjects to guess when they did not know the colas. He was quite explicit about this in his instructions when he pointed out to the subjects that even if they were not sure of the brand they should tell him what brand they thought it was.

Note also that Thumin has (in a sense) corrected for guessing in his statistical analysis (see footnote to Table 11.1). There were three choices: One choice was correct and two were incorrect. Thus, the probability of being correct by chance was one out of three. The statistic tested to see if the subjects' identifications were more correct than would be expected by guessing alone (chance).

Subject variables. Chapter 6 presented a lengthy discussion of subject variables. Much of this discussion centered around the problem of having people in one experimental treatment who were different as to some subject characteristics from people in another treatment. Thumin avoided this problem by having all subjects participate in all treatments—a *within-subject design*. Thumin controlled for some of the problems of the within-subject design by controlling for order effects. However, another subject variable problem could appear in this experiment. Suppose that only heavy cola drinkers can distinguish between the different brands. Suppose further that the sample of subjects Thumin picked had a small proportion of heavy cola drinkers. Because the sample would contain only a few heavy drinkers (who can make correct identifications), then the majority of light drinkers might make it appear as if the identification of cola beverages is little better than chance. Thumin accounted for this subject variable problem by finding out how much cola each subject drank per week. It is apparent that this subject variable did not influence the data (see Table 11.2); however, Thumin took this possibility into account and tested to see what effect this subject variable had on the dependent variable.

This discussion of design and control procedures should make it evident to the student that careful planning is needed before an experiment can be executed. We pointed out previously that control problems vary with each experimental area. The Thumin experiment illustrates some of the considerations that must be taken into account in experimenting with the identification of the taste of substances.

Identification of Cola Beverages

FREDERICK J. THUMIN
WASHINGTON UNIVERSITY

An attempt was made to overcome certain methodological inadequacies of earlier studies in determining whether cola beverages can be identified on the basis of taste. Some 79 Ss completed questionnaires on their cola drinking habits and brand preferences, then were tested individually on samples of cola beverages presented under methods of paired comparisons. Significant chi-square values were obtained for Coca-Cola and Pepsi Cola, due to the large number of correct identification for these brands. Correct identification of Royal Crown, however, did not differ from chance expectancy. No significant relationship was found between ability to identify cola beverages and degree of cola consumption; nor were Ss any better at identifying their "regular" brand than they were other brands.

Earlier studies attempting to determine whether cola beverages can be identified on the basis of taste[1] have, in the main, obtained negative results (Bowles & Pronko, 1948; Pronko & Bowles, 1948; Pronko & Bowles, 1949; Pronko & Herman, 1950; Prothro, 1953). These results may, in part, be attributed to certain methodological difficulties. For example, in the majority of these studies, **the subjects were not informed as to what brands they were attempting to identify. This lack of restriction encouraged guessing behavior, which resulted in the naming of irrelevant beverages (e.g., Dr. Pepper),** as well as

a

[1]In this report, the word "taste" is used in the broad sense—that is, to include gustation, olfaction, and possible tactual qualities as well.

SOURCE: Reprinted by permission from *Journal of Applied Psychology*, 1962, *46* (5), 358–360. Published by the American Psychological Association.

ANALYSIS

REVIEW OF LITERATURE, STATEMENT OF PROBLEM, AND HYPOTHESIS

The first paragraph attempts briefly to describe the previous research on cola identification and to note the scope of the present research. Although it is not common to define terms in research papers, the author uses an acceptable method to define *taste* in a footnote.

Three methodological factors (**a, b,** and **c**) were not controlled effectively in the previous literature, which may have had an effect on the results. These factors were (**a**) not informing the subjects of the colas to be identified, (**b**) not determining

relatively frequent mentions of the more heavily advertised brands such as Coca-Cola.

b Moreover, the subjects were expected to identify the various colas on the basis of past experience, yet apparently **no attempt was made to determine whether the subjects had ever tasted these beverages, or to relate identification to degree of cola consumption.**

c Each of these previous studies used essentially the same method of stimulus presentation; namely, **all beverages were presented simultaneously to the subject,** and only one such presentation was made. This technique, while satisfactory, would appear to be somewhat less sensitive than the method of paired comparisons, which requires the subject to identify each brand a number of times under various experimental conditions.

d Thus, the purpose of the present study was to determine whether methodological inadequacies in the earlier studies may have contributed to the subjects' relative inability to identify brands. The primary modifications in experimental design were as follows: an indication of cola consumption habits was obtained, subjects were told in advance what beverages they were attempting to identify, and the method of paired comparisons was used for presentation of stimuli.

subjects' experience with cola beverages, and **(c)** presenting the colas simultaneously. The author tentatively identifies the possible biasing (or confounding) result of each of these uncontrolled factors.

In **(d)** a succinct statement is made about the methodological differences between this and previous studies.

In this experiment, paired comparison was used. This is a standard technique in which a subject is given two stimuli and asked to judge them. The judgment in this study was a qualitative one; subjects were asked to identify specific colas by taste. (It should be noted that this method of comparison also permits quantitative judgments. For example, in this experiment, the subjects could be asked to identify greater or lesser amounts of a certain quality.) A related procedure, which Thumin mentions in **(c)** as the predominant procedure used in cola tasting, is *multiple comparison.* In this method, subjects are asked to make judgments on a variety of different stimuli.

This study does not clearly state a hypothesis to be tested, yet the reader can provide his or her own statement of a hypothesis with the material presented. How would you state the hypothesis?

METHOD

The method section of this report is fairly brief, but it contains the necessary information (including the exact instructions given to the subjects) to allow the study

METHOD

e

Seventy-nine subjects were employed, all of whom were either college students or college graduates between the ages of 18 and 37 years. The subjects were first asked to fill out a questionnaire on their cola consumption habits and brand preferences. The cola beverages were presented to the subjects individually in an experimental room which was kept dimly lighted to eliminate possible visual cues. Instructions were as follows:

> I would like to have you taste and identify some cola drinks. I will place two cups at a time in front of you—one on your left and one on your right. Taste these two colas in any order you wish; then tell me what brand you think each one is. Be careful not to change the position of the cups while you are tasting them; that is, keep the left cup on the left, and the right cup on the right. Each time you finish with one pair of cups, rinse your mouth well by taking a few swallows of water from the water cup. When you have done this, I will give you the next pair.

> There are three colas involved in this study—Coca-Cola, Pepsi Cola, and Royal Crown. Even if you are not sure of the brand in some cases, I still want you to tell me what brand you think it is. The two members of a pair are always different brands; that is, a brand is never compared with itself. Are there any questions?

Using the method of paired comparisons, six pairs of beverages were presented to the subject, one pair at a time. The subjects were exposed to each brand four times for a total of 12 judgments. The order of presentation of stimulus pairs was randomly determined. Stimulus cups contained 2 ounces of the beverages at an approximate temperature of 5° centigrade.

to be replicated **(e)**. However, there are some features of the design that the author does not make explicit, which we will discuss here.

Type of Design

This design is quite similar to the design used in Chapter 10 for investigating the relationship between the frequency of a tone and the subject's sensitivity to that tone and is typical of the type of design found in these areas. In this experiment the independent variable is three types of colas, whereas in the audition experiment the independent variable was tones of different frequencies. In the audition experiment the dependent variable was the absolute threshold of the tone, that is, some hypothetical point above which the subject could hear the tone and below which he or she could not hear the tone. In the cola experiment the dependent variable is the correctness of the identification of the cola. In both experiments there were multiple presentations of the same stimulus; in the audition experiment there were six attempts to determine the threshold for each tone, three using the descending method and three using the ascending method. In the cola experiment the subjects

RESULTS

f   **The chi-square was used to determine whether ability to identify brands differed significantly from chance expectancy.** As Table 11.1 shows, the chi-square values for both Coca-Cola and Pepsi Cola were significant at the

g   **0.01 level of confidence,** while that for Royal Crown was not significant. Inspection of the data indicates that the significant divergencies obtained with Coca-Cola and Pepsi Cola are due to the large number of correct identifications of these brands; for example, more than twice the expected number of subjects were able to identify these brands correctly at least three times out of four.

were exposed to each brand four times. By using several presentations of the same stimulus, the experimenter may get a more stable or reliable measure of each subject's judgments. The basic difference in the two experiments is that in the audition experiment the experimenter presented one tone at a time. This was an appropriate method for that problem because the experimenter was asking, "When can you hear the tone?" In the Thumin experiment, the problem is identifying the cola: "Which cola is it?" To get an answer to this question, Thumin argues that presenting two stimuli at the same time (paired comparison method) is more effective than the methods previously used.

RESULTS

The basic findings and the summary of the statistical analysis are the principle parts of the results section.

The χ^2 (chi-square) statistical procedure was used to treat the data **(f).** This is a relatively simple procedure in which the results are compared with what one would expect by chance alone. *Chance* is defined as the variation in the results that is due to uncontrolled factors such as guessing, experimental error, failure to perfectly mask stimuli, failure to achieve a perfect matching of subjects or randomization in experiments using different groups of subjects, and so on. In this experiment subjects who had no knowledge of the cola but simply guessed would be correct sometimes. But if the experimental results vary greatly from what would be expected by chance, then some experimental conditions are probably responsible for this diversity.

The level of confidence or "level of significance" **(g)** is a reflection of the probability that the results would be obtained by chance alone. In this study, the author establishes a level of confidence of $p = 0.01$; that is, the probability that such results would occur by chance alone is 1 in 100. Frequently, the level of confidence is set at 0.05 (or 5 in 100), but in some research it may be useful to demand lower levels.

Table 11.1   CHI-SQUARE FOR OBSERVED AND EXPECTED FREQUENCIES OF BRAND IDENTIFICATION

BRAND OF COLA	OBSERVED AND EXPECTED FREQUENCIES	NUMBER OF CORRECT IDENTIFICATIONS				χ^2
		0	1	2	3 OR 4	
Coca-Cola	(f_o)	13	23	24	19	14.57**
Pepsi Cola	(f_o)	12	20	26	21	22.14**
Royal Crown	(f_o)	18	28	19	14	4.60*
All brands	$(f_c)^a$	15.6	31.2	23.4	8.8	

Note: For each comparison, $df = 3$.
[a]Expected values were obtained from the expression $N(\frac{1}{3} + \frac{2}{3})^4$. Each sample had one chance in three of being identified correctly, and each brand was presented to the subject four times.
*$p > .05$.
**$p < .01$.

The results presented in Table 11.2 indicate that ability to identify cola beverages correctly was unrelated to degree of consumption; that is, correct identifications were essentially the same for heavy, medium, and light cola drinkers. Further analysis of the data showed that ability to identify a given brand was also unrelated to whether that brand was considered by the subject to be his or her "regular" brand.

By telling the subjects in advance what brands they were attempting to identify, irrelevant brand naming was eliminated as well as excessive naming of heavily advertised brands. Specifically, Coca-Cola was mentioned 317 times, Pepsi Cola 321 times, and Royal Crown 310 times.

DISCUSSION

h

i

The present study clearly demonstrated that certain brands of cola can be identified on the basis of taste. The significant chi-square values obtained with Coca-Cola and Pepsi Cola were due to the large number of correct identifications for these brands. **The subjects' inability to identify Royal Crown Cola can probably be attributed to a lack of *recent* experience with this brand. Some 58 percent of the subjects said they had not had a Royal Crown for at least 6 months prior to the experiment.**

No relationship was found between ability to identify cola beverages and degree of cola consumption (i.e., number of colas consumed in an average

The author makes a clear summary statement and also provides the reader with two tables that contain the experimental data. It is noted that two of the obtained χ^2 values exceed the 0.01 level.

Table 11.2   CHI-SQUARE FOR BRAND IDENTIFICATION AS RELATED TO CONSUMPTION

NUMBER OF COLAS CONSUMED PER WEEK	NUMBER OF CORRECT IDENTIFICATIONS		
	0–3	4–6	7–12
Heavy (7 or more)	10 (8.5)	14 (12.0)	3 (6.5)
Medium (3–6)	7 (8.2)	9 (11.5)	10 (6.3)
Light (0–2)	8 (8.2)	12 (11.5)	6 (6.3)

Note: Expected values appear in parentheses.
$\chi^2 = 5.44$; $df = 4$; $p < .05$.

week). Moreover, the subjects were no better at identifying their regular brand than they were at identifying other brands. Thus, it would appear that the subjects needed a certain minimal amount of recent experience with a brand in order to identify it, but beyond this minimal amount, additional experience (i.e., heavier consumption) did not help.

j   Within the framework of this study, **the method of paired comparisons proved to be sufficiently sensitive to detect small but significant abilities to identify cola beverages.** There appeared to be no problem with the development of sensory adaptation as successive pairs of stimuli were presented. Analysis of the data revealed that, as trials progressed, the subjects showed small (though nonsignificant) increases in ability to identify brands. ◆

REFERENCES

Bowles, J. W., Jr., & Pronko, N. H. (1948). Identification of cola beverages: II. A further study. *Journal of Applied Psychology, 32,* 559–564.

Pronko, N. H., & Bowles, J. W., Jr. (1948). Identification of cola beverages: I. First study. *Journal of Applied Psychology, 32,* 304–312.

Pronko, N. H., & Bowles, J. W., Jr. (1949). Identification of cola beverages: III. A final study. *Journal of Applied Psychology, 33,* 605–608.

Pronko, N. H., & Herman, D. T. (1950). Identification of cola beverages: IV. Postscript. *Journal of Applied Psychology, 34,* 68–69.

Prothro, E. T. (1953). Identification of cola beverages overseas. *Journal of Applied Psychology, 37,* 494–495.

The author wishes to express his appreciation to A. Barclay who served as critical reader for earlier drafts of this paper.

DISCUSSION

In the discussion section the author reviews the major findings (**h**) and offers a plausible explanation for the lack of statistically significant findings regarding the Royal Crown condition (**i**).

Finally, the sensitivity of the paired-comparison technique in detecting abilities to identify colas is mentioned **(j)**.

QUESTIONS

1. In the design of this experiment each cola was paired with the other colas an equal number of times, and the presentation of stimulus pairs was randomly determined. Using a within-subject design, lay out exactly how one sequence of 12 pairs might have been presented to a subject.
2. Why was a dimly lighted room used **(e)**? What effects would a red illuminated room have? A green illuminated room? A nonilluminated room? Would these conditions be worthy of research?
3. Why did the subjects rinse their mouths? Should they eat something neutral (e.g., a cracker) between tests? Do you think this is a critical variable?
4. How would you interpret the fact that some subjects could not correctly identify the beverages? What factors could determine a person's preference for one cola over another?
5. No sex differences were noted in the selection of subjects. Why? Subjects were not asked to identify their favorite cola. What influence might this have had on the results?
6. Design an experiment to test whether or not subjects can correctly identify whole, skim, and powdered milk.
7. Design an experiment to test whether whole, skim, or powdered milk tastes better.

Chapter 12

Distance and Rank

INTRODUCTION

How close do you stand to a person with whom you are talking? If you think about the answer, you will probably say, "It depends on several factors." One factor is familiarization. In general, we tend to stand closer to people we know than to those we do not know. But other factors also influence our behavior. An important consideration is the influence of our cultural background. Foreign diplomats know that misunderstandings can develop because some people tend to stand close to someone during casual conversations, while others tend to stand far away. A member of the U.S. foreign service once told of a diplomatic reception in a South American country in which he was "chased" around the room. While the diplomat was talking with his South American colleague, the colleague would edge a bit closer and the service officer would withdraw slightly. The colleague would again move closer, the diplomat would withdraw, and so on. Throughout the reception the two could be seen advancing, retreating, advancing, retreating. If we could have interviewed the two, the diplomat probably would have said, "They are very 'pushy' people, those South Americans," while the South American would have probably complained that the Americans are "stand-offish, distant, and aloof."

In an experiment by Dean, Willis, and Hewitt (1975), the influence of military rank on distance between officers was studied empirically. This article may be interesting in that the results may tell you something about your own behavior, but the purpose of including this article is to demonstrate two special issues: *field based studies* and *unobtrusive measures*. Pay close attention to these issues as you read the article.

SPECIAL ISSUES

Field-Based Studies

Psychological research can be divided into two categories: laboratory experiments and field experiments. Most research takes place in controlled laboratory settings. The reason psychologists tend to favor a laboratory setting is for methodological control. In the experimental laboratory the researcher can isolate the subject from the noisy world and precisely control the type of stimulation he or she receives. In effect, the experimental laboratory allows the researcher to create a microcosm in which the only factors that are allowed in are those that the experimenter wishes to influence the subject; other cues can be either eliminated or brought under experimental control. Paradoxically, in this very strength of laboratory experimentation lies a weakness: the artificial nature of the laboratory setting. By removing the subject from his or her natural setting, as is done in laboratory experiments, the subject is deprived of the forces that are necessarily part of his or her normal life, and these forces may help determine the subject's reactions. In technical language, some stimuli that are eliminated or controlled by the laboratory setting may be critical independent variables that significantly affect the dependent variable.

In choosing between a laboratory experiment and a field experiment, one critical question should be answered: Are the desired experimental results significantly related to the social situation? Many types of social and educational problems fall into this category. Consider mob behavior. The very presence of other members of a mob undoubtedly influences the behavior of any given individual, and his or her behavior may in turn influence the others. How can an experimenter control for these pressures and other independent variables (e.g., the presence of an inflammatory stimulus, such as a lynched body)? It is simply impossible to recreate such a complex situation in a laboratory. Therefore, a researcher interested in mob behavior must either isolate some hypothesized factors of the larger issue for laboratory investigation, or turn to the field study. A large number of worthy research projects must be studied in the realistic field in which they occur. In addition to the previous mob example, many educational problems are studied in the context of field experimentation. This article and the research article by Fisher and Harris on note taking (Chapter 24) are field studies.

Unobtrusive Measures

When subjects know they are being observed, they frequently behave differently than when they are not being observed. This generalization is true not only for human subjects in a psychological experiment but also in some cases of biological research in which animals are being observed. A partial explanation of this phenomenon is that the experimenter becomes part of the field whose influence he or she is attempting to measure. By way of analogy, consider the problem of making critical temperature measurements. A thermometer inserted into a substance not only reacts to the temperature of the substance but also to its own temperature. In a similar way, an experimental psychologist is not only an observer of behavior but also a part of the environment to which the subject reacts.

Since precise control over stimulus variables in field-based experiments is sometimes impossible, their effect on response variables is also sometimes questionable. As you may recall from our previous discussion of scientific theory and methodology, a cardinal principle of experimental research is to identify the precise cause of a special effect. If the causes of behavior (in experimental terms, the independent variables) are doubtful, then the behavior (the dependent variable) may or may not be related to the cause. The resolution of this profound dilemma is not easy; however, the issue has been reviewed in two influential books: Webb, Campbell, Schwartz, and Sechrest, *Unobtrusive Measures: Nonreactive Research in the Social Sciences* (1966), and Cook and Campbell, *Quasi-Experimentation: Design & Analysis Issues for Field Settings* (1979). In these sources (and others) experimental psychologists grapple with the problem of the intrusion of an observer into the psychological field. To reduce the influence of an observer on a psychological experiment—especially on a field-based experiment—the authors suggest a number of unobtrusive means experimenters can use to collect information. The interested student is referred to these sources for a more complete discussion of the issue.

But when unobtrusive measures are made in field-based experiments this raises questions of experimental ethics. Is it proper or ethical to skulk around with a clipboard, collecting data on unsuspecting people? Although questions may be raised about the propriety of field-based experiments, it would be erroneous to conclude that such experiments are unethical. We do, however, strongly recommend that in the planning of all experiments, including field experiments, the experimenter consult the APA ethical standards reprinted in Chapter 8.

INITIAL INTERACTION DISTANCE AMONG INDIVIDUALS EQUAL AND UNEQUAL IN MILITARY RANK

LARRY M. DEAN, FRANK N. WILLIS, AND JAY HEWITT
UNIVERSITY OF MISSOURI AT KANSAS CITY

Interaction distances for 562 pairs of individuals in military settings were recorded at the moment a conversation was begun. It was found that interactions initiated with superiors were characterized by greater distance and, further, that this distance increased with the discrepancy in rank. When interactions were initiated with subordinates, interaction distance was unrelated to rank discrepancy. It was hypothesized that superiors are free to approach subordinates at distances that indicate intimacy, while subordinates are not at liberty to do so.

The increased interest in research in personal space is evidenced by three reviews of the literature in this area (Evans & Howard, 1973; Linder, 1974; Pederson & Shears, 1973). Research has been reported for both animal and

a   human subjects. **Studies have been conducted using personal space as both an independent and a dependent variable.** As a dependent variable, personal

SOURCE: Reprinted by permission from *Journal of Personality and Social Psychology*, 1975, *32* (2), 294–299. Some portions of the original article have been omitted.

ANALYSIS

In this study, as in most articles, we suggest that you survey the entire article, not getting bogged down in the specific methodological issues but instead trying to gain an overview of the purpose, the technique, the results, and the conclusion. Having established this overview, you will see that the article is more than a simple experiment studying how close people of unequal rank stand, but that it deals with the broader issues of personal space and perceived status.

REVIEW OF LITERATURE

This article follows logically from a well-defined sequence of related studies of personal space. Of particular note in the introduction is the authors' statement that personal space has been treated as both an independent *and* a dependent variable (a). Can you think of another variable that could be treated as both an independent and a dependent variable?

Dean et al. point out that studies of personal space have been done with

space has been related to the personal characteristics of initiators and recipients and to the relationships between initiators and recipients. The personal characteristics that have been related to their interaction distances have included the following: handicaps or stigma (Kleck, Buck, Goller, London, Pfeiffer, & Vukcevic, 1968), psychopathology (Horowitz, 1968), extroversion-introversion (Patterson & Sechrest, 1970), sex (Pellegrini & Empey, 1970), race (Jones & Aiello, 1973), and age (Willis, 1966).

Hall (1966) suggested four zones for social interaction: intimate, personal, social, and public. He asserted that the selection of a zone for a particular transaction was dependent upon the degree of intimacy of the relationship for the individuals involved.

In research involving personal space as a dependent variable, one characteristic of the relationship between subjects that has received considerable interest in animal studies and little interest in human studies is that of status or dominance. King (1965) reported that the distance a subordinate chicken remained from a stationary dominant bird was linearly related to the frequency with which the dominant had pecked the subordinate in the home coop. King (1966) also found a similar relationship with human children as subjects. Lott and Sommer (1967) reported that college upperclassmen sat closer to peers than to freshmen "doing poorly in school" and sat nearer to peers than to professors. It is Sommer's (1969) opinion that studies relating spacing and dominance in humans have been far more infrequent than those with infrahumans "probably because this type of experimentation requires conditions that are uncommon in naturalistic settings" (p. 24). The lack of information in this area may be due to the difficulty in specifying observable dominance in humans. **The present study was designed to overcome this difficulty by observing interaction distances in military settings where dominance relations are indicated on the clothing of the individual.** If, as has been indicated in infrahuman studies, superiors are capable of inflicting aversive consequences, whereas peers and subordinates are not, then subjects might prefer greater interaction distances with superiors. If this is true, the relative status of the interaction initiator and recipient would be an important variable in determining their interaction distance. The following questions were of interest in the present study: (a) Would interaction distances be greater when people initiated an interaction with superiors than when they initiated them with peers? (b) Would interaction distances be less when people initiated an

animals and humans and at least hint (**b**) that the critical factor in both human and animal studies might be related to dominance.

In section (**c**) the problem is stated, followed by the hypothesis in section (**d**). The hypothesis is presented in terms of an "if . . . then" proposition, which is supported by a series of specific questions (**e**) the research purports to answer.

interaction with subordinates than when they initiated them with peers? (c) Would interaction distance increase monotonically with status discrepancy when people initiated interaction with superiors?

METHOD

Subjects

Subjects were 562 uniformed active duty United States naval personnel who were observed as they initiated an interaction with another uniformed member of the U.S. Navy.

Procedure

f_1

g
h

i
f_2

Observations were made at the U.S. Naval Station, Long Beach, California, by three uniformed naval personnel who were assigned to duties at that station. Subjects were observed during duty hours but in nonworking settings. These settings were chosen for the study in order to minimize the effects of regular working relationships upon the interaction distances. **The settings were as follows: the Navy exchange, the cafeteria, the lobby of the station hospital, and a station recreation center.** All subjects were standing when observed. Interactions were selected when one person was standing and was approached by another **who began a conversation.** At the moment the first words were spoken, observers **recorded the number of floor tiles between their nearest feet to the closest half-tile.** All settings had standard 22.86-cm. square floor tiles. Also recorded was the rank of the initiator and of the recipient of the interaction. The subjects were an incidental sample of all personnel in the settings meeting the above requirements during the hours of observation. The observers attempted to use each subject only once in the role of initiator. Interjudge reliability for the distances was so high in a pilot study that only one observer was used for each dyad. **Observers were not informed as to the hypotheses** of the study. **Notations were made unobtrusively,** and no subject was observed to react to the observers.

METHOD

There are several important features in the procedure section. First, the data were collected in a natural setting (f_1), and the notations were made "unobtrusively" (f_2). Second, the data (i.e., the distance between subjects) were collected at the moment conversation started (g). Third, the researchers used an inventive measure of distance: the number of floor tiles between the subjects (h). Finally, the observers were not told of the hypothesis (i), which made this a double-blind design in that both the experimenters and the subjects were unaware of the purpose of the experi-

RESULTS

Peers Versus Subordinates and Superiors

For naval personnel of ranks of commander through third-class petty officer, information was available for interaction distance between these individuals (as initiators) and others of higher, equal, or lower status. For captains this information was available for others of equal or lower status. For seamen, this information was available for others of equal or greater status. The mean interaction distances are presented in Table 12.1. Combined across all ranks, mean interaction distance was greater when people initiated an interaction

j_1 with a superior than with a peer, $t(289) = 3.18, p < .01$. When people initiated an interaction with a subordinate, however, mean initial interaction distance was not significantly different from that which existed for peer interactions,

j_2 $t(304) = .37, p < .05$. Examining interaction distances for people at each rank

j_3 in Table 12.1, it can be seen that in **10 of the 11 ranks** for which information was available, the mean interaction distance was greater when the recipient was higher in status than when the recipient was equal in status. Finding such a pattern in 10 of 11 cases is significantly higher than one would expect on the

j_4 basis of chance $\chi^2(1) = 5.82, p < .05$.

Interaction Distance and Status Discrepancy Collapsed Across Ranks

[Another . . .] of investigating the relationship between status discrepancy and interaction distance is to compute the mean distance for people who initiate an interaction with someone one step above them in rank (e.g., a commander with a

ment. (Subjects in most field experiments are not aware they are participating in an experiment.) This procedure section is particularly rich in methodological issues associated with field research and deserves careful attention and reflection.

RESULTS

The results are initially presented in tabular form (Table 12.1). But we think the results could also be effectively presented in graphic form. From the data in Table 12.1 we have composed Figure 12.1, where we can clearly see the essential results, which indicate that the distance between military personnel tended to be greater when a person of lower rank approached a person of higher rank than when peers approached each other or when a person of higher rank approached a person of lower rank. This observation is confirmed by statistical analysis (j_1, j_2, j_3, and j_4). These data answer the first and second questions raised in section (e).

Table 12.1 MEAN INITIAL INTERACTION DISTANCE AND NUMBER OF INTERACTIONS OBSERVED AS A FUNCTION OF STATUS OF INITIATOR OF INTERACTION

	STATUS OF RECIPIENT					
	HIGHER		SAME		LOWER	
STATUS OF INITIATOR	$\bar{\chi}$	n	$\bar{\chi}$	n	$\bar{\chi}$	n
Captain	—	—	4.00	1	3.45	38
Commander	3.50	4	3.17	3	3.03	15
Lieutenant commander	3.60	5	3.00	2	3.54	27
Lieutenant	4.71	7	2.81	8	3.51	35
Lieutenant junior grade	4.50	8	3.40	5	3.12	17
Ensign	4.01	12	3.00	2	3.35	10
Warrant officer	3.79	21	3.33	3	3.15	20
Chief	3.98	27	3.55	11	3.57	36
1st-class petty officer	3.76	21	3.79	14	3.38	17
2nd-class petty officer	3.58	20	3.45	11	3.63	31
3rd-class petty officer	3.95	39	3.69	8	3.40	5
Seaman	4.06	45	3.46	14	—	0

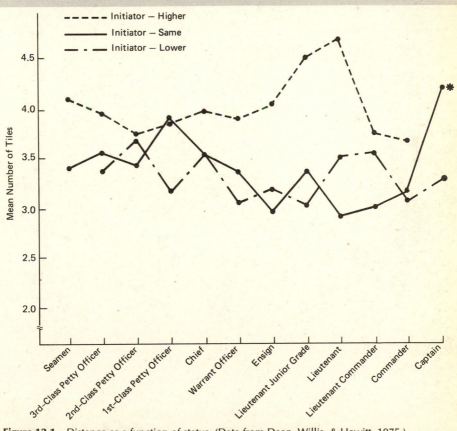

Figure 12.1 Distance as a function of status. (Data from Dean, Willis, & Hewitt, 1975.)

captain), the mean distance for people who initiate an interaction with someone two steps above them in rank, etc. As status discrepancy increases, interaction distance should also increase for interactions initiated upward.

The results of a one-way analysis of variance carried out on the means in the top half of Table 12.2 (interactions initiated upward) indicated that there was

Table 12.2   MEAN INITIAL INTERACTION DISTANCE BETWEEN INDIVIDUALS OF UNEQUAL RANK AS A FUNCTION OF NUMBER OF STEPS IN RANK SEPARATING INDIVIDUALS

	STEPS IN RANK SEPARATING INDIVIDUALS						
INITIATOR	1	2	3	4	5	6–7	8+
Subordinate							
$\bar{X}$	3.54	3.88	4.38	3.82	3.97	3.96	4.59
n	55	36	21	28	16	26	27
Dominant							
$\bar{X}$	3.55	3.29	3.53	3.59	3.21	3.14	3.33
n	84	36	36	38	17	35	26

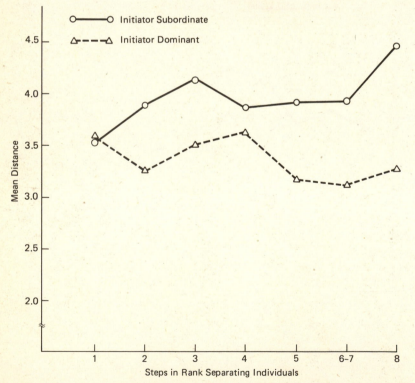

Figure 12.2   Distance as measured by mean number of tiles between subjects of varying military rank. (Data from Dean, Willis, & Hewitt, 1975.)

k_1
k_2
k_3

a significant effect of status discrepancy, $F(6,202) = 2.90$, $p < .05$. The linear trend was significant, $F(1,202) = y.84$, $p < .001$, as was the cubic, $F(1,202) = 5.76$, $p < .01$. It would thus appear that interaction distance tends to increase as a function of status discrepancy for interactions initiated upward. A rank-order correlation was carried out between the obtained rank order of the seven means and the rank order predicted by the model (increasing distance as a function of increasing status discrepancy). The rank-order correlation coefficient was .71, which is not significant with an n of 7.

DISCUSSION

l

The results of the present study support the proposition that interaction directed toward superiors is characterized by greater interaction distance than those directed toward peers and further that the distance is greater when the difference in rank is greater. The relationship does not obtain when interactions are directed toward individuals lower in rank. Previous research has shown the closer interaction distances are characteristic of intimacy (e.g., Willis, 1966). We might argue that a superior has the option of a formal (more distant) or more intimate (less distant) interaction, while a subordinate is usually required to initiate a formal interaction. Brown and Ford (1961) have stated this clearly:

m

If the person of lesser value were to initiate associative acts, he would run the risk of rebuff; if the person of higher value initiates such acts there is no risk. The superior then must be the pacesetter in progression to intimacy; the person of higher status is, we believe, the pacesetter not in linguistic address alone but in all acts that increase intimacy. (p. 383)

The results are also analyzed to compare the distance between ranks. A reader may ask, "As the difference in rank increases, is there also a measurable increase in the physical distance between individuals?" We have presented the results graphically in Figure 12.1. The results were statistically treated (k_1, k_2, and k_3). These tests suggest that distance tends to increase as the difference in rank between subjects increases. This tendency is shown in Figure 12.2 and answers the third question raised in (e).

DISCUSSION

The researchers briefly review their findings (l), relate the results to previous experiments (m), and suggest an alternate hypothesis (n).

Previous reports of studies of spatial relation have frequently failed to specify the initiated and recipient of the interaction. The results of the present study strongly suggest that these are important sources of information. If the interaction distance is great in a superior-subordinate interaction, our results indicate that it is the subordinate who is responsible for this distance.

Although our study indicated that interaction distances could not be predicted by status discrepancy when interactions were initiated by a superior, the average distance in one of the relationships in this setting may be of interest. The greatest interaction distance for junior officers initiating conversation with military subordinates was that for senior enlisted men. The average age and length of service is much greater for senior enlisted men than for junior officers. It is possible that dominance relationships even in a highly structured organization like the Navy are affected not only by the formal organization but also by norms regarding age and experience provided by society at large. ◆

n

REFERENCES

Brown, R., & Ford, M. (1961). Address in American English. *Journal of Abnormal and Social Psychology, 62,* 375–385.

Evans, G. W., & Howard, R. B. (1973). Personal space. *Psychological Bulletin, 80,* 334–344.

Hall, E. T. (1966). *The hidden dimension.* New York: Doubleday.

QUESTIONS

1. In this article personal space was treated as a dependent variable. Briefly outline a design in which personal space would be an independent variable. Also, be sure to identify the dependent variable in your design.
2. The authors suggest one alternate hypothesis. Can you think of any others?
3. Briefly define a procedure in which personal space could be empirically observed.
4. Review the basic problems associated with field-based experiments.
5. Because the experiment called for a judgment by the experimenter (e.g., counting the number of tiles), wouldn't it have been better to use two or more observers? Why or why not? What influence might multiple observers have on the results?
6. Does this experiment violate any ethical principle? Which principles apply?

Horowitz, M. J. (1968). Spatial behavior and psychopathology. *Journal of Nervous and Mental Disease, 146,* 24–35.

Jones, S. E., & Aiello, J. R. (1973). Proxemic behavior of black and white first-, third-, and fifth-grade children. *Journal of Personality and Social Psychology, 25,* 21–27.

King, M. G. (1965). Peck frequency and minimal approach distance in domestic fowl. *Journal of Genetic Psychology, 106,* 25–28.

King, M. G. (1966). Interpersonal relations in preschool children and average approach distance. *Journal of Genetic Psychology, 109,* 109–116.

Kleck, R., Buck, P. L., Goller, W. L., London, R. S., Pfeiffer, J. R., & Vukcevic, D. P. (1968). Effect of stigmatizing conditions on the use of personal space. *Psychological Reports, 23,* 111–118.

Linder, D. E. (1974). *Personal space.* Morristown, N.J.: General Learning Press.

Lott, D. F., & Sommer, R. (1967). Seating arrangements and status. *Journal of Personality and Social Psychology, 7,* 90–94.

Patterson, M. L., & Sechrest, L. B. (1970). Interpersonal distance and impression formation. *Journal of Personality, 38,* 161–166.

Pederson, D. M., & Shears, L. M. (1973). A review of personal space research in the framework of general systems theory. *Psychological Bulletin, 80,* 367–388.

Pellegrini, R. J., & Empey, J. (1970). Interpersonal spatial orientation in dyads. *Journal of Psychology, 76,* 67–70.

Sommer, R. (1969). *Personal space: The behavioral basis of design.* Englewood Cliffs, N.J.: Prentice-Hall.

Willis, F. N. (1966). Initial speaking distance as a function of the speaker's relationship. *Psychonomic Science, 5,* 221–222.

Requests for reprints should be sent to Frank N. Willis, Department of Psychology, University of Missouri, Kansas City, Missouri 64110.

Larry M. Dean is a lieutenant j.g. of the United States Navy and is presently head of the Fleet Problems Branch of the Operational Psychiatry Division of the Navy Medical Neuropsychiatric Research Unit, San Diego, California 92152.

Chapter 13

Fanning Old Flames

INTRODUCTION

The next two papers are from poster sessions at psychological meetings. The first, a paper by Gold and Wegner (1991) presented at an APA meeting, is a paper about "fanning old flames" and examines the effect of thoughts for former lovers on a subject's physiological reactivity. It is accompanied by our analysis. The second paper deals with bruxism (teeth grinding) and its relationship to stress-related variables. You should analyze this article.

SPECIAL ISSUES

Poster Sessions at Professional Meetings

An increasingly popular forum for presenting current research is in poster sessions at professional meetings. Here, both seasoned researchers and students display their research on placards in a format that allows them to interact directly with their audience. Typically, after one has conducted a research project, he or she will perform the necessary statistical tests, write the report, and submit it to a professional organization, which will then display the research at a meeting. Most organizations, including the American Psychological Association, the American Psychological Society, Psychonomic Science, and regional associations such as the Eastern Psychological Association, the Midwestern Psychological Association, and the Western Psychological Association, have poster sessions at their annual meetings.

Before a paper is accepted for presentation as a poster, an abstract is reviewed by a panel of experts in the field. But the standards for accepting a poster are far less rigorous than the standards for accepting a full research paper in a leading psycho-

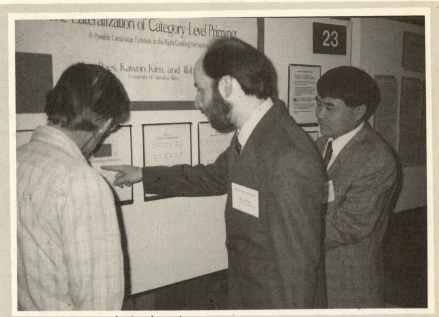

Poster session at a professional meeting.

logical journal. Reviewers normally have only an abstract of the paper and cannot evaluate in detail the method and results, for example. As such, a researcher often will present posters that (1) show only a part of the research findings, (2) reach a tentative conclusion, or (3) invite criticism from others.

The idea of a poster session is to provide a venue for new research. The audience mills around the displayed work, reads the contents, and asks questions or makes comments to the researcher. It is usually a lively venue for the exchange of ideas.

Fanning Old Flames: Arousing Romantic Obsession Through Thought Suppression[1]

DANIEL B. GOLD and DANIEL M. WEGNER
University of Virginia

We examined whether thought suppression intensifies physiological reactivity to thoughts of old flames. We asked subjects to think about an old flame, and found that those who thought about a still-desired relationship showed higher skin conductance levels (SCLs) than those who focused on one no longer desired. Then, the subjects were instructed either not to think about their old flame, or not to think about the Statue of Liberty. In a subsequent second thinking period, subjects who were asked to think of the still-desired relationship and had just suppressed that thought showed elevated SCLs compared to the others. These results suggest that trying not to think about a past relationship may prolong one's emotional responsiveness to thoughts of the relationship.

Ruminations about past loves can be painful and persistent. Hank Williams (1947) captured the reflections of despondent lovers everywhere (as well as those of generations of other songwriters) when he wrote "I can't get you off of my mind, when I try I'm just wasting my time. Lord I've tried and I've tried, and all night long I've cried—but I can't get you off of my mind." Unwanted thoughts about lost loves can be distressing, with thoughts of "old flames" sometimes turning into obsessions that interrupt daily lives and hinder the ability to form new relationships.

1. Presented at the meeting of the American Psychological Association, August, 1991, San Francisco, CA. This research was supported by National Science Foundation Grant BNS 90-96263. Correspondence should be addressed to Daniel M. Wegner, Department of Psychology, Gilmer Hall, University of Virginia, Charlottesville, VA 22903 (e-mail DMW2M@VIRGINIA.EDU).

ANALYSIS

REVIEW OF LITERATURE

Gold and Wegner capture the human pathos involved in the painful thought of a lost lover by quoting country and western songwriter Hank Williams: "I can't get you off my mind." And how true that is! Anyone who has left a relationship has recurrent thoughts of the other person and what might have been. But where do these thoughts come from and is there a physiological component that accompanies them?

a   There are several theories of how these thoughts begin. Martin and Tesser (1989) have argued that unsatisfied desires generally promote ruminations. The goal of a romantic relationship that is unfulfilled results in obsessive thinking about the lost love. It has also been hypothesized that the unwanted thoughts themselves are serving a beneficial purpose. Epstein (1983) believes that ruminations are coping strategies that promote habituation to the stressful situation. In this view, romantic ruminations help the person get over the loss.

b   A different way of viewing ruminations has been proposed by Wegner (e.g., Wegner, 1989; Wegner, Schneider, Carter, & White, 1987; Wegner, Shortt, Blake, & Page, 1990). Ruminations may develop not from unsatisfied desires or from needs to habituate, but rather from the common response people have to unwanted thoughts—**the suppression of the thought. By suppressing an unwanted thought, people may gain some immediate relief from psychological distress.** But at the same time, they prevent themselves from habituating to the bothersome thought. The distress associated with that thought will be

c   novel each time the thought reenters the stream of consciousness. **The continued distress fuels further rumination, which in turn prompts suppression and renews the problem.**

d   This study was designed to examine this reasoning by testing whether thought suppression can intensify physiological reactivity to thoughts of an old flame. We asked subjects first to think about an old flame for a period of time, expecting that those who focused on a relationship that was still desired (a hot flame) would show higher skin conductance levels (SCLs) than those who focused on a relationship no longer desired (a cold flame). Then, subjects were instructed either not to think about their old flame, or not to think about the Statue of Liberty. In a second expression period, we invited subjects again to think about their old flame, reasoning that those who had just suppressed the thought of a hot flame would continue to show elevated SCLs. Those suppressing thoughts of the cold flame, in turn, or those who suppressed the irrelevant thought, were expected to show reduced SCL in this second expression period.

In section (a) several of the past theories of the origin of these thoughts are mentioned, and an alternate hypothesis (b) suggests that suppression of the thought of a lost love may give some immediate relief. But this does little to help the person adjust and may cause further anguish (c).

In a poster session, the researcher gets to the point directly. The next paragraph (d) gives a brief but informative synopsis of the research plan.

METHOD

Procedure and Design

University of Virginia undergraduates (38 men and 32 women) were tested individually. Each completed a questionnaire about a "meaningful past relationship," identifying this person by initials. SCL measurement electrodes were explained while the subject was being attached. The subject was then asked to think aloud for tape recording.

From an adjacent room, the experimenter prompted subjects by intercom to perform three tasks for ten minutes each. **Two-minute baseline think aloud periods,** during which subjects were instructed to think about anything, began and ended each task session. In the three task sessions, subjects were asked first to express **(". . . think of the person"),** then suppress **(". . . try NOT to think of the person, however, mention it if you do"),** and then express again their thoughts of the past relationship. **Control subjects in the suppression period were asked to suppress an irrelevant target (the Statue of Liberty) instead of their thoughts of the past relationship.**

Physiological Measurement

The measurement of SCL was accomplished as recommended by Fowles et al. (1981) for finger electrode placement. Ag/AgCl electrodes were adhered to the second phalanges of the first and third fingers of the subject's right hand and attached to a J & J Electronics I-330 PC Interface System which continuously recorded the results. Each suppression or expression period was divided into three 3-minute periods, and the deviation each minute from one minute of baseline directly preceding that period was used for analysis.

METHOD

Of particular interest in this experiment is the establishment of a baseline task in which subjects were asked to think about anything that came to mind (**e**). After the baseline measures had been established the subjects were asked to think of a person (**f**) and then to *not* think of that person (**f**); the suppression part of the experiment is the independent variable. The control subjects were given the same instructions as the experimental group except that the control group was asked to suppress (presumably) an irrelevant "person" such as the Statue of Liberty during the final stage (**g**). The design is classic in form and straightforward.

The dependent variable is the measurement of skin conductance level (SCL). The exact procedures for making these measures are described in (**h**).

RESULTS

i   An initial median split of subjects was made into **hot and cold flame** groups according to their questionnaire responses about their degree of longing for their old flame. The numbers of subjects were hot = 36 (16 in the suppress "old flame" group, 20 in the suppress "Statue of Liberty" group) and cold = 34 (18 suppressing "old flame," 16 suppressing "Statue"). Analyses for each task period used a 2 (Hot vs. Cold Flame) × 2 (Suppression Target: Flame vs. Statue of Liberty) × 3 (Time: first, middle or last 3 minutes) ANOVA, with

j   repeated measures on the last variable. No effect for subject gender was found in any of the task periods, therefore subject gender was collapsed in further analyses.

In all three task periods, a significant main effect was found for hot vs. cold flame. Subjects thinking of hot flames significantly differed from those thinking of cold flames in the first expression period $F(1,66) = 4.05$, $p < .05$, in the

k   suppression period, $F(1,66) = 7.78$, $p < .01$, and in the second expression period, $F(1,66) = 5.59$, $p < .05$, respectively, with the subjects in the hot flame condition becoming more physiologically aroused than those in the cold flame condition.

In the second expression period, however, subjects who were in the hot flame condition who had previously suppressed the thought of the past relationship continued to show an increase in physiological responsiveness (see Figure 13.1). Subjects who suppressed the comparison target showed decreased SCL as time went by, while subjects in the cold flame conditions showed a similar decline in SCL across the expression period. This interaction effect was significant $F(2,132) = 5.91$, $p < .01$.

RESULTS

Here the researchers recognize the potent impact of an old relationship that one still desires (a "hot flame") and the less acute impact of an old relationship one does not desire (a "cold flame"). Section (**d**) explains the rationale. An overall analysis of variance (ANOVA) was also conducted (**j**).

The results from each group are reviewed in section (**k**). The results of the appropriate statistical analysis are disclosed with the probability values. These data are also presented in Figure 13.1, which shows that the skin conductance level was lowest for neutral thoughts, was higher for thoughts of a hot old flame, and was highest when the thoughts of the hot old flame were suppressed. Other measures of skin conductance seemed to decline over time.

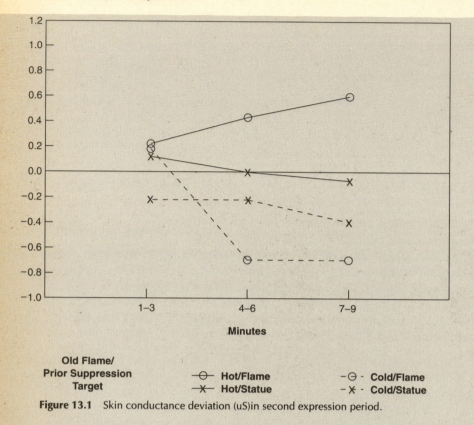

Old Flame/
Prior Suppression   —○— **Hot/Flame**   –⊝ – **Cold/Flame**
Target   —✕— **Hot/Statue**   – ✕ – **Cold/Statue**

Figure 13.1   Skin conductance deviation (uS)in second expression period.

DISCUSSION

These results indicate that the process of suppression may be responsible for extended physiological responsiveness to a lost love. Subjects who were prompted to suppress the thought of their still-desired old flame subsequently showed enhanced SCL on thinking about the flame. Those who were not initially responsive to the old flame—who indicated initially that the flame was "cold," were not prompted to become more responsive by suppression. It seems that suppression may prevent normal habituation to a stimulus that is already exciting, thus making the electrodermal reaction to that stimulus similar to if it was novel each time. **In choosing to suppress emotional**

DISCUSSION

In the final section **(l)** of the poster the authors restate their findings and offer a further explanation that may have important applications **(m)**. Finally, they add poignancy to the paper in **(n)**.

m **thoughts rather than express them** (see Pennebaker, 1988), **people may create the very emotional reactions they seek through suppression to eliminate.**
It seems that if people follow Hank Williams' example and try to get a past
n love off their minds, they may come to suffer the same fate: **"All night long I've
cried." ◆**

REFERENCES

Epstein, S. (1983). Natural healing processes of the mind: Graded stress inoculation as an
inherent coping mechanism. In D. Meichenbaum & M. E. Jaremko (Eds.), *Stress reduction and prevention* (pp. 39–66). New York: Plenum Press.

Fowles, D. C., Christie, M. J., Edelberg, R., Grings, W. W., Lykken, D. T., & Venables,
P. H. (1981). Publication recommendations for electrodermal measurements. *Psycho-
physiology, 18,* 232–239.

Martin, L. L., & Tesser, A. (1989). Toward a motivational and structural theory of rumina-
tive thought. In J. S. Uleman & J. A. Bargh (Eds.), *Unintended Thought* (pp. 306–326).
New York: Guilford Press.

Pennebaker, J. W. (1988). Confession, inhibition, and disease. In L. Berkowitz (Ed.),
Advances in experimental psychology (Vol. 22, pp. 211–242). Orlando, FL: Academic Press.

Wegner, D. M. (1989). *White bears and other unwanted thoughts: Suppression, obsession, and the psychology of mental control.* New York: Penguin.

Wegner, D. M., Schneider, D. J., Carter, S. R. III, & White, L. (1987). Paradoxical effects
of thought suppression. *Journal of Personality and Social Psychology, 53,* 5–13.

Wegner, D. M., Shortt, J. W., Blake, A. W., & Page, M. S. (1990). The suppression of
exciting thoughts. *Journal of Personality and Social Psychology, 58,* 409–418.

QUESTIONS

1. What was the purpose of the baseline data in this experiment?
2. What follow-up experiments are suggested by these data?
3. Based on this study, suggest how the use of SCL data might be used to
   distinguish deeply disturbing events from less disturbing events.
4. Some people have been known to "carry a flame" for someone for years.
   Design a study that could distinguish people who have carried a flame for a
   long time from people who have just recently broken up.
5. Design an experiment that could measure other emotional factors (e.g., the
   loss of a loved one).
6. Search the literature for examples of skin conductivity and emotional
   reactions. Prepare a brief report on the use of this technique in gathering
   experimental data.

Chapter 14

Teeth Grinding and Arousal Tendencies

INTRODUCTION

In this section you will find an interesting article that examines whether teeth grinding (bruxism) is influenced by stress-related variables in college students. This paper was presented during a poster session at a recent meeting of the Western Psychological Association.

This article is the first of four articles in Part Two that we have left for you to analyze. We suggest that you follow the format used in the previous three chapters to write your analysis. Try to identify the essential features of experimental design in this paper. In your analysis address the following issues:

1. What is bruxism?
2. What is the cause of bruxism and, given the authors' questionnaire, how widespread is it?
3. What is the "problem" the authors address?
4. What is the dependent variable in this study? What is the independent variable?
5. Describe the Arousal Predisposition Scale. (The library may have a better description of this scale.)
6. What are the results of this study?
7. What statistical tests are applied?
8. What conclusions do the authors offer? What additional conclusions could be drawn?
9. Is the experiment well controlled? Why or why not? What alternative hypothesis would you suggest?
10. Suggest further research in this area.

BRUXISM AND AROUSAL PREDISPOSITION IN COLLEGE STUDENTS

DARRAH A. WESTRUP, STEPHEN R. KELLER, TERRY A. NELLIS, AND ROBERT A. HICKS
SAN JOSE STATE UNIVERSITY

Bruxism, also known as teeth grinding, is a significant health problem. This painful, dental illness is characterized by the non functional gritting, grinding and clenching of the teeth, usually occurring during sleep. If not checked, this disorder can result in excessive irreversible toothwear to the point where actual fractures can occur. In addition, bruxing can lead to deterioration of the alveolar bone which holds the teeth, resulting in periodontal disease, and is frequently accompanied by chronic severe headaches, facial pain and sensitive teeth.

The fact that the incidence of bruxism is widespread further underscores the need for research on this disorder. A recent study, based upon three large samples of college students, showed a four-fold increase (i.e., from 5.1% to 20.9%) of self-reported nocturnal bruxism over a 23 year period (Hicks & Conti, 1989).

Despite the prevalence of this disorder, the etiology of bruxism is currently unknown. At present, theories range from the mechanical, such as the idea that occlusal or biting discrepancies interfere with jaw movement, to the belief that a disturbance in the central nervous system is to blame. Increasingly, research points to links between stressful experience and bruxing behavior (e.g., Hicks, Conti & Bragg, 1990; Hicks & Conti, 1991).

The purpose of this study was to further examine the relationships between stress-related variables and bruxism by measuring the relationships between arousability and incidence of self reported bruxing behavior. We felt that arousability might be implicated in bruxism, in that this variable seems to reflect the consequences of exposure to stressful experience (e.g., Matthews, Weiss, Detre, Dembroski, Falkner, Manuck & Williams, 1986).

METHOD

To locate bruxers, we administered a questionnaire to 780 university undergraduates during regularly scheduled classes. Embedded within this questionnaire was the critical item: "Do you experience bruxism (teeth grinding)?"

From this, 116 bruxers, 41 men and 75 women, were identified by their positive (yes) response to this question. Further, to ensure that this group would not be confounded by students who may brux and not realize it, 116 nonbruxers were additionally qualified by a score of zero on a bruxing symptoms checklist also embedded with the survey. This checklist was comprised of questions describing common symptoms of bruxing (e.g., Upon waking, are your cheek muscles sore or tired?) to which students responded with degree of severity ranging from "never" to "always."

To measure arousability, we used a scale that Coren (1988; 1990) has recently developed and standardized, i.e., the Arousal Predisposition Scale (APS). Of relevance to this study, score on the APS has been shown to be positively correlated with the degree to which students report experience with stress-related physical symptoms (Hicks, Conti & Nellis, 1992). Students are asked to select the response to each item which best describes them and their behaviors (e.g., I startle easily, and I am a calm person) and in choosing from five alternatives that range from "never" to "always." These responses are then scored in an ascending sequence on a five point scale and then summed for a total arousability score. For the purposes of our study, the items of the APS were embedded in the questionnaire that we administered to our sample of university undergraduates.

RESULTS

After tabulating the APS totals for the group of bruxers and the group of nonbruxers, we computed their means and standard deviations. We found a significant difference in APS scores between the bruxers and nonbruxers.

As shown in Table 14.1, the bruxing group showed a mean score of 35.82 ± 8.24, with the nonbruxers at a mean of 32.83 ± 7.20. Using a two-tailed t test for independent groups, we found a statistically significant difference with $t(116) = 2.94 \ p < .004$.

Interestingly, when the subjects were separated by gender, we found that the statistical difference found between bruxers and nonbruxers was attributable only to the female subjects. While there was no statistical difference between male bruxers and nonbruxers, female bruxers showed a mean APS score of 37.40 ± 8.17 compared with a mean of 33.12 ± 6.91 for female nonbruxers. A two-tailed t test showed a significant difference between these two groups of $t(116) = 3.47 \ p < .001$ (see Tables 14.2 and 14.3).

Table 14.1   MEANS AND STANDARD DEVIATIONS FOR APS SCORES AND
t STATISTICS COMPUTED BETWEEN BRUXERS AND NONBRUXERS

GROUPS	N	MEAN	SD	t BRUXER/NONBRUXER	p
Bruxers	116	35.82	8.24	2.94	.004
Nonbruxers	116	32.83	7.20		

Table 14.2   MEANS AND STANDARD DEVIATIONS FOR APS SCORES AND
t STATISTICS COMPUTED BETWEEN MALE BRUXERS AND MALE NONBRUXERS

GROUPS	N	MEAN	SD	t BRUXER/NONBRUXER	p
Male Bruxers	41	32.93	7.65	.37	.711
Male Nonbruxers	41	32.29	7.78		

Table 14.3   MEANS AND STANDARD DEVIATIONS FOR APS SCORES AND
t STATISTICS COMPUTED BETWEEN FEMALE BRUXERS AND FEMALE
NONBRUXERS

GROUPS	N	MEAN	SD	t BRUXER/NONBRUXER	p
Female Bruxers	75	37.40	8.17	3.47	.001
Female Nonbruxers	75	33.12	6.91		

DISCUSSION

Our prediction that the incidence of self-reported bruxing behavior among college students would be linked with one's arousal predisposition was supported only as far as female students were concerned. The data from this study, while supporting the theory that psychological factors are possible etiological factors of bruxism, suggests that the various factors leading to bruxing behavior in women may differ from those found in men. This research and that previously conducted, as well as the unquestionable severity and prevalence of this dental disorder points to the need for further research in this area.

REFERENCES

Coren, S. (1988) Prediction of insomnia from arousability predisposition scores: Scale development and cross-validation. *Behavior Research & Therapy, 26,* 415–420.

Hicks, R.A. & Chancellor, C. (1987) Nocturnal bruxism and type A-B behavior in college students. *Psychological Reports, 60,* 1211–1214.

Hicks, R.A. & Conti, P.A. (1989) Changes in the incidence of nocturnal bruxism in college students 1966–1989. *Perceptual and Motor Skills, 69,* 481–482.

Hicks, R.A., Conti, P. A. & Bragg, H. R. (1990) Increases in nocturnal bruxism among college students implicate stress. *Medical Hypotheses, 33,* 239–240.

Hicks, R.A. & Conti, P. A. (1991) Nocturnal bruxism and self reports of stress-related symptoms. *Perceptual and Motor Skills, 72,* 1182.

Hicks, R.A., Marical, C. M. & Conti, P. A. (1991) Coping with a major stressor: Differences between habitual short and longer sleepers. *Perceptual and Motor Skills, 72,* 631–636.

Hicks, R.A., Conti, P. A. & Nellis, T. A. (1992) Arousability and stress-related physical symptoms: A validation study of Coren's Arousal Predisposition Scale. *Perceptual and Motor Skills, 74,* 659–662.

Chapter 15

Picture Memory

INTRODUCTION

Of all our perceptual attributes, identification of visual scenes occupies a central position in our cognitive domain. In this experiment, Bower, Karlin, and Dueck demonstrate that contextual cues greatly facilitate the way we learn and place simple line pictures in memory.

Enhancing memory by "priming" a subject with mnemonic devices has often been used in the study of linguistic material. Consider the following syntactically correct but hard to understand and memorize sentence: The notes were sour because the seams split. Nonsense, you might think, but suppose you put this sentence into the context of "bagpipe." Suddenly you easily understand why the notes were sour, and your ability to memorize that sentence is significantly enhanced. But do we memorize pictures, especially ambiguous pictures, in a similar way? Two well-designed and interesting experiments by Bower et al. indicate that we do.

COMPREHENSION AND MEMORY FOR PICTURES

GORDON H. BOWER, MARTIN B. KARLIN, AND ALVIN DUECK
STANFORD UNIVERSITY

The thesis advanced is that people remember nonsensical pictures much better if they comprehend what they are about. Two experiments supported this thesis. In the first, nonsensical "droodles" were studied by subjects with or without an accompanying verbal interpretation of the pictures. Free recall was much better for subjects receiving the interpretation during study. Also, a later recognition test showed that subjects receiving the interpretation rate as more similar to the original picture, a distractor, which was close to the prototype of the interpreted category. In Experiment II, subjects studied pairs of nonsensical pictures, with or without a linking interpretation provided. Subjects who heard a phrase identifying and interrelating the pictures of a pair showed greater associative recall and matching than subjects who received no interpretation. The results suggest that memory is aided whenever contextual clues arouse appropriate schemata into which the material to be learned can be fitted.

The following experiments address the question of how people remember pictures. We may begin with the observation that pictures (drawings, diagrams, photographs) comprise a two-dimensional notational system which, like language, has both a "surface structure" (the medium) and a meaningful "deep structure" (the message). Like language, pictures have a terminal vocabulary (of strokes, shadings, etc.), sets of combination rules, often a referential field, and conventional rules for interpreting what a picture is about (see Gombrich, 1960; Goodman, 1968). Pictures, especially "realistic" ones, denote objects or scenes in a manner that parallels the symbolic way that words and sentences do. And just as language appears to be acquired as a perceptual motor skill, so also does it appear that children learn the conventional rules for interpreting the notational symbolism of pictures. These rules guide our construction of what a picture is about—what conceptualizations it expresses

SOURCE: Reprinted by permission from *Memory and Cognition,* 1975, 3, 216–220.

ANALYSIS

REVIEW OF LITERATURE

In this experiment, pictures were conceptualized as a "two-dimensional notational system" with features similar to those of language structure and usage. As such, pictures, like language, are a communicative medium. But the rules that gov-

or what objects it symbolizes. That we learn to interpret drawings is illustrated clearly by the difficulty novices have acquiring the symbolic system of their profession, such as ballet Labanotation, musical scoring, molecular structure, and the like.

a

We are interested in memory for pictures. **The hypothesis to be tested is that a major determinant of how well a person can remember a picture is whether or not he "understands" it at the time he studies it.** If he comprehends the picture—achieves a compact interpretation of it—then he should remember it much better than if he fails to comprehend it.

b

This hypothesis was suggested by the work of Bransford and Johnson (1972), Bransford and McCarrell (1974), and Doll and Lapinski (1974) on memory for *linguistic* material. They showed convincingly that a person's ability to recall a sentence depends on whether the sentence causes him to call to mind an appropriate referential situation. For example, consider causal sentences such as: (1) The notes were sour because the seams split; (2) The voyage wasn't delayed because the bottle shattered; (3) The haystack was important because the cloth ripped. Though simple in syntax and word meanings, such sentences prove difficult to understand and recall. The mind boggles because a causal connection is asserted to hold between two apparently unrelated events; the subject cannot call to mind an appropriate schemata (known scenario) into which the events can be substituted and thus related causally. But all difficulties dissolve if the subject is provided with a clue as to an appropriate causal schemata: The clue is a simple "thematic prompt" (for the three sentences above, *bagpipe, ship christening, parachutist*). The clue calls to mind a known scenario (see the "frames" theory of Minsky, 1974) into which the events mentioned in the sentence can then be fitted. The sentences then become comprehensible and memorable.

c

We wish to advance here a parallel argument for the role of comprehension in memory for pictures. **Our experiments will, therefore, expose subjects to pictures which are very difficult to "understand" unless one is given a the-**

ern pictorial memory and understanding may or may not be similar to the linguistic rules that govern its understanding. On an intuitive level our "understanding" of a picture at the time we see it ought to influence how well we remember that picture. The researchers in this experiment set out to see if that was true, stating their hypothesis in (**a**).

That hypothesis was suggested by contemporary research on memory for linguistic material (**b**). This experiment asks, If memory for pictorial material were similarly primed, would the effect be similar? An abbreviated description of the experimental design is found in (**c**). Notice how clearly these psychologists set the stage for these experiments. First, they must be convinced that memory is facilitated

matic clue; we then later test memory for pictures that had been shown with or without the clues. Of course, this means that we are investigating memory for "nonsensical" pictures, one for which subjects usually have no interpretation. But what makes a picture nonsensical or meaningless? There are doubtless several kinds of nonsense, but included would be pictures for which the viewer (1) does not know the conventions for interpretation (e.g., a musical score for a musician who only "plays by ear"), (2) does not know the conceptual denotations of the symbols (e.g., the step sequences corresponding to ballet Labanotation), or (3) knows both of the above but still can achieve no coherent understanding by applying the standard conventions because the picture does not supply enough interpretive clues. Examples of the latter kind, which we shall use in our research, occur with "impoverished" pictures: These are pictures which reduce or eliminate the salient clues or distinctive features of objects which typically guide our selection of their schemata from memory. Such pictures present fragments of hidden figures which may be seen only by suggestion. They appear uninterpretable until a clue retrieves from memory an appropriate conceptual frame which can then be fit onto the line fragments.

d   A curious side effect results from finally finding a conceptual schema which fits: The tension of "What is it?" dissolves with laughter into "Oh, now I get it!" Many of us became familiar with such visual jokes in the early 1960s with the "droodles" rage in America: A droodle was an uninterpretable drawing that turned out to have a funny interpretation (see Price, 1972). Figure 15.1 shows two of the examples used in our experiment.

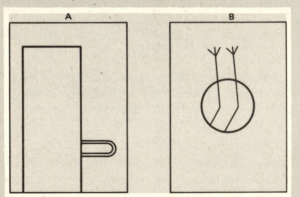

Figure 15.1   Droodles of Experiment I. Panel A: A midget playing a trombone in a telephone booth. Panel B: An early bird who caught a very strong worm.

by presenting a cue that assists understanding. And then they illustrate the memory technique they use in terms of verbally interpreted "droodles" **(d)**. One droodle we remember from years ago is shown below. It is either a used lollipop or the side view of a postcard.

EXPERIMENT I

The first experiment used free recall to assess the effects of comprehension on picture memory. Subjects studied a series of droodles pictures that they were to recall. Some subjects heard the interpretation as they saw each picture; other subjects, the controls, simply viewed the pictures without hearing any interpretive comment. The session ended with subjects drawing copies of those pictures they could recall.

e   A second, minor hypothesis tested was that subjects might distort their memory of the picture in a direction which provided a better fit to the prototype of the category used to interpret the picture upon original viewing. This "assimilation hypothesis" is an old one (see Carmichael, Hogan, & Walter, 1932; Riley, 1963), but the evidence regarding it has been equivocal. To test the

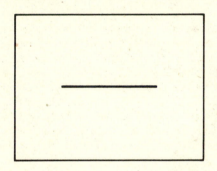

Once you "see" it, your memory for the object is significantly altered. Let's return to the experiments and see how the authors use these simple line drawings to describe empirically how cueing pictures affects memory.

EXPERIMENT I

From (c) we already have a good idea what the experimental design will be. The first experiment's dependent measures were a free-recall task that was given immediately after the presentation of the stimuli and a recognition memory test that was given one week later. Subjects were assigned either to a group that heard an interpretation with each picture or to a group that saw only the ambiguous pictures.

In addition to the general hypothesis, the experimenters were also trying to find support for the assimilation hypothesis (e), which states that subjects distort their memory for a picture to make it more representative of the interpretation than the actual picture itself.

hypothesis, we had subjects return after a week for a recognition memory test. Besides the correct picture, the multiple-choice set for each item contained two distractor pictures that were equally similar to the correct picture in terms of a line-overlap measure. One of these distractors exemplified a minor variation on the target which made it look even more like the interpreted prototype than did the original target (call this the "prototype" distractor). The other distractor involved a similarly minor line alteration but was done in such a manner as to violate the fit of the interpretive schema to the picture (call this the physically "similar" distractor). The expectation of the assimilation hypothesis is that, when recognition errors are made, subjects who learned the picture with a suggested interpretation will make relatively more errors on the prototype distractor than on the physically similar distractor. On the other hand, subjects who do not achieve the appropriate interpretation should tend to divide their errors evenly between the prototype and similar distractors.

Method

f   **The subjects were 18 undergraduates fulfilling a service requirement for their introductory psychology course. They were tested individually, assigned in random order to the "label" or "no-label" condition of the experiment.** All subjects studied a series of 28 simple droodles pictures shown on 3 × 5 in. cards at a rate of one every 10 seconds. As each picture was shown, its appropriate interpretation was given by the experimenter to the subjects in the label group but not to subjects in the no-label group. **Following presentation of the list, subjects had 10 minutes**

g   **to draw all the pictures they could remember in any order they wished.** The recall sheets were 8 × 11 in. papers marked off into a 3 by 3 matrix; the subject was

Method

In the experimental procedure all subjects were shown a series of 28 simple droodles at the rate of one every 10 seconds. However, one group heard the appropriate interpretation after each picture (label group), whereas the other group did not (no-label group). In all other respects the two groups were treated exactly alike. Thus, we have a simple two-group experiment.

The authors note that the 18 subjects were randomly assigned to either the label or the no-label group (**f**), but they do not report the order in which they presented the 28 pictures to the subjects. However, that really doesn't matter in this experiment because we are not concerned with any possible order effects of the cards. The important point is that the order of presentation was exactly the same in both the label and the no-label groups.

After the subjects had seen the 28 cards they were asked to recall as many of the pictures as they could in a free-recall situation (**g**) where the subject had to

instructed to recall by quickly sketching a recalled picture in one of the nine boxes on his recall sheet and to use as many sheets as necessary to complete his list recall. Before recall commenced, it was emphasized that the subject should aim for sketching the "gist" of the pictures recalled rather than for providing a lot of artistic detail of each picture. (The pictures could in fact be drawn very simply.) Following completion of the recall task, the subject was dismissed with an appointment to return the next week **"for other experiments."**

h

Upon returning the next week, subjects received the three-alternative multiple-choice test over 24 of the 28 pictures of the originally learned list (for four of the original pictures, we were unable to think up two similar distractors which met our criteria). The subject received a six-page booklet, with four multiple-choice triplets arranged in rows down each page. He was told that each triplet (row) contained one picture he had seen the week before as well as two closely similar pictures. **He was asked to rank order the three alternatives in each row, placing a 1 beside that test figure he considered most like the one he remembered seeing, a 2 beside the next most similar one, and a 3 beside that picture he considered least similar to the one he remembered.** The test was self-paced. Upon completing the test, the subject was debriefed and dismissed. One subject of the no-label group failed to return for the 1-week test, leaving eight subjects in that group at that point.

i

sketch the pictures in any order from memory. Each subject had 10 minutes for this task. Experimenters have to set a time limit on tasks such as these because otherwise the subjects could sit for hours trying to recall another droodle. Usually researchers choose a time limit by "pretesting" the material to find out how long it takes most subjects to recall most of the pictures that they know.

Note that when the subjects were dismissed they were asked to return the next week "for other experiments" (**h**). If the subjects had been told that they were coming back to recall the droodles, they could have practiced during the week. To eliminate these practice effects, the subjects were led to believe they were coming back for a different experiment.

Each subject returned the next week and took a recognition test. Three versions of each droodle were shown to the subject—one was the original droodle, one looked more like the interpretation given to the droodle in the label group (prototype distractor), and one was similar to the original but did not correspond to the interpretation. The subject was asked to rank order the figures (**i**). We assume the arrangement of the three pictures for each droodle was random across each row. Note that the subjects for both groups were treated exactly alike—the instructions and materials were the same for both groups. Also note that enough of the experimental procedure is included to allow the replication of the experiment.

Results

Free Recall

A first noteworthy fact is that we had relatively few problems in scoring for "gist recall" of the sketches. We had anticipated severe problems produced by interfering or confused combinations of several pictures, or at least deletions causing the sketch to be unidentifiable. But subjects tended to recall (sketch) the pictures either relatively accurately or not at all. The primary result of interest is that an average of 19.6 pictures out of 28 (70%) were accurately recalled by the label group [standard error of the mean (SEM) = 1.25], whereas only 14.2 pictures (51%) were recalled by the no-label group (SEM = .92). **The means differ reliably in the predicted direction [$t(16)$ = 3.43, $p < .01$]. Thus, we have clear confirmation that "picture understanding" enhances picture recall.**

Recognition Memory

Despite the closeness of the distractors to the target, recognition of the correct target at the 1-week retention interval was very high. Subjects who received labels during study correctly recognized (gave a 1 rating to) a mean of 22.0 out of the 24 test triplets (92%); subjects receiving no labels during study correctly

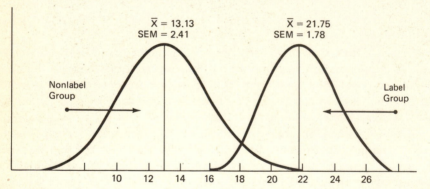

Figure 15.2   Number of pictures recalled (theoretical distribution based on data by Bower, Karlin, & Dueck, 1975).

Results

In this experiment two groups were tested in the free-recall task. The results are portrayed in Figure 15.2. From these data the authors used the t statistic to test for statistical significance and found **(j)** that the two means differed in the predicted direction for the 0.01 level of probability. Therefore, the authors concluded that labeling enhances picture recall.

recognized a mean of 20.1 pictures (84%). With standard errors of .83 and 1.08, respectively, the means do not differ reliably.

Even noting the high levels of recognition accuracy, we may still ask whether the label and no-label subjects react differently to the prototype vs. the physically similar distractor. There are several indications that the label subjects considered the prototype distractor much closer subjectively to the target. First, considering only cases when the correct picture was not ranked first, the conditional probability that the prototype (rather than the similar distractor) was ranked first was .75 for the label subjects but only .38 for the no-label subjects. However, these conditional probabilities are based on very few observations. A more stable measure of differentiation is provided by the difference in rankings (on similarity to the remembered target) between the prototype distractor and the physically similar distractor. For the label group the mean rank assigned to the prototype was 2.17, whereas that for the similar distractor was 2.76, a difference of .59. In contrast, for the no-label group, the rankings of the two distractors was closer: 2.34 were for the prototype and 2.48 for the similar distractor, a difference of only .14. The difference in rankings is reliably larger for the label group than for the no-label group [$t(15)$ = 2.79, $p < .02$]. **This result accords with the assimilation hypothesis: Subjects receiving the picture interpretation during study later reported that the distractor which moved in the direction of the interpreted prototype was closer to the target than was the distractor which involved a similarly minimal physical alteration but one which violated the interpretation given to the original target.** In contrast, the no-label subjects showed no comparable differentiation between the two kinds of distractors.

k

Experiment II

The initial experiment demonstrated the role of semantic comprehension in facilitating free recall of pictures. Having a meaningful name for a picture may

The second part of the experiment involved recognition memory for the pictures after a one-week interval. Both groups recognized the original figures with a high degree of accuracy. The authors also speculate on the nature of the recognition task and the assimilation hypothesis (k).

Experiment II

The first experiment established that a meaningful name helped the recall of a picture. But what about more complex learning, as when two ambiguous pictures are paired together? Would a label that cleverly tied the two pictures together facilitate the association between them?

facilitate recall because it provides a memorable summary or cue for later free recall. But an interpretation does more than provide a meaningful mnemonic label for a picture; it also causes unification or knitting together of the disparate parts of the picture into a coherent whole or schema.

l

m

The second experiment sets out to test more directly the influence of the unifying coherence of an interpretation upon picture memory. The subject was asked to study *pairs* of nonsensical pictures and was later tested by cueing with one member of each pair for recall (in drawing) of the other member of the pair. Again, half the subjects received no interpretation of the pictures, whereas half heard a phrase that made both pictures and their pairing a meaningful sequence. Examples of picture pairs are shown in the three rows of Figure 15.3. Their interpretations are (from left to right panels in each pair): (1) rear end of a pig disappearing into a fog bank, and his nose coming out the other side of the fog; (2) piles of dirty clothes, then pouring detergent into the washing machine to wash the clothes; and (3) uncooked

n

spaghetti, then cooked spaghetti and meatballs. **The hypothesis is that subjects hearing such interpretations during study will show much higher associative recall than will subjects who study the pictures without the interpretations.**

Method

The subjects were 16 university students attending summer school. They were recruited by an advertisement and paid $1.50 for their participation. They were tested individually, assigned in random alteration to the label and no-label conditions ($n = 8$ per group). The subject was told to learn 30 picture pairs that were shown to him at a rate of one every 12 sec. The pairs were drawn and shown by means of 3×5 flashcards, one picture on a white card and its mate on a pink card. The subject had been told that in the later recall test he would be shown the picture on the white card and would have to recall (draw) its mate from the pink card. During presentation of each pair, the label subjects heard the experimenter supply an interactive interpretation to bind the picture-pair together. These were descriptions like those given for the three panels of Figure 15.3. Following presentation of the 30 pairs, the white deck was shuffled and presented as recall cues at a 20-second rate. "Gist" sketching of the recalled pictures was emphasized. If the subject had begun his drawing before 20 sec, he was allowed to complete it; other-

The second experiment was designed to answer these questions (**l**). In (**m**) the experimenters give an abbreviated description of what they are going to do. Such a statement puts the reader "in the ball park" so that he or she can read the technical method section with a clear frame of reference. The hypothesis is stated in (**n**).

The method in this experiment involved presenting pairs of nonsense pictures. As in the first experiment, half of the subjects also received an interpretation,

o wise, the next cue was presented after 20 seconds. **Subjects drew their recall sketches in numbered boxes, nine to a page;** they left blank any numbered box for which they could recall nothing to the corresponding cue.

p After the cued recall test (conducted without feedback regarding the correctness of subject's recall), **the subject received an associative matching test.** The 30 white and 30 pink cards were spread out in a random array over the table top. The subject was instructed to scan over the array, looking for the pairs of white and pink pictures he had studied. As pairs were recognized, the two cards were picked up by the subject and handed to the experimenter. The subject continued this pairing

Figure 15.3   Pairs of nonsensical pictures used in Experiment II. See the text for an explanation of their contents.

which in this case linked the pictures conceptually, while half of the subjects received no interpretation. The authors hypothesized that subjects who had the interpretation would have better associative recall of the pairs than subjects who did not receive the interpretation.

The authors supply the pertinent details for this experiment: subjects, independent variables, description of stimulus materials, and procedures of the experiment. The dependent measures were an associative recall test **(o)** and an associative matching test **(p)**.

until he had selected all pairs he could remember: because they were asked not to guess, many subjects stopped short of pairing off all members. The subject's associative matching score was simply the number of correct pairs he selected from the array before terminating. (The expected correct pairs obtainable by guessing in an associative matching test is about one, regardless of the number of pairs to be guessed at. See Feller, 1957, p. 97.) After completing the matching test, the subject was debriefed and dismissed.

Results

Associative Recall

Again no problem was encountered in scoring correct gist recall of the cued picture. **Cued recall averaged 21.75 pictures out of 30 (73%) for the label subjects (SEM = 1.78) and 13.13 (44%) for the no-label subjects (SEM = 2.41).** These percentages differ reliably [$t(14) = 2.87$, $p < .02$]. The effect is remarkably consistent over items, too: For 22 items the label subjects recalled more than the no-label subjects, for three items the no-label subjects recalled more, and there were five ties. The 22/25 predominance of items with more recalls by label subjects exceeds chance of 50% ($z = 6.55$, $p < .01$). Thus, the label group is uniformly superior in recall to the no-label group.

Associative Matching

The number of correct matches (of pairs) averaged 27.50 for the label subjects (SEM = .98) compared to 16.63 for the no-label subjects (SEM = 2.80). These differ reliably [$t(14) = 3.66$, $p < .01$]. Moreover, the relative gain in recognition performance above what could be recalled was much larger for the label subjects (70%) than for the no-label subjects (21%). The data show that the label subjects still exhibit superior associative coherence even when all the pictures are available and do not need to be recalled.

In the results section we see that the influence on memory of labeling ambiguous pictures was statistically tested by means of a t-test, which indicated that labeling strongly facilitates associative learning. The average number of correctly cued recall items was 21.75 for the label group and 13.13 for the no-label group (**q**). The hypothetical distribution of scores is shown in Figure 15.4.

The number of correct matches is reported in (**r**).

DISCUSSION

The results of these experiments are succinctly summarized (**s**), and the authors suggest that their results have empirically demonstrated something that may have been suspected about human nature for some time.

DISCUSSION

It has been argued that memory for a picture depends upon the subject achieving a conceptual interpretation of the picture as he views it. The hypothesis is the pictorial analog of that relating sentence recall to comprehension (e.g., Bransford & McCarrell, 1974). The point is intuitively obvious once it has been noticed (as are many other "facts" of psychology), and the experiments above are primarily demonstrational in nature. **Subjects provided with meaningful interpretations of single droodles show superior free recall. Subjects who hear an interpretation identifying and relating two pictures together show greater coherence of the pictures on later association tests.** Although control subjects were probably trying to come up with some sensible interpretation of the pictures, the difficulty of the task precluded much success. Presumably, if we had collected control subjects' attempts at interpretations (recording their "thinking aloud"), those pictures for which they achieved a meaningful interpretation would more likely have been recalled (see, e.g., Montague, 1972). The likelihood of this being the case remains to be checked.

One might question whether the associative coherence found in Experiment II is a result merely of identifying the objects in each picture or whether it depends in addition upon providing the meaningful linking relationship between the two pictures of a pair. For some pairs the linking relation was that the pictures in Figure 11.3 denoted different parts of the same object (the pig), different states of an object as it underwent changes (the spaghetti), or different objects associated with a common process (the clothes and washer). We feel these relations are very important for promoting associative coherence of the elements of the pair. To illustrate this point, four further subjects from the same source were tested under the same procedure as Experiment II, except that new pairs were constructed by re-pairing the old pictures in a random manner. As the pair was shown, each picture was separately interpreted (e.g., a pig's tail and detergent pouring into a washing machine). No linking relation other than contiguity was stated for connecting the two contents. These four subjects averaged only 7.75 correct in cued recall (SEM = 1.89) and 8.25 correct in associative matching (SEM = 1.11). If anything, the scores are lower than those for the controls who studied the original pairs without hearing the objects or relation identified. Quite possibly, this "mispairing" list was so difficult because semantically related objects appeared in different pairs, creating intrapair interference. Teasing out the several contributors to this poor learning would be a task for further experimentation. The significant fact we wish to glean from this poor recall of mispaired items is that associative

The authors are thorough in their discussion and are careful not to come to wild conclusions that are not supported by the data in the experiments. In one

coherence depends heavily upon relating the two identified pictures and relatively little upon identifications per se that do not call to mind a known relationship between the two pictures.

How are our results to be related to previous work on picture memory? Previous work on learning of nonsense figures typically used recognition rather than reproduction measures and have been largely concerned with testing hypotheses of acquired distinctiveness or acquired equivalence of forms induced by learning different or the same arbitrary labels for the forms. Of more direct relevance to our results are those by Ellis (reviewed by Ellis, 1973), who found that the learning of "representative labels" to complex polygons enhanced their later recognition. Of course, the "representative labels" were simply a plausible name or interpretation of the figures. The fact that pairing with a representative label makes the pictures more memorable seems quite consistent with our hypothesis relating picture comprehension to memory. Since "association value" or "codability" has been a common variable in research on pictorial memory, one may ask whether our notion of a "semantic interpretation" of a picture is just a fancy name for an association to it. We think not. We intend "semantic interpretation" to be much more specific than the concept of "picture association" suggests. Associations may occur to many surface features of a picture or to fragments of it, all without improving memory for it. Presumably, picture memory would improve with greater "depth of processing," as does memory for words (Craik, 1973) or faces (Bower & Karlin, 1974). But this implies comprehending the picture, figuring out what conceptualization it expresses or what object it denotes: It means getting the "message" behind the "medium."

u

instance they worry about the results of Experiment II (t) and check their hunch by varying the procedure slightly for a small group of subjects.

In the final paragraph (u) Bower et al. fit their newly found information into the larger structure of memory. We urge the reader to study this section for its clarity of writing and thought.

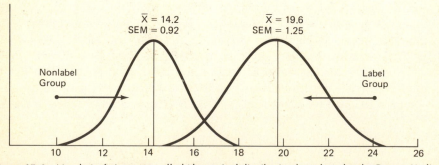

Figure 15.4   Number of pictures recalled (theoretical distribution based on data by Bower, Karlin, & Dueck, 1975).

REFERENCES

Bransford, J. D., & Johnson, M. K. (1972). Contextual prerequisites for understanding: Some investigations of comprehension and recall. *Journal of Verbal Learning and Verbal Behavior, 11*, 717–726.

Bransford, J. D., & McCarrell, N. S. (1974). In W. Weimer & D. Palermo (Eds.), *Cognition and the symbolic processes*. Hillside, N. J.: Lawrence Erlbaum Associates.

Bower, G. H., & Karlin, M. B. (1974). Depth of processing pictures of faces and recognition memory. *Journal of Experimental Psychology, 103*, 751–757.

Carmichael, L., Hogan, H. P., & Walter, A. A. (1932). An experimental study of the effect of language on the reproduction of visually perceived form. *Journal of Experimental Psychology, 15*, 73–86.

Craik, F. I. M. (1973). "Levels of analysis" view of memory. In P. Pliner, L. Krames, & T. Alloway (Eds.), *Communication and affect: Language and thought*. New York: Academic Press.

Doll, T. J., & Lapinski, R. H. (1974). Context effects in speeded comprehension and recall of sentences. *Bulletin of the Psychonomic Society, 3*, 342–345.

Ellis, H. C. (1973). Stimulus encoding processes in human learning and memory. In G. H. Bower (Ed.), *The psychology of learning and motivation* (Vol. 7). New York: Academic Press.

Feller, W. (1957). *An introduction to probability theory and its applications.* (Vol. 1. 2nd ed.) New York: Wiley.

Gombrich, E. (1960). *Art and illusion.* New York: Pantheon.

Goodman, N. (1968). *Languages of art* (2nd ed.). Indianapolis, Ind: Bobbs-Merrill.

Minsky, M. (1974). *Frame systems.* Unpublished manuscript. M.I.T. AI Project.

Montague, W. E. (1972). Elaborative strategies in verbal learning and memory. In G. H. Bower (Ed.), *The psychology of learning and motivation: In research and theory* (Vol. 6). New York: Academic Press. Pp. 225–302.

Price, R. (1972). *Droodles.* Los Angeles: Price/Stern/Sloan.

Riley, D. A. (1963). Memory for form. In L. Postman (Ed.), *Psychology in the making.* New York: Knopf.

NOTE: The authors thank Susan L. Karlin for creating most of the stimulus materials for Experiment II. The research was supported by Grant MH 13905-07 from the National Institute of Mental Health to Gordon H. Bower.

QUESTIONS

1. Devise another experiment to test the assimilation hypothesis mentioned in this experiment.
2. Why did the experimenters use two dependent measures in Experiment II: (1) a free-recall test and (2) a recognition test one week later?
3. The authors state that it is intuitively obvious that providing a conceptual interpretation for ambiguous pictures facilitates recall of the pictures. Why, then, must it be experimentally tested?
4. Design an experiment that tests whether groups of musical notes are easier to remember if they are accompanied by a verbal explanation.
5. Describe some practical and creative applications of this study.
6. The authors don't state specifically how they randomly assigned the 18 subjects into two groups (such that there were 9 subjects per group). How would you do it?

Chapter 16

Hormones and Toy Preferences

INTRODUCTION

"Little boys play with dump trucks and little girls play with dolls." Once in a while, though, the preference for play objects reverses; some girls prefer dump trucks and some boys prefer dolls. At an early age, however, these role reversals are often discouraged—girls may be taunted and called tomboys and boys may be called sissies, and sometimes the expressions are even more explicit and harsh. But sooner or later, most young children learn to play with toys normally associated with their own gender. Some psychologists have developed elaborate theories of social learning that suggest that gender role behavior is essentially a result of the socialization process and that genetics and/or hormones are of little consequence in the selection of gender specific toys. But in a striking contrast to these theories, Berenbaum and Hines present experimentally supported evidence that hormones (specifically, androgen levels in young girls) have a pronounced effect on what types of toys girls play with.

SPECIAL ISSUES

Selection of Subjects as a Source of Independent Variable

In many experiments it is possible to use subject characteristics as an independent variable. This is sometimes a practical route in studies measuring the differences between two or more groups of subjects who may react differently to some test or experience. Perhaps the most commonly used subject variable is gender. In this study by Berenbaum and Hines, the researchers ask the question, "Are sex-typed toy preferences related to androgens?" To answer this they used male and female children who had a particular genetic disorder that caused elevated levels of androgens. Other researchers may choose to study schizophrenics, gays, Hispanics, bruxers (see Chapter 14), military people of differing ranks (see Chapter 12), good students, football players, people living in Connecticut, poor people, stutterers, dyslexic readers, and so on. Although some experimental psychologists argue that the selection of a group of subjects who exhibit certain characteristics introduces special types of control problems, many times this technique is the only one available. Furthermore, if used prudently, this technique can yield reliable data that researchers can use to make valid conclusions.

Early Androgens Are Related to Childhood Sex-Typed Toy Preferences

SHERI A. BERENBAUM
University of Health Sciences

MELISSA HINES
University of California at Los Angeles

Girls with congenital adrenal hyperplasia (CAH) who were exposed to high levels of androgen in the prenatal and early postnatal periods showed increased play with boys' toys and reduced play with girls' toys compared with their unexposed female relatives at ages 3 to 8. Boys with CAH did not differ from their male relatives in play with boys' or girls' toys. These results suggest that early hormone exposure in females has a masculinizing effect on sex-typed toy preferences.

a   Sex differences in children's toy preferences have been demonstrated repeatedly. Boys prefer construction and transportation toys, whereas girls prefer dolls, doll furnishings, and kitchen supplies (Connor & Serbin, 1977; Liss, 1981). **These preferences appear to be partially learned, through modeling and reinforcement.** For example, children at various stages of development emulate the behavior of same-sex models in preference to opposite-sex ones (Bussey & Bandura, 1984; Huston, 1983). We present evidence that these sex-typed toy preferences are also related to prenatal or neonatal hormones (androgens).

b   Gonadal hormones play a major role in the development of sex differences in behavior and the brain in a variety of species, including rodents, songbirds, and primates (Arnold & Gorski, 1984; Beatty, 1979; Goy & McEwen, 1980;

SOURCE: Reprinted with permission of *Psychological Science,* 1992, *3,* 203–206.

ANALYSIS

Review of Literature

The authors begin by stating a commonly accepted supposition regarding the etiology of boys' preference for dump trucks and girls' preference for dolls: "These preferences appear to be partially learned, through modeling and reinforcement." (See (a).) On the other hand, in (b) Berenbaum and Hines suggest that hormones also appear to play an important role in sex differences, especially in animals (e.g., c   MacLusky & Naftolin, 1981). **A unique opportunity to study hormonal influences on human sex-typed behavior is provided by the genetic disorder congenital adrenal hyperplasia (CAH). Because of an enzymatic defect, individuals with CAH produce high levels of adrenal androgens** beginning *in utero.* Postnatal treatment with corticosteroids (and mineralocorticoids for the 75% who are also salt-losers) normalizes hormone levels (White, New, & Dupont, 1987).

d   **If early hormone exposure affects the development of sex-typed behavior in human beings as it does in other species, then CAH girls should show behavior similar to that of normal boys.** Previous studies have suggested that CAH females show intense physical energy expenditure, "tomboyism," rough outdoor play, preference for traditionally masculine toys and activities (Ehrhardt

d_1 & Baker, 1974; Ehrhardt, Epstein, & Money, 1968), and greater spatial ability than their female relatives (Resnick, Berenbaum, Gottesman, & Bouchard, 1986). These findings parallel those reported in other samples with prenatal exposure to masculinizing hormones due to maternal ingestion during pregnancy (Ehrhardt & Meyer-Bahlburg, 1979; Hines, 1982), and are especially interesting in light of recent reports of sex differences in human brain structure (Allen, Hines, Shryne, & Gorski, 1989; Hines & Green, 1991; Swaab & Fliers, 1985).

e   Although prior studies of CAH girls suggest that hormones can influence childhood behavior, methodological limitations encouraged us to pursue this issue. Specifically, in prior studies, (a) behavior was assessed from interviews rather than direct observation; (b) data were collected with knowledge of the patient or control status of the subject; (c) behaviors were usually rated as present or absent, rather than as continuous traits; and (d) masculine and feminine behaviors were often not assessed separately, but instead treated as opposite ends of a single continuum. Therefore, the present study of individ-

rats, birds, and primates). In (**c**) we find that it is possible to study the effect of adrenal androgens on behavior by studying children who are born with a congenital disorder called CAH or congenital adrenal hyperplasia. This condition produces high levels of androgens (the male hormone) in patients with the disorder.

In (**d**) a working hypothesis is presented (a formal hypothesis (**e**) is also shown) in which the authors let us know the general scheme of the investigation. In section (**d_1**) the authors provide a review of literature, which, in general, supports the idea that CAH leads to male related behavior.

In section (**e**) the problems are stated, and in section (**f**) Berenbaum and Hines formally state their hypothesis.

uals with CAH focused on objective, quantifiable measures of sex-typed toy preferences, with assessments made by raters blind to the patient or control status of subjects. **We hypothesized that CAH girls would show greater preference for boys' toys than their unaffected female relatives, and reduced preference for girls' toys.** We did not predict effects for CAH boys because androgen treatment has inconsistent effects in male experimental animals (Baum & Schretlen, 1975; Diamond, Llacuna, & Wong, 1973).

f

METHOD

Subjects

We recruited 3- to 8-year-old children with CAH from pediatric endocrine clinics at eight hospitals in the Midwest and California and tested 26 girls and 11 boys. Because of the restricted age range studied and family constellations, only 16 patients (43%) had a same-sex control. Therefore, we combined relatives of male and female patients to obtain our control groups. **The control groups consisted of 15 unaffected female relatives** (10 sisters and 5 first cousins) **and 18 unaffected male relatives** (14 brothers and 4 first cousins). Patients and controls did not differ significantly in birth order or age. Mean ages in months were as follows: **female patients, 66.54 (range: 36–99); female controls, 61.80 (range: 36–93); male patients, 64.18 (range: 41–101); and male controls, 69.78 (range: 33–99).**

g_1

g_2

g_3

Illness Characteristics

Although behavioral changes in CAH girls may be caused by androgen influences on the developing brain, it has been suggested that these changes might instead result from social or illness factors (Quadagno, Briscoe, & Quadagno, 1977; Slipjer, 1984). For example, because androgen levels are high *in utero,* females have masculinized genitalia. Surgical reconstruction is often necessary, although postnatal treatment prevents further virilization. Parents may treat their CAH daughters in a masculine fashion as a response to this masculine appearance at birth. Therefore, we asked parents to complete a questionnaire which included the item "I encourage my child to act as a girl should." The questionnaire was available for 24 CAH and 11 control girls.

METHOD

The method section begins with a description of the subjects used in the study (g_1), the control groups (g_2), and the children's ages (g_3). Note that the researchers used close relatives of the experimental group to form the control groups. Thus, these groups approximated the genetic and environmental features of the experimental group. The "Illness Characteristics" section gives a further description of CAH and its features.

We also examined characteristics of the child's disease: age at diagnosis, degree of genital virilization at diagnosis, and salt-losing status. This information was obtained from medical records, available on 25 girls and 10 boys. Medical ratings were made by two research assistants who had no knowledge of the child's behavioral scores. Age at diagnosis and salt-losing status were recorded with perfect reliability. Most patients were diagnosed in the early neonatal period: Median age at diagnosis was 11 days for girls (range: 0 days to 64 months) and 30 days for boys (range: 14 days to 54 months). Degree of virilization was rated on a scale ranging from 0 (normal female) to 6 (normal male), with intermediate values reflecting varying degrees of clitoral enlargement and fusion of the labia (Prader, 1954). Interrater reliability was .82; mean ratings were used. All girls had some degree of genital virilization: The mean Prader score was 3.0 (range: 1–5). One girl had been raised as a boy for the 1st month.

Toy Preference Materials and Procedure

h
Toy preference was measured by the amount of time the child played with toys shown by others to be preferred by girls, by boys, or equally by the sexes (neutral). The boys' toys included transportation toys (a helicopter, two cars, and a fire engine) and construction toys (blocks and Lincoln Logs). The girls' toys included three dolls, kitchen supplies, a toy telephone, and crayons and paper. Neutral toys, used as a control, included books, two board games, and a jigsaw puzzle. The toys were arranged in a standard order on the floor of the pediatric clinic or the child's home, in an area approximately 8 ft by 10 ft surrounded by screens.

i
The child was brought into the play area individually and told to play with the toys however he or she wanted. **The 12-min session was videotaped for later scoring.** The first 10 min of each session were usually scored; the additional 2 min were scored only if sections of the initial 10 min were unscorable (e.g., if the child attempted interaction with the videotaper). Order of testing the patient and control was random.

We scored the amount of time the child played with each toy and then summed the time spent in play with the toys in each of the three categories, to produce total scores for play with boys' toys, girls' toys, and neutral toys. **All tapes were rated by**

The dependent variable is operationally stated in (**h**). In (**i**) we learn that the sessions were videotaped for later scoring. The videotapes also also might provide important documentation of the actual play sessions should other investigators or the authors want to reexamine the data for other behavioral characteristics.

j   the same two raters, who had not tested the children and who were blind to their patient or control status. Interrater reliability was very high: The median correlation was .99 for individual toys and .99 for total scores. Data reported here represent the mean scores of the two raters.[1]

RESULTS

Statistical tests are one-tailed when hypotheses are directional (sex differences and female CAH–control comparisons for time spent in play with boys' and girls' toys) and two-tailed when no difference is hypothesized (all comparisons for play with neutral toys) or when the direction of the difference is not specified (male CAH–control comparisons).

Sex Differences

k   As expected, control boys and girls differed in the amount of time they played with boys' toys and girls' toys, but not neutral toys (see Fig. 16.1). The magnitudes of the differences (in standard deviation units, or d; Cohen, 1977) are consistent with those reported by other investigators. Further, control boys preferred boys' toys to girls' toys, $t(17) = 4.58$, $p < .001$. Control girls had the opposite preference, but the difference was not significant, $t(14) = -1.17$.[2]

[1]Two children (both CAH girls) who played with no toys during their sessions were excluded from all analyses. These girls are similar to the other subjects in disease characteristics. Analyses including these subjects do not change the interpretation of the results.

[2]Because the distributions of play scores are skewed, we conducted all analyses on data transformed in various ways (e.g., arcsine, square root) and with nonparametric procedures. All analytic procedures produced similar results.

In (j) the authors tell us that two people who were blind as to which children were in which group independently scored the behavior of the children. Although reason would suggest that there is little ambiguity in scoring the behavior—time playing with a specific type of toy leaves little room for mistakes—the authors exhibit meticulous attention to potential experimenter bias that might creep into the observations.

In footnote (k) the authors mention another issue: subject attrition. In experimental work in psychology, subjects are often lost for one reason or another. It is prudent to anticipate these losses and have a systematic way to deal with them *before* the experiment begins (insofar a such farsighted planning is possible). The researcher should have a plan for replacing or in some way compensating (e.g., statistically) for lost data. In this experiment, two little girls simply did not play with the toys; in effect, they did not produce any data on the dependent variable (h). The data from these two subjects were excluded from all analyses. In this passage the word *analysis* is left undefined; it could mean statistical analysis (likely) or inspection of the data.

CAH Patient–Control Comparisons

Females

1

m

As hypothesized, CAH girls spent significantly more time playing with boys' toys than did control girls ($d = .89$), $t(37) = 2.66$, $p < .01$, and about as much time as the control boys (see Figure 16.1). **Further, like control boys, CAH girls played significantly more with boys' toys then with girls' toys, $t(23) =**

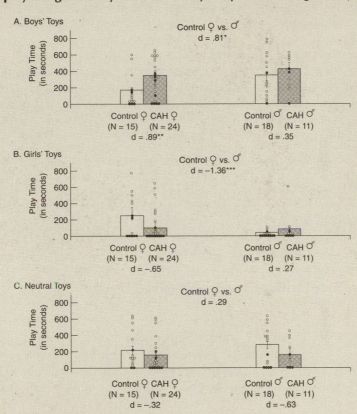

Figure 16.1 Time spent in play with sex-typed toys by female and male CAH patients and controls during 10 min of play. Scores are the sum of play with individual toys and therefore may exceed 600 s. Bars represent group means; lines represent standard errors; points represent individual subjects; d = difference between group means/average standard deviation. Group differences were evaluated by t test; $*p < .05$, $**p < .01$, $***p < .001$, one-tailed.

RESULTS

In section **(l)** the main results of the two groups of girls are presented first. The probability (p) value is also presented. In **(m)** the authors statistically test the results showing how much the CAH girls played with girls' toys and how much they played with boys' toys. Further tests are conducted on the data and these details are presented in the next two paragraphs.

2.93, $p < .01$. This effect is also seen in the subsample of patients with a same-sex relative control ($N = 11$ pairs): The mean time in play with boys' toys was 327.23 s for CAH girls and 166.55 s for matched girls, $t(10) = 2.20$, $p < .05$.

CAH girls also played less with girls' toys than did the control girls, $t(37) = -2.02$, $p < .05$, but this effect is smaller than that for boys' toys ($d = .65$). Perhaps because of low statistical power, the difference is not significant in matched-pairs analysis of the 11 patient–control pairs, $t(10) = -1.09$. Because total play time was limited, play with one set of toys tended to reduce play with another set, so reduced play with girls' toys may have partly resulted from increased play with boys' toys. In fact, the correlations among play with boys', girls', and neutral toys are all moderately negative, ranging from $-.30$ to $-.69$ (all $ps < .01$, one-tailed). In this context, however, it is important to note that there were no differences between CAH and control girls in play with neutral toys (both matched and unmatched t-values < 1.0).

There were no significant differences between CAH and control girls in parents' responses to the question "I encourage my child to act as a girl should," with 59.1% and 63.6% responding "yes," respectively. Amount of time spent playing with sex-typed toys was not significantly related to any disease characteristic. For play with boys' toys, rs are $-.10$ for virilization, $-.29$ for age at diagnosis, and $.25$ for salt-losing status ($1 = $ salt-loser, $0 = $ simple virilizer). For play with girls' toys, corresponding rs are $-.09$, $-.14$, and $-.06$. Note, however, that the small sample size and restricted range (e.g., only 2 girls were simple virilizers) reduce the power of these analyses.

Males

In contrast to the differences observed between CAH and control girls, there were no significant differences between CAH boys and control boys on sex-typed play ($ps > .20$ for play with boys' and girls' toys). Examination of medical characteristics in relation to sex-typed play in boys is not meaningful because of the lack of patient–control differences and limited variability in the sample.

DISCUSSION

n   **Large differences between CAH girls and unaffected female relatives indicate masculinization of toy preferences in girls exposed to high levels of androgen during early development.** These results are consistent with data

DISCUSSION

The discussion section begins with a restatement of the principal findings and conclusion (**n**). Then the results are related to previous findings (**o**). The authors

o from animal models and with prior studies of CAH and other females exposed to masculinizing hormones *in utero*. **Our findings strengthen the conclusions of previous play studies in CAH girls because we used objective, quantifiable measures and behavioral ratings made without knowledge of patient or control status.** The data also suggest a relative defeminization of toy preferences, although this result was less robust. Other studies of hormone-exposed samples have also been more likely to observe "masculinizing" than "defeminizing" effects of prenatal hormones, perhaps in part because the studies have concentrated on the former (Berenbaum, 1990; Ehrhardt & Meyer-Bahlburg, 1979; Hines, 1982). Although other hormones, such as 17-hydroxyprogesterone and corticosteroids, are also abnormal prenatally in CAH, it is unlikely that they account for the masculinized behavior in CAH girls, because these hormones have smaller and less consistent behavioral effects than androgen and may actually prevent masculinization (Erpino, 1975; Hull, Franz, Snyder, & Nishita, 1980).

It appears unlikely that the behavioral changes are due to social or illness factors. Our failure to find relationships between sex-typed play and physical virilization is consistent with other data in CAH patients indicating no relationship between degree of virilization and gender-role behavior (Slipjer, 1984). Although it is possible that any virilization results in different parental treatment, this argument is weakened by parents' retrospective reports that they did not treat CAH girls in a "masculine" fashion, here and in other studies (Ehrhardt & Baker, 1974; Resnick, 1982), and by data from rhesus macaques indicating that masculinization of juvenile play is unrelated to genital virilization or maternal behavior (Goy, Bercovitch, & McBrair, 1988). Nevertheless, given the disproportionate rate of return of questionnaires for our CAH and control girls and other problems inherent in questionnaire measures, direct observation of parental behavior would be valuable.

also state the rigor involved in the present studies, thus addressing some issues raised in the statement of the problem (see **e**). Further findings are presented throughout the next several paragraphs. In section (**p**) the authors offer an expanded view of how their experimental results fit into the larger issue of hormones' influence on behavior. Of particular interest is that the researchers do not go beyond their data—they confine their conclusions to the specific data reported in this study. The unseasoned researcher might be tempted to make sweeping generalizations regarding adult women who exhibit masculine tendencies or even social phenomena such as the women's movement. While such matters are of great interest, this research addresses a specific problem (which may ultimately be part of a larger matter) that yielded clear-cut results with direct implications for the authors' hypothesis. You should also note that this example is especially clear in exhibiting many of the features of precise experimental design mentioned throughout this book.

Our data indicate no changes in sex-typed toy preferences in CAH males, consistent with previous reports in nonhuman animals and people. CAH males have generally been reported not to differ from controls in sex-typed activities or abilities (Ehrhardt & Baker, 1974; Resnick et al., 1986).

P | Although the data are consistent with an androgen influence on sex-typed toy choices, it is not necessary that hormones have a direct influence on these choices. Hormones may affect toy choices indirectly, perhaps through an influence on activity level, motor skills, abilities, or temperament. For example, CAH girls may be more active than control girls, and boys' toys may facilitate active play (O'Brien & Huston, 1985).

Results from this study may also be relevant to understanding the development of sex-typed toy preferences in normal children. Normal males have higher testosterone levels than normal females from approximately week 8 to week 24 of gestation and from approximately the 1st to the 5th month of life (Smail, Reyes, Winter, & Faiman, 1981). These sex differences in hormones may contribute to subsequent sex differences in behavior, such as those in toy preferences. Further, natural variations in levels or availability of testosterone among normal males and females might contribute to individual differences in sex-typed behavior.

Our data may also be relevant to evidence that early sex-typed toy preferences predict later behavior, including sexual orientation (Green, 1987) and spatial ability (Newcombe, Bandura, & Taylor, 1983; Sherman, 1967), and to evidence that spatial ability (Resnick et al., 1986) and sexual orientation (Money, Schwartz, & Lewis, 1984) are masculinized in CAH females (for discussion, see Berenbaum, 1990). Specifically, it is possible that hormonal influences on adult behaviors are mediated by childhood sex-typed toy preferences, although it is also possible that various sex-typed behaviors are influenced separately by hormones (Arnold & Gorski, 1984; Goy & McEwen, 1980; Goy et al., 1988). Further studies of CAH and other endocrine syndromes should help us to understand not only whether gonadal hormones influence human behavior, but also how they do so.

ACKNOWLEDGMENTS

This study was supported by National Institutes of Health Grants HD19644, HD24542, and NS20687. We thank the following people who contributed to this project: Drs. Stephen Duck, Orville Green, Ora Pescovitz, Julio Santiago, Jo Anne Brasel, Robert Clemons, Gertrude Costin, Richard Fefferman, Lynda Fisher, Francine Kaufman, and Thomas Roe generously provided access to their patients and answered medical questions; Brenda Henderson, André

Black, Ruth Estes, Erin Foy, Kim Kerns, Deena Krumdick, Jennifer Lawrence, Naomi Lester, Anne Maxwell, Kristie Nies, Ellen Rochman, Robyn Reed, Elizabeth Snyder, and Christopher Verbin helped with data collection, scoring, entry, and analysis; Kim Kerns established a superb data-entry and tracking system. We are particularly grateful to the subjects and their parents for their enthusiasm and cooperation. We also thank Michael Bailey, Susan Resnick, Michael Taylor, and Gary Oltmans for helpful comments on this paper. Portions of this paper were presented at the meeting of the Society for Research in Child Development, Kansas City, April 1989.

REFERENCES

Allen, L.S., Hines, M., Shryne, J.E., & Gorski, R.A. (1989). Two sexually dimorphic cell groups in the human brain. *Journal of Neuroscience, 9*, 497–506.

Arnold, A.P., & Gorski, R.A. (1984). Gonadal steroid induction of structural sex differences in the central nervous system. *Annual Review of Neuroscience, 7*, 413–442.

Baum, M.J., & Schretlen, P. (1975). Neuroendocrine effects of perinatal androgenization in the male ferret. *Progress in Brain Research, 42*, 343–355.

Beatty, W.W. (1979). Gonadal hormones and sex differences in nonreproductive behaviors in rodents: Organizational and activational influences. *Hormones and Behavior, 12*, 112–163.

Berenbaum, S.A. (1990). Congenital adrenal hyperplasia: Intellectual and psychosexual functioning. In C. Holmes (Ed.), *Psychoneuroendocrinology: Brain, behavior, and hormonal interactions* (pp. 227–260). New York: Springer-Verlag.

Bussey, K., & Bandura, A. (1984). Influence of gender constancy and social power on sex-linked modeling. *Journal of Personality and Social Psychology, 47*, 1292–1302.

Cohen, J. (1977). *Statistical power analysis for the behavioral sciences* (rev. ed.). New York: Academic Press.

Connor, J.M., & Serbin, L.A. (1977). Behaviorally based masculine- and feminine-activity-preference scales for preschoolers: Correlates with other classroom behaviors and cognitive tests. *Child Development, 48*, 1411–1416.

Diamond, M., Llacuna, A., & Wong, C.L. (1973). Sex behavior after neonatal progesterone, testosterone, estrogen, or antiandrogens. *Hormones and Behavior, 4*, 73–88.

Ehrhardt, A.A., & Baker, S.W. (1974). Fetal androgens, human central nervous system differentiation and behavior sex differences. In R.C. Friedman, R.M. Richart, & R.L. Vande Wiele (Eds.), *Sex differences in behavior* (pp. 33–51). New York: Wiley.

Ehrhardt, A.A., Epstein, R., & Money, J. (1968). Fetal androgens and female gender identity in the early-treated adrenogenital syndrome. *Johns Hopkins Medical Journal, 122*, 160–167.

Ehrhardt, A.A., & Meyer-Bahlburg, H.F.L. (1979). Prenatal sex hormones and the developing brain: Effects on psychosexual differentiation and cognitive function. *Annual Review of Medicine, 30*, 417–430.

Erpino, M.J. (1975). Androgen-induced aggression in neonatally androgenized female mice, inhibition by progesterone. *Hormones and Behavior, 6*, 149–158.

Goy, R.W., Bercovitch, F.B., & McBrair, M.C. (1988). Behavioral masculinization is independent of genital masculinization in prenatally androgenized female rhesus macaques. *Hormones and Behavior, 22,* 552–571.

Goy, R.W., & McEwen, B.S. (1980). *Sexual differentiation of the brain.* Cambridge, MA: MIT Press.

Green, R. (1987). *The "sissy boy syndrome" and the development of homosexuality.* New Haven, CT: Yale University Press.

Hines, M. (1982). Prenatal gonadal hormones and sex differences in human behavior. *Psychological Bulletin, 92,* 56–80.

Hines, M., & Green, R. (1991). Human hormonal and neural correlates of sex-typed behaviors. *Review of Psychiatry, 10,* 536–555.

Hull, E.M., Franz, J.R., Snyder, A.M., & Nishita, J.K. (1980). Perinatal progesterone and learning, social and reproductive behavior in rats. *Physiology and Behavior, 24,* 251–256.

Huston, A.C. (1983). Sex-typing. In P.H. Mussen (Ed.), *Handbook of child psychology* (Vol. IV, 4th ed., pp. 387–467). New York: Wiley.

Liss, M.B. (1981). Patterns of toy play: An analysis of sex differences. *Sex Roles, 7,* 1143–1150.

MacLusky, N.J., & Naftolin, F. (1981). Sexual differentiation of the central nervous system. *Science, 211,* 1294–1303.

Money, J., Schwartz, M., & Lewis, V.G. (1984). Adult erotosexual status and fetal hormonal masculinization and demasculinization: 46, XX congenital virilizing adrenal hyperplasia and 46, XY androgen-insensitivity syndrome compared. *Psychoneuroendocrinology, 9,* 405–414.

Newcombe, N., Bandura, M.M., & Taylor, D.G. (1983). Sex differences in spatial ability and spatial activities. *Sex Roles, 9,* 377–386.

O'Brien, M., & Huston, A.C. (1985). Activity level and sex stereotyped toy choice in toddler boys and girls. *Journal of Genetic Psychology, 146,* 527–534.

Prader, A. (1954). Der genitalbefund beim Pseudohermaphroditismus femininus des kongenitalen adrenogenitalen Syndroms. *Helvetica Paediatrica Acta, 3,* 231–248.

Quadagno, D.M., Briscoe, R., & Quadagno, J.S. (1977). Effects of perinatal gonadal hormones on selected nonsexual behavior patterns: A critical assessment of the nonhuman and human literature. *Psychological Bulletin, 84,* 62–80.

Resnick, S.M. (1982). *Psychological functioning in individuals with congenital adrenal hyperplasia: Early hormonal influences on cognition and personality.* Unpublished doctoral dissertation, University of Minnesota, Minneapolis.

Resnick, S.M., Berenbaum, S.A., Gottesman, I.I., & Bouchard, T.J. (1986). Early hormonal influences on cognitive functioning in congenital adrenal hyperplasia. *Developmental Psychology, 22,* 191–198.

Sherman, J. (1967). Problem of sex differences in space perception and aspects of intellectual functioning. *Psychological Review, 74,* 290–299.

Slipjer, F.M.E. (1984). Androgens and gender role behavior in girls with congenital adrenal hyperplasia (CAH). In G.J. DeVries, J.P.C. DeBruin, H.B.M. Uylings, & M.A. Corner (Eds.), *Progress in brain research* (Vol. 61, pp. 417–422). Amsterdam: Elsevier.

Smail, P.J., Reyes, F.I., Winter, J.S.D., & Faiman, C. (1981). The fetal hormone environment and its effect on the morphogenesis of the genital system. In S.J. Kogan & E.S.E. Hafez (Eds.), *Pediatric andrology* (pp. 9–19). The Hague: Martinus Nijhoff.

Swaab, D.F., & Fliers, E. (1985). A sexually dimorphic nucleus in the human brain. *Science, 228,* 1112–1115.

White, P.C., New, M.I., & Dupont, B. (1987). Congenital adrenal hyperplasia. *New England Journal of Medicine, 316,* 1519–1524.

Chapter 17

Maternal Behavior

INTRODUCTION

Maternal behavior (i.e., nest building, care of the young, retrieving, nursing, etc.) has long been considered an instinct. However, in recent years it has become apparent that calling a behavior an instinct tells us little about its causes. So psychologists have begun to study the factors underlying such behavior. These factors include stimuli coming from the nest or the young, brain mechanisms, and past experience. Some suggest that hormones, too, may play a role in controlling such behavior. This idea derives from the fact that maternal behavior develops very gradually and fades away in the same manner; that is, nest building and the bodily changes accompanying maternity take place prior to birth and disappear gradually as the young grow older. This suggests that the behavior may be related to the gradual buildup and decline of some chemical substance in the blood. One way to demonstrate this is to inject the hormone and see whether it can induce the behavior. Previous studies of this type have failed to produce clear-cut results, and this study attempts to approach the problem in a somewhat different fashion.

This article illustrates several important issues in experimental design. Among these is the problem of control. Several types of control were considered. The experimental animals were divided into equal-sized groups (presumably in a random fashion). Some groups were defined as control groups, and other groups were designated the experimental groups. Also, all animals were treated identically except for the type of chemical injection they received. Another type of control involved the preparation of substances used in the injection.

This experiment has a distinctive independent variable and a control group. The experimenters measured the reaction of an experimental group against the reaction of a control group. You should note that something was done *to* the subjects—they were injected with various chemicals.

SPECIAL ISSUES

The Use of Animals in Psychological Research

Some of the earliest experiments in psychology were done with animals. Ivan Pavlov used dogs; E. L. Thorndike used cats; William James used chickens; Harry Harlow used monkeys; the Gardners used chimpanzees; and B. F. Skinner used rats. Other psychologists have used guinea pigs, seals, porpoises, monkeys, elephants, whales, planera, bees, pigeons, sheep, pigs, horses, rabbits, and so on. Given this, one might think psychologists were more interested in the behavior of nonhuman creatures than in the behavior of *homo sapiens*. But in most instances experimental psychologists—even those who restrict their studies to animals—strongly assert that their experiments are designed to lead to a better understanding of humans.

The basic premise on which psychological experimentation is predicated is that conclusions reached through the study of nonhuman subjects are in some way applicable to an understanding of behavior. There are many reasons for using animals in psychological research. In general, animals are more available than human subjects. It is also possible to keep an animal under observation 24 hours a day, seven days a week for months, if necessary. In addition, it is possible to perform procedures with animals that are impossible with humans. Such research could involve the use of noxious stimuli, prolonged periods of deprivation, psychosurgery, the use of experimental drugs, and so on. It would be almost impossible to get a college sophomore to volunteer for these experiments, and even if subjects were available, many of these experiments would constitute a serious breach of ethics. But the maintenance of ethical restraints is also important for nonhuman subjects. The experimental psychologist who uses animals is honorbound to conduct his or her research within the guidelines described in Chapter 8.

This experiment by Joseph Terkel and Jay Rosenblatt on maternal behavior induced by injecting maternal blood plasma into virgin rats was selected for inclusion in this section (along with the article on the creative porpoise) because it illustrates an experiment in which animals serve as the experimental subjects. A similar experiment with humans would not be possible, and yet the conclusions reached may give us a clue as to the source of maternal behavior in humans.

Maternal Behavior Induced by Maternal Blood Plasma Injected into Virgin Rats

JOSEPH TERKEL
INSTITUTE OF ANIMAL BEHAVIOR

JAY S. ROSENBLATT
RUTGERS UNIVERSITY

Induction of maternal behavior (i.e., retrieving) in virgins by exposure to young pups was studied to investigate effects of blood plasma from a postparturient female. Control groups were injected with blood plasma from nonmaternal females in proestrus and diestrus phases of the vaginal estrous cycle and with saline solution. Virgins injected with maternal blood plasma had significantly shorter latencies of maternal behavior than other groups. Injections of saline and proestrus blood plasma had no effect on maternal behavior. Virgins injected with diestrus blood plasma were significantly delayed in displaying maternal behavior. The findings indicate that there is a humoral basis for the appearance of maternal behavior after parturition.

Several recent attempts to induce maternal behavior in the rat by means of various hormones (i.e., estrogen, progesterone, and prolactin) injected directly into virgin or experienced females have not yielded results that would increase our understanding of the hormonal basis of this behavior (Beach & Wilson, 1963; Lott, 1962; Lott & Fuchs, 1962). In this laboratory injected hormones (prolactin and oxytocin) have failed also to maintain maternal behavior in females that have become maternal after parturition or have been made maternal by Caesarean-section delivery of their fetuses several days before parturition. However, the conviction that maternal behavior in the rat is based upon hormones is supported by the success in inducing nest building in the mouse with progesterone (Koller, 1952, 1955) and in the hamster with estrogen and progesterone (Richards, 1965). Some success has been reported

SOURCE: Reprinted by permission from *Journal of Comparative and Physiological Psychology*, 1968, *65* (3), 479–482. Published by the American Psychological Association.

ANALYSIS

REVIEW OF LITERATURE

In the review of literature, Terkel and Rosenblatt carefully present two sides of a controversy surrounding the hormonal basis of maternal behavior: While some

in inducing maternal nest building in rabbits using a combination of hormones (e.g., stilbestrol, progesterone, and prolactin were used by Zarrow, Sawin, Ross, & Denenberg, 1962).

a

In view of the difficulty of inducing maternal behavior in female rats with injected hormones, a difficulty no doubt based upon failure to introduce either the proper hormone or hormones in the proper order and dose level, we have attempted a different approach to the problem. Remaining close to the natural conditions under which maternal behavior normally appears at and shortly after parturition, we have attempted to transfer blood plasma from postparturient females that have become maternal within the past 48 hr to virgins, hoping thereby to induce maternal behavior in the latter. **Establishing**

b

that maternal blood plasma carries a substance or substances capable of inducing maternal behavior in virgins would be a first step in identifying the humoral basis of maternal behavior.

We have shown recently that virgin females can be induced to show maternal behavior when they are exposed to young pups continuously for about 5 days (Rosenblatt, 1967). Cosnier (1963) reported a similar finding using shorter daily exposures. Both studies confirm an early suggestion by Wiesner and Sheard (1933) which was only partially verified in their own studies. Maternal behavior induced under these conditions appears to be of nonhormonal origin since ovariectomizing or hypohysectomizing virgins before exposing them to pups did not prevent the appearance of maternal behavior or alter, significantly, latencies for the onset of a major item of maternal behavior, namely retrieving (Cosnier & Couterier, 1966; Rosenblatt, 1967). In this study, therefore, we observed whether the latency for the appearance of retrieving (and other items

c

of maternal behavior) by virgins exposed to young pups was significantly reduced by prior injection of maternal blood plasma as compared to prior injection of blood plasma taken from virgins in the proestrus or diestrus phases of the vaginal estrous cycle, or prior injection of saline solution.

researchers have found strong evidence for the hormonal basis of maternal behavior, others have not. In **(a)** the authors speculate that the ambiguity of previous experimental results may be due to the fact that such studies have not employed the proper hormones in the proper order and at proper dose levels. But they suggest that if the blood conditions in virgin rats (presumably behaviorally nonmaternal) were nearly identical to the blood conditions of rats who had recently delivered a litter of pups (presumably behaviorally maternal), then a reliable measure of the serological basis of maternal behavior would be possible. The authors suggest in **(b)** that we should identify a general hormonal basis of maternal behavior as a first step that would presumably lead to a more discrete analysis of specific chemicals that might be responsible for this behavior.

METHOD

Subjects

Thirty-two virgin females, 60 days of age at the start of the experiment, were obtained from Charles River Breeding Farm, Dover, Mass. Twenty-four additional *S*s of the same age provided blood plasma for the injections. Other rats provided the pups used in the maternal behavior tests. The *S*s were housed individually in 45 × 50 × 40 cm. rectangular cages, each with transparent Plexiglas walls, grid floor, wall feeder, water bottle, and two bins containing hay and coarse wood shavings for nesting material. They were fed Purina chow and water ad lib supplemented twice weekly with vitamin-enriched bread, carrots, and lettuce.

The *S*s were divided into four equal-sized groups. One group received plasma taken from maternal *S*s; injections were given when *S*s were in various unspecified phases of the vaginal estrous cycle. One control group consisted of *S*s in proestrus that received plasma taken from females that were also in proestrus, and a second control group consisted of *S*s in diestrus that received plasma taken from females that were also in diestrus. The fourth group of *S*s received an injection of 0.9 percent saline solution at various unspecified phases of the vaginal estrous cycle.

The last sentence of the review (**c**) contains the hypothesis to be tested and a brief version of the research plan. In effect, the researchers state the dependent variable (latency of retrieving pups) and the independent variable (injection of different groups with (1) maternal blood plasma, (2) blood from animals in the proestrus phase, (3) blood from animals in the diestrus phase, and (4) saline solution). One group was not injected and served as a control. The proestrus phase in rats is the period immediately prior to estrus ("heat"), and the diestrus period follows estrus. Presumably, these phases were selected to reduce the possibility that hormones present during the estrus phase would cause a general increase in the activity level, which could have been falsely interpreted as maternal behavior.

METHOD

Three parts are included in the method section: The subjects are identified, the procedures are described, and the tests of maternal behavior are specified. In this study, enough detail is present to allow its replication. For example, in describing the subjects the authors identify the breeding farm, age, sex, cage size and construction, feed, water schedule, and nesting material. There is a practical limit to the amount of methodological detail that can be presented, so some items, such as temperature, humidity, lighting, and so on, are deleted, but one could logically infer that these items were controlled.

Procedures

Blood was withdrawn from the donors within 48 hr after parturition, after it was clearly established that these Ss were performing maternal behavior normally. Blood taken from estrous-cycling donors was withdrawn within 1 hr after the vaginal smear indicated either proestrus or diestrus. Between 6 and 8 cc of blood was withdrawn from the heart. To withdraw the blood, the donors were anesthetized with ether, and the heart was surgically exposed by a chest incision to one side of the midline. The blood was withdrawn using a 40 mm. 16-gauge needle with a 20-cc syringe containing 15–20 units of Heparin Sodium to prevent blood clotting.

About 4 min elapsed from the time the donor was judged to be completely anesthetized until the blood was first transferred from the syringe to a test tube and centrifuging was started. When a zone of clear plasma, free of blood cells, appeared 3–4 cc of plasma was drawn into a 5-cc syringe and injected into a subject with a 20 mm. 27-gauge needle. Plasma injection was completed in 4 min. Each experimental S received all of its plasma from a single donor female.

The S was lightly anesthetized with ether, a small incision was made on the inner surface of the upper thigh, and the right femoral vein was exposed. The needle was inserted into the vein, a small amount of blood was withdrawn, and then the plasma (or saline solution) was injected slowly over a period of 2½ min. The incision was closed with wound clips. In this inbred strain of rats, plasma transfer between any two strain mates does not result in anaphylactic shock.

Maternal Behavior Tests

Each animal was given a 15-min retrieving test 1 hr before blood was withdrawn from the proestrus and diestrus donors and plasma or saline was injected into the recipient Ss. Since no S retrieved during this test it was not necessary to eliminate any of them from the experiment.

Tests following the injection of either plasma or saline were begun after it was judged that Ss were fully recovered from the anesthetic used during the injection. Each S was judged to be recovered if it was able to walk around the edge of a bell

In the procedure section, Terkel and Rosenblatt demonstrate their considerable talent for unambiguous writing. They explain in detail the method used for blood transfusion so that a researcher with some surgical skill could exactly replicate their procedure.

A critical aspect of this study is the definition of the dependent variable: maternal behavior. The validity of scientific inquiry rests to a large extent on the operational definition of dependent variables and the testing of these variables. Terkel and Rosenblatt suggest several behavioral characteristics that typify maternal behavior, including latency in retrieval of pups, crouching over young, licking young, and nest building.

jar maintaining its balance; if it fell from the edge it was retested at a later time. It was possible for the first postinjection test to begin in all Ss 1 hr after the injection since all Ss were fully recovered from the ether about 5 min after the injection.

Five pups, 5–10 days of age, were placed at the front of each Ss cage. Retrieving was observed for 15 min, following which observations for 1-min periods at 20-min intervals were made over the next 2 hr. During the 1-min period of observation the occurrence of retrieving, crouching over the young, licking the young, nest building, and other maternal and nonmaternal items of behavior were recorded. Nesting material had previously been spread over the floor. At the end of the 2-hr test the pups were left with Ss until the next morning at which time they were removed and replaced by a fresh litter of five pups in the same age range. The test procedure was repeated daily until an S retrieved pups in two consecutive daily tests. Since retrieving is usually the last item of maternal behavior to appear when pups are used to induce it, all Ss had already shown the other main items of maternal behavior (i.e., crouching over young, licking young, nest building) by the termination of testing.

Several Ss that had been injected with maternal plasma were observed continually on the first day following the first test to see if times of maternal behavior would appear between the first and the second test, 22 hr later.

RESULTS

d   Mean latencies in days for the onset of retrieving for the various groups, shown in Table 17.1, indicate that plasma taken from a lactating mother within 48 hr after delivery is capable of inducing a more rapid maternal response to pups than saline and either proestrus or diestrus plasma ($F = 9.79$, $df = 3/28$, $p < .01$; data transformed to square roots). Under the combined influence of maternal plasma and stimulation from pups, retrieving appeared in an average of 2 days. This time was significantly shorter than when saline or proestrus plasma was injected or when diestrus plasma was injected (Duncan's New Multiple Range test at the .05 level). Proestrus plasma combined with the proestrus condition of the recipient virgin was similar in its effect on maternal behavior to saline but diestrus plasma given to females in diestrus produced a significant delay in the mean latency for the onset of retrieving (Duncan's New Multiple Range test at the .05 level).

RESULTS

The essence of this research paper is typified in the first sentence (**d**) of the results section. The main result of this experiment is plainly stated: Plasma of lactating rats is capable of inducing more rapid maternal behavior than other substances. Statistical evidence in the form of the F statistic is then presented. In addition, the authors employ Duncan's New Multiple Range test and the Mann-Whitney U statis-

In a previous study (Rosenblatt, 1967) it was established that maternal behavior (i.e., retrieving and other items) can be induced in virgins by exposure to pups, without any prior injection, with an average latency of 5.79 ± 2.69 days. The saline-injected proestrus plasma- and diestrus plasma-injected Ss of the present study had average latencies which did not differ significantly from Ss in the earlier study (Mann-Whitney $U = 40$–42, $p > .10$). The average latency of the maternal plasma-injected Ss was, however, significantly shorter than that of the Ss that were only exposed to pups (Mann-Whitney $U = 20$, $p = .02$).

The onset of retrieving was accompanied in all groups by the occurrence of the three other main items of maternal behavior (i.e., crouching over the young, licking, and nest building). With pups continuously present, the onset of an item of maternal behavior, and particularly the onset of retrieving, was followed by its appearance from then on in each of the subsequent daily tests. Observations of the Ss between the first and second tests led us to believe that the maternal plasma induced maternal behavior more rapidly than our formal test procedure was capable of detecting. Several Ss that were injected with maternal plasma began to show maternal behavior in attenuated fashion within 4–8 hr after the injection and the beginning of exposure to pups, although, several hours earlier, during the first scheduled test of maternal behavior, they were indifferent to the pups. Two of these were fully maternal, according to our criteria for the virgins, by the second test, which was begun 1 day after the injection. Others were not fully maternal until the third test was begun 2 days after the injection.

Table 17.1 MEAN LATENCIES IN DAYS FOR THE ONSET OF RETRIEVING

GROUP	N	MEAN	SE
Maternal plasma	8	2.25	.97
Proestrus plasma	8	4.62	1.21
Diestrus plasma	8	7.00	2.96
Saline	8	4.00	1.41
Untreated[a]	14	5.79	2.69

[a]Taken from Rosenblatt (1967).

tic. Duncan's test ascertains whether significant differences exist between each of the means, while the Mann-Whitney statistic is a special type of analysis for ranking data with two classes of information.

In the second paragraph of the results section the authors introduce an untreated control group used in a previous study. This is obviously not as desirable as if the authors had used their own control group. There may be differences between the two experiments in the sample of rats, the handling of the rats, or the experimen-

DISCUSSION

e | Our study established for the first time that substances carried in the plasma of the newly maternal rat are capable of increasing the readiness of virgins to respond maternally to pups. We have, therefore, finally found a way of accomplishing what Stone (1925) set out to do when he joined in parabiosis a maternal and nonmaternal rat hoping that blood-borne substances responsible for maternal behavior in the former would induce maternal behavior in the latter. Were it not for the failure of these substances to cross from the maternal to the nonmaternal animal, because of selective transmission across the parabiotic union, Stone would have demonstrated what we have found and perhaps the effect would have been stronger with the continuous exchange of blood that he attempted.

f | The present study does not enable us to identify the substance or substances that are responsible for increasing the maternal responsiveness of the virgins or to determine whether these substances act on the virgin via the endocrine system or directly upon the nervous system. Initially we thought that dividing our plasma control group into two groups, one receiving proestrus plasma

tal procedures. On the other hand, the procedure used in this series of studies is fairly standardized, which makes a cross-experiment comparison somewhat feasible (although not totally desirable). Data from one experiment should not be analyzed with data from a second experiment unless the experimenter is quite sure that the subjects and procedures of the two experiments are quite similar.

In the third paragraph of this section the authors introduce some observational data that their experimental procedures were not sensitive enough to detect. Observational data are often quite valuable in helping to clarify or further interpret results, but it should be noted that data of this type are usually not collected as systematically or precisely as the data we have just discussed. Therefore, these data should be treated as supplementary to the results found in the formal testing of the dependent variable, which is how the authors present them.

DISCUSSION

The discussion section begins with a bold statement (**e**) that is a combination of empirically validated evidence and a logically inferred statement. Terkel and Rosenblatt quickly point out that their hypothesis is not a new one and suggest a reason for previous failures to validate the hypothesis. They also state a limitation (**f**) by noting that their study did not specifically identify the substance(s) within the blood responsible for increasing maternal behavior. Nonetheless, one can feel the excitement in this research, and even the newcomer to psychology can anticipate the next development.

and the other diestrus plasma, the virgins themselves being in the corresponding phases of the estrous cycle, would enable us to make a first step in identifying the active substances. To the extent that the diestrus blood plasma combined with the diestrus condition of the recipient virgin produced a delay in the onset of maternal behavior we have been partially successful. However, any identification of ovarian hormones or pituitary secretions as the active substances would be highly speculative and incapable of substantiation at this time. Our findings therefore await further analysis of hormonal secretions during the estrous cycle, pregnancy, and parturition.

An added finding of importance does emerge from this study which was surprising to us. Our previous work indicated that maternal responsiveness increases gradually during pregnancy (Lott & Rosenblatt, 1967), and we interpreted this as indicating that the hormonal conditions during pregnancy gradually sensitized the neural substrate of maternal behavior thereby preparing for the appearance of maternal behavior at parturition. The present study suggests that there need be no prolonged period (i.e., 22 days of pregnancy) of sensitization for substances contained in maternal plasma to have their effect on maternal behavior. It would appear that the gradual increase in maternal responsiveness which we found during pregnancy after Caesarean-section deliveries (Lott & Rosenblatt, 1967) need not be built up by a continual addition of "units" of maternal responsiveness. Rather the level of maternal responsiveness at each period of pregnancy reflects for that particular moment the current capability of the blood to stimulate maternal behavior and this capability presumably undergoes a continuous increase until it is fully established around parturition. In this respect then our findings agree with those of Moltz and Weiner (1966) and Denenberg, Grota, and Zarrow (1963) that hormonal secretions at parturition are likely to be important for the induction of maternal behavior.

REFERENCES

Beach, F. A., & Wilson, J. R. (1962). Effects of prolactin, progesterone, and estrogen on reactions of nonpregnant rats to foster young. *Psychological Report, 13,* 231–239.

Cosnier, J. (1963). Quelques problèmes posés par le "comportement maternel provoqué" chez la ratte. *CR Soc. Biol.,* Paris, *157,* 1611–1613.

Cosnier, J., & Couterier, C. (1966). Comportement maternel provoqué chez les rattes adultes castrées. *CR Soc. Biol.,* Paris, *160,* 789–791.

Denenberg, V. H., Grota, L. J., & Zarrow, M. X. (1963). Maternal behavior in the rat: Analysis of cross-fostering. *Journal of Reproduction and Fertility, 5,* 133–141.

Koller, G. (1952). Der Nestbau der Weiber Mause und seine hormonale Auslosung. *Verh. dtsch. zool. Ges.,* Freiburg, 160–168.

Koller, G. (1955). Hormonale und psychische Steuerung beim Nestbau Weiber Mause. *Zool. Anz., (Suppl.), 19,* 125–132.

Lott, D. F. (1962). The role of progesterone in the maternal behavior of rodents. *Journal of Comparative and Physiological Psychology, 55,* 610–613.

Lott, D. F., & Fuchs, S. S. (1962). Failure to induce retrieving by sensitization or the injection of prolactin. *Journal of Comparative and Physiological Psychology, 55,* 1111–1113.

Lott, D. F., & Rosenblatt, J. S. (1967). Development of maternal responsiveness during pregnancy in the rat. In B. M. Foss (Ed.), *Determinants of infant behavior IV.* London: Methuen.

Moltz, H., & Weiner, E. (1966). Effects of ovariectomy on maternal behavior of primiparous and multiparous rats. *Journal of Comparative and Physiological Psychology, 62,* 382–387.

Richards, M. P. M. (1965). *Aspects of maternal behaviour in the golden hamster.* Unpublished doctoral dissertation, Cambridge University, 1965.

Rosenblatt, J. S. (1967). Non-hormonal basis of maternal behavior in the rat. *Science, 156,* 1512–1514.

Stone, C. P. (1925). Preliminary note on maternal behavior of rats living in parabiosis. *Endocrinology, 9,* 505–512.

Wiesner, B. P., & Shard, N. M. (1933). *Maternal behaviour in the rat.* London: Oliver and Boyd.

Zarrow, M. X., Sawin, P. B., Ross, S., & Denenberg, V. H. (1962). Maternal behavior and its endocrine bases in the rabbit. In E. L. Bliss (Ed.), *Roots of behavior.* New York: Harper & Row.

This research was supported by National Institute of Mental Health Research Grant MH-08604 to J. S. R. and Biological Medicine Grant FR-7059 to J. T. We wish to thank D. S. Lehrman and B. Sachs for reading the manuscript. Publication No. 50 from the Institute of Animal Behavior, Rutgers University, Newark.

QUESTIONS

1. Speculate about what might happen if greater quantities of blood had been transfused.
2. Based on the results of this experiment, would you make a generalization to human transfusion? Is this generalization warranted? Why or why not? Do you see any practical application of this study?
3. Why did the authors include a saline group? A proestrus group? A diestrus group? An untreated group?
4. What significance do you attribute to the results of the diestrus group?
5. What group(s) would you like to add?
6. In addition to the behavioral indices of maternal behavior, suggest several physiological measures of maternal tendencies. Could these be quantifiable? Would evaluation of these changes be important to this study? Why or why not?
7. This paper could just as easily have been published in a physiological journal. What is the relationship between psychology and physiology?
8. Why are rats used in psychological studies?

Chapter 18

Long-Term Memory

INTRODUCTION

The next article, "Does It Pay to Be 'Bashful'? The Seven Dwarfs and Long-Term Memory" by Meyer and Hilterbrand, is a type of controlled "Trivial Pursuit" in which long-term memory for common events is examined.

You are to analyze this article. Some things you should look for include the following:

1. What is the theoretical motivation behind this experiment?
2. Who were the subjects?
3. What is the independent variable?
4. How would you classify this design?
5. What was the dependent variable?
6. What controls (e.g., subjects) were employed?
7. How could long-term memory be assessed?

The original article had three experiments. For simplicity only the first experiment is presented here.

DOES IT PAY TO BE "BASHFUL"? THE SEVEN DWARFS AND LONG-TERM MEMORY

GLENN E. MEYER AND **KATHIE HILTERBRAND**
LEWIS AND CLARK COLLEGE

Recall of the Seven Dwarfs of the Disney film was measured. The name Bashful was at a disadvantage in both percentage recall and position of recall. An analysis of errors from a "tip-of-the-tongue" viewpoint suggests this is due to retrieval difficulties.

In a classic review of literature, Miller (1956) pointed out that subjects were able to discriminate between seven plus-or-minus two different stimuli without making an error in various kinds of memory and discrimination tasks. He referred to the number seven as the "magical number seven" and questioned whether it is more than coincidence that there are seven objects in the span of attention, seven categories for absolute judgment, seven digits in the span of immediate memory, and the classical groups of seven, such as the seven wonders of the world, the seven levels of hell, the seven daughters of Atlas in the Pleiades, and the seven ages of man. Because most students (and probably faculty) are unaware of Atlas's daughters, we used as a class example a contemporary group of seven, the Seven Dwarfs, from the Walt Disney Studio film *Snow White and the Seven Dwarfs*. Upon asking people to name the Seven Dwarfs, we noted an interesting effect. The name Bashful appeared to be at a disadvantage and was forgotten most often. Similar experiments have examined recall for lists assumed to be common in memory such as the presidents (Roediger & Crowder, 1976) and popular songs (Bartlett & Snelus, 1980). In Experiment 1, we decided to verify and quantify the "Bashful effect" using a free-recall task.

METHOD

We gave 141 college students a form with seven spaces and told them to write the names of the Seven Dwarfs in the order that they were remembered.

SOURCE: Reprinted with permission of *American Journal of Psychology*, 1984, *97*(1), 47–55.

RESULTS

The percentage recall and average position of recall are presented in Figure 18.1 for each dwarf. We first analyzed the percent correctly identified (Cochran's $Q = 138.03$, $df = 6,986$; $p < .0001$). Bashful was named by only 35 percent of the subjects. Sleepy was remembered by 86 percent of the subjects.

The second analysis was performed on the average recall-position data because position might give another indication of long-term memory (LTM) accessibility. Because only 22 subjects correctly identified all seven of the dwarfs, we conducted several different analyses of position. In the first, all subjects were used, and we assigned average positions for the missing or nonrecalled dwarfs for each subject. When a dwarf was left out of a subject's list, he was assigned an average position on the blank spaces for that subject. If a subject correctly identified four dwarfs, the unnamed dwarfs would each be assigned an average position of six. Friedman's chi square was found to be significant ($\chi^2 = 36.62$; $df = 6$; $p < .001$). Again, Sleepy was strongest with an average recall position of 2.6, whereas Bashful was weakest with 5.39. Pearson's correlation between the average position of the dwarfs and the percentage correctly identified was $r = -.94$ ($df = 5$; $p < .01$). These data are presented in Figure 18.1. Also calculated was the average position for the 22 subjects who correctly recalled all dwarfs (Sleepy = 2.682, Grumpy = 3.318, Dopey = 3.409, Sneezy = 3.545, Doc = 4.791, Bashful = 4.909, Happy = 5.045). Again, Friedman's chi square was significant ($\chi^2 = 36.67$, $df = 6$, $p < .001$). We also calculated the average position of recall for all the subjects, excluding all those who had perfect recall ($141 - 22 = 119$). The mean positions were Sleepy

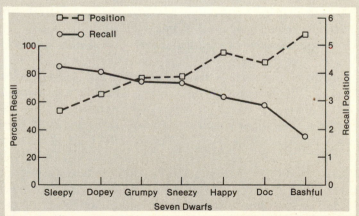

Figure 18.1 Recall percentage and average position of recall for the Seven Dwarfs.

= 2.150, Dopey = 2.768, Grumpy = 3.250, Sneezy = 3.253, Doc = 3.271, Happy = 3.687, Bashful = 4.643. The positions data scored by all three methods were highly correlated ($r > .81$, $p < .02$). Most interesting was that the data for percentage of correct recall was highly correlated with the position data from the 22 subjects who recalled all the dwarfs ($r = -.85$, $p < .01$). Thus, high recall is linked with early position of recall and vice versa. There were no differences in recall between those who had seen or not seen the movie ($p > .10$).

DISCUSSION

The two analyses verify that Bashful is the poorest recalled of the Seven Dwarfs. Fewest remember him, and if they do he is usually recalled toward the end of the list. The major reason for this may lie in retrieval problems. Several factors support this view. First, we found that the dwarfs recalled best were cited first, because the position of recall by subjects having all correct correlated strongly with percentage correct. Thus, both measures indicate that these dwarfs are easily accessible. Second, our analysis of the incorrect guesses also suggested retrieval difficulties. We recorded 88 incorrect responses. Six were rejected as obviously facetious (i.e., the reporting of Fred and Ethel as one dwarf's name).

Our review of the incorrect guesses of the dwarfs' names suggests that many of the subjects were in a "tip-of-the-tongue" state. In an experiment designed to induce "tip-of-the-tongue" phenomena, Brown and McNeill (1966) found that recall for the ending and beginning letters of a word was better than that for middle letters. Our data parallel those findings. Of the Seven Dwarfs, the initial letters are B, D, H, S, and G, and 65.8 percent of the guesses started with these letters. We correlated the percentage recall of the dwarfs versus the number of guesses that had the starting letters of the dwarfs' names (percentages were averaged for names with the same starting letter) and found the relationship strong (Spearman's rho = .87; $df = 5$; $p = .027$). Thus, a letter with a strong recall of its dwarfs produced many guesses. Poor recall produced few guesses. Our subjects also kept the length of their guesses close to the length of the dwarfs' names. Mean length of the seven names is 5.2 letters and of the guesses it was 5.4, a nonsignificant difference ($t = 1.52$; $df = 81$; $p > .05$). Similarly, 5 of 7 (71 percent) of the dwarfs' names end in y and 74 percent of the guesses ended in y. That Doc and Bashful, the two least remembered dwarfs, do not end in y tends to support our position. Bousfield and Wicklund (1969) found that rhyming words tended to be recalled in a cluster in a free-recall task.

Semantics also seems to play a role. We divided the dwarfs' names into three categories: names and "neutral" adjectives, "good" adjectives, and "bad" adjectives (defined as a personal trait). Colleagues thought the categories reasonable. Of the dwarfs, the names fell into three categories: neutral, 1 (14.3 percent); bad, 4 (57.1 percent); good, 2 (28.6 percent). Of the guesses, 25 percent were name and neutral, 52 percent bad, and 23 percent good. Happy, Doc, and Bashful were the weakest dwarfs in recall and position. That the other four were negative adjectives might have aided their recall. Guesses therefore seemed concentrated around negative adjectives with similar phonetic structures. For example, the two most popular guesses were Dumpy (17 percent) and Grouchy (12 percent).

Apparently forces seen in the laboratory can be demonstrated in the internal structuring, acquisition, and recall of material in the real world (if the Seven Dwarfs represent the "real world"). The magical number seven seems prominent in contemporary as well as classical mythology, and retrieval factors operate in the magic number and the magic kingdom. Further, these data are readily replicable in a classroom and can be used for an easy demonstration of basic processes in memory. Asking a class to recall the dwarfs makes the differences between recall and recognition, TOT, organization, and so on readily apparent.

REFERENCES

Bartlett, J. C., & Snelus, P. (1980). Lifespan memory for popular songs. *American Journal of Psychology, 93,* 551–560.

Bousfield, W. A., & Wicklund, D. A. (1969). Rhyme as a determinant of clustering. *Psychonomic Science, 16,* 183–184.

Brown, R., & McNeill, D. (1966). The "tip-of-the-tongue" phenomenon. *Journal of Verbal Learning and Verbal Behavior, 5,* 325–337.

Miller, G. A. (1956). The magical number seven, plus or minus two: Some limits on our capacity for processing information. *Psychological Review, 63*(2), 81–96.

Roediger, H. L., III, & Crowder, R. G. (1976). A serial position effect in recall of United States presidents. *Bulletin of the Psychonomic Society, 8,* 275–278.

Sidey, P. (1980). Introduction. *Walt Disney's Snow White and the Seven Dwarfs.* New York: Harmony Books.

NOTE: Parts of these data were presented at the 1982 meeting of the Western Psychological Association. Requests for offprints should be sent to Glenn E. Meyer, Department of Psychology, Lewis and Clark College, Portland, OR 97219.

ADDITIONAL QUESTIONS

1. Offer an alternate hypothesis for the results.
2. Search the literature for other examples of long-term memory. How did other researchers investigate this phenomenon?
3. In what other ways could one measure the dependent variable in this experiment? (Hint: Look up the original published experiment and look at experiments 2 and 3.)
4. How general are these results? For memory research? For cross-cultural research?
5. What additional research is suggested by this experiment?
6. Replicate this experiment with different materials using classmates as subjects. Analyze and report your results.
7. What individual personality factors might be operating?
8. Do a literature search for the relationship between knowledge for common events and intelligence.

Chapter 19

Creative Porpoise

INTRODUCTION

Two models of learning dominated the research activities of learning psychologists during the early part of this century. One model was developed by Ivan Pavlov and is commonly called *classical conditioning;* the other model was suggested by E. L. Thorndike and refined by B. F. Skinner and is referred to as *operant conditioning.* The initial experiments of both groups attempted to identify the conditions for learning using infrahuman subjects. Pavlov used his famous salivating dogs, while Thorndike studied the effects of reward on the behavior of cats. Skinner employed rats in his early experiments, then pigeons, and finally, among other species, humans. The contemporary period has seen a great proliferation in the use of different species in the learning laboratory, all with the basic purpose of establishing laws of behavior. The use of porpoises in this study is a logical step to illustrate the effectiveness of operant conditioning in yet another species.

The authors of the "The Creative Porpoise" have skillfully applied several basic components of the model developed by B. F. Skinner. As you read this experiment try to identify some of the following principles of operant behavior.

An important principle of operant conditioning is that responses that are followed by a reward or positive reinforcement tend to increase, while responses that are not rewarded tend to decrease. Skinner demonstrated this principle in rats by measuring whether bar-pressing responses increased after a reward. The apparatus developed by Skinner has been described previously. The initial behavior of a rat placed in a "Skinner box" is normally exploratory in nature; it sniffs the corners, moves from one side to the other, examines the walls, and washes its face. Only a small number of these responses have anything to do with a bar-pressing response, but the skilled researcher can identify preparatory bar-pressing responses and reinforce them. The process of selectively reinforcing successive approximations of the principal response (bar pressing) is called *shaping.* Gradually the animal moves closer to the bar, then places its paw close to the bar, touches it, and eventually depresses it.

Much information has been collected regarding the specific conditions that facilitate operant conditioning. For example, if the reward is presented immediately after the appropriate response, then conditioning is more rapid than if the reward is delayed. In reading the article by Pryor, Haag, and O'Reilly, notice how they try to present the reinforcement as soon after the appropriate response as is practical.

Other researchers have studied the role that secondary reinforcers (stimuli that have been associated with the primary reward) have on behavior. The general conclusion of these studies is that secondary reinforcers have strong rewarding properties. Consider the reward properties of the secondary reinforcement of money in many societies. In this study, Pryor et al. use a distinguishable signal (a whistle) as a conditioned reinforcer.

The results of operating conditioning experiments lend themselves to graphic representation in the form of a cumulative frequency record. On such a graph, the subject's responses are accumulated and scaled on the ordinate, while time is recorded on the abscissa. Since the responses are accumulative, the response curve never goes down; a nonresponse is depicted as a line parallel to the abscissa.

This article was selected not only to illustrate the principles of operant behavior but also to show you an experiment that is largely descriptive in nature. Note that only one subject was used, and the statistical portion of the paper is largely descriptive. The researchers show how data from psychological experiments can be effectively portrayed in graphic form. From the graphs we can quickly grasp their results.

Another feature of this experiment is that it combines a laboratory setting with a natural setting. The experiment was conducted in a large artificial pool, but the habitat is designed to simulate, as close as is practical, the natural environment of the subject. To do this research in the natural setting of the experimental animal (i.e., the ocean) is probably impractical (although not impossible, as the researchers cite research conducted in the open field), but to conduct the research in a confined and artificial environment may inhibit the responses of the porpoise.

Most research in psychology depends on large samples; however, to infer that psychological research must study large samples in order to be valid is an unwarranted conclusion. Pryor et al. describe the learning process with clarity by thoroughly examining the behavior of a single subject.

Special Issues

Small *n* Designs

Research that is based on a single subject or small sample is called a small *n* or small number of subjects design. Most research in psychology is based on a large sample of subjects that generates a large amount of data. These designs are sometimes called large *n* designs. Data gathered from these experiments are amenable to statistical analysis in which probabilistic conclusions about the source of the outcome are based. Because of the enormous popularity of large *n* experiments and their compatability with modern statistical techniques, it may seem that small *n* designs are methodologically inferior. However, three branches of psychology have resisted the trend toward large *n* experiments. The first of these groups is led by psychophysicists, another by clinical researchers (see Chapter 21, "Special Issues: Single-Subject Design in Clinical Studies"), and a third by operant learning psychologists.

Psychophysical experiments generally deal with the relationship between physical stimuli and a subject's perception of those stimuli. In many psychophysical designs each subject may be exposed to dozens or even hundreds of different stimuli in a single experiment. Thus, the number of observations in psychophysical experiments, although based on only a few subjects or even a single subject, may exceed the number of observations in a large *n* experiment.

Studies of operant conditioning frequently use small *n* design and in some instances a single subject. In these experiments a single subject (or only a few subjects) is intensely studied. As in psychophysical experiments and clinical studies, the design is frequently a within-subject design, where subjects serve in each condition. A chief proponent of small *n* designs has been B. F. Skinner, who has observed the development of behavior in a single subject. The research presented in this chapter is a clear example of a single-subject experiment. As you study the experiment note that the authors do not use inferential statistics but instead use descriptive statistics. This is a general feature of small *n* experiments, but very sophisticated statistics are available to analyze the results obtained in single-subject experiments.

The Creative Porpoise: Training for Novel Behavior

KAREN W. PRYOR
Oceanic Institute

RICHARD HAAG
Makapuu Oceanic Center

JOSEPH O'REILLY
University of Hawaii

Two rough-toothed porpoises *(Steno bredanensis)* were individually trained to emit novel responses, which were not developed by shaping and which were not previously known to occur in the species, by reinforcing a different response to the same set of stimuli in each of a series of training sessions. A technique was developed for transcribing a complex series of behaviors onto a single cumulative record so that the training sessions of the second animal could be fully recorded. Cumulative records are presented for a session in which the criterion that only novel behaviors would be reinforced was abruptly met with four new types of responses, and for typical preceding and subsequent sessions. Some analogous techniques in the training of pigeons, horses, and humans are discussed.

a   The shaping of novel behavior, that is, behavior that does not occur, or perhaps cannot occur, in an animal's normal activity, has been a preoccupation of animal trainers for centuries. The fox terrier turning back somersaults, the elephant balancing on one front foot, or Ping-Pong playing pigeons (Skinner, 1962) are produced by techniques of successive approximation, or shaping. However, novel or original behavior that is not apparently produced by shaping or differential reinforcement is occasionally seen in animals. Originality is a fundamental aspect of behavior but one that is rather difficult to induce in the laboratory.

SOURCE: Reprinted by permission from *Journal of the Experimental Analysis of Behavior*, 1969, *12*, 653–661. Copyright 1969 by the Society for the Experimental Analysis of Behavior, Inc. Some portions of the original article have been omitted.

ANALYSIS

Other articles in this book began with a review of the literature, but Pryor, Haag, and O'Reilly introduce their paper with a general statement about the use of operant conditioning in a variety of species (**a**). In the following two paragraphs they de-

In the fall of 1965, at Sea Life Park at the Makapuu Oceanic Center in Hawaii, the senior author introduced into the five daily public performances at the Ocean Science Theater a demonstration of reinforcement of previously unconditioned behavior. The subject animal was a female rough-toothed porpoise, *Steno bredanensis,* named Malia.

Since behavior that had been reinforced previously could no longer be used to demonstrate this first step in conditioning, it was necessary to select a new behavior for reinforcement in each demonstration session. Within a few days, Malia began emitting an unprecedented range of behaviors, including aerial flips, gliding with the tail out of the water, and "skidding" on the tank floor, some of which were as complex as responses normally produced by shaping techniques, and many of which were quite unlike anything seen by Sea Life Park staff in Malia or any other porpoise.

b

c
To see if the training situation used with Malia could again produce a "creative" animal, the authors repeated Malia's training, as far as possible, with another animal, one that was not being used for public demonstrations or any other work at the time. **A technique of record keeping was developed to pinpoint if possible the events leading up to repeated emissions of novel behaviors.**

METHOD

d
A porpoise named Hou, of the same species and sex as Malia, was chosen. Hou had been trained to wear harness and instruments and to participate in physiological experiments in the open sea (Norris, 1965). This individual had a large repertoire of shaped responses but its "spontaneous activity" had never been reinforced. Hou was considered by Sea Life Park trainers to be "a docile, timid individual with little initiative."

scribe some natural observations that led to the hypothesis (**b**), and they provide an abbreviated statement of their method (**c**). It is important to note that the development of a hypothesis may emanate from many sources. Although the most frequent source of a hypothesis is previous research, another significant source is the observation of behavior in a natural setting.

METHOD

Paragraphs (**d**), (**e**), and (**f**) identify the subject, describe how conditioning was done, and explain the technique used to record the data. In paragraph (**e**) the authors describe the use of a bell that was rung at the beginning of the training session. The use of a signal at the outset of an experiment is called a *discriminative*

Training sessions were arranged to simulate as nearly as possible Malia's five brief daily sessions. Two to four sessions were held daily, lasting from 5 to 20 min each, with rest periods of about half an hour between sessions. Hou was given normal rations; it is not generally necessary to reduce food intake or body weight in cetaceans to make food effective as a reinforcer. Any food not earned in training sessions was given freely to the animal at the end of the day, and it was fed normal rations, without being required to work, on weekends. During the experimental period, no work was required of Hou other than that in the experiment itself. **A bell was rung at the beginning and end of sessions to serve as a context marker.** The appearance and positioning of the trainers served as an additional stimulus that the opportunity for reinforcement was now present.

To record the events of each session, the trainer and two observers, one above water and one watching the underwater area through the glass tank walls, wore microphones and made a verbal commentary; earphones allowed the experimenters to hear each other. The three commentaries and the sound of the conditioned reinforcer, the whistle, were recorded on a single tape. A typed transcript was made of each tape; then, by comparing transcript to tape, the transcript was marked at 15-sec intervals. Each response of the animal was then graphed on a cumulative record, with a separate curve to indicate each type of response in a given session (Figures 19.2 to 19.5).

stimulus and lets the conditioned operant know the experiment is beginning. In this experiment the bell signified to Hou that it was time for it to do its thing.

This paper may appear to be an observational study rather than an experiment. But upon closer examination it is apparent that there are independent and dependent variables, operational definitions of terms, hypothesis testing, and controlled conditions, to name but a few of the characteristics of experimentation.

This experiment differs from many of the experiments reviewed in this book in that the dependent variable (novel behavior) is followed by the independent variable (reinforcement). Compare this sequence with Terkel and Rosenblatt's article, "Maternal Behavior Induced by Maternal Blood Plasma Injected into Virgin Rats" (Chapter 17), in which the dependent variable (maternal behavior) is *preceded* by the independent variable (maternal blood plasma). But you may have observed that in this article the porpoise's behavior and reinforcement was ongoing, so it could be argued that a chain of responses—reinforcement, response; reinforcement, response—would develop. The sequence of independent and dependent variable then becomes a matter of deciding which came first: the reinforcement or the behavior.

Several statements in paragraph (**f**) require comment. The researchers have the two observers tape-record their observations, which are then transcribed into a typed manuscript. They also keep a precise record of the time and frequency of the conditioned reinforcer (a whistle).

g
It was necessary to make a relatively arbitrary decision about what constituted a reinforceable or recordable act. In general, a reinforceable act consisted of **any movement that was not part of the normal swimming action of the animal, and which was sufficiently extended through space and time to be reported by two or more observers.** Such behavior as eye-rolling, inaudible whistling, and gradual changes in direction may have occurred, but they could not be distinguished by the trainers and therefore could not be reinforced, except coincidentally. This unavoidable contingency probably had the effect of increasing the incidence of gross motor responses. Position and sequence of responses were not considered. An additional criterion, which had been a contingency in much of Hou's previous training, was that only one type of response would be reinforced per session.

The experimental plan of reinforcing a new type of response in each session was not fully met. Sometimes a previously reinforced response was again chosen for reinforcement, to strengthen the response, to increase the general level of responding, or to film a given behavior. Whether the "reviewing" of responses was helpful or detrimental to the animal's progress is open to speculation.

h
Interobserver reliability was judged from the transcripts of the taped sessions, in which a new behavior was generally recognized in concert by the observers. Furthermore, each new behavior chosen for reinforcement was later diagrammed in a series of position sketches. At no time did any of the three observers fail to agree that the drawings represented the behaviors witnessed. These behavior diagrams were matched, at the end of the experiment, with film of each behavior, and were found to represent adequately the topography of those behaviors that had been reinforced (see Figure 19.1).

In (g) an operational definition of a reinforceable act is stated. They do this because in shaping behavior the act to be reinforced is sometimes ambiguous, and some researchers will "play it by ear"; that is, they will make a decision to reinforce or not reinforce on the basis of ongoing behavior. Since scientific procedure must be specified in sufficient detail to allow the experiment to be exactly replicated, Pryor et al. attempt to reduce the ambiguity of shaping by operationally defining reinforceable responses. In (h) they explain the measure of experimental reliability that they used.

A final procedural question is raised in (i), and some evidence is presented to answer the question. Complexity of responses rather than novelty of responses may have been conditioned, so to resolve the issue the experimenters used a form of validation in which observations made by one experimenter are compared with observations of another experimenter.

RESULTS

The results reported in this article are generally self-explanatory, and we suggest that you thoroughly read the original transcript. Several general remarks may guide your reading.

After 32 training sessions, the topography of Hou's aerial behavs became so complex that, while undoubtedly novel, the behaviors exceeded the powers of the observers to discriminate and describe them. This breakdown in observer reliability was one factor in the termination of the experiment.

To corroborate the experimenters' observation that certain of Hou's responses were not in the normal repertoire of the species, and constituted genuine novelties, the diagrams of each reinforced behavior were shown or sent to the past and present staff members who had had occasion to work with animals of this species. Each trainer was asked to rank the 16 behaviors in order of frequency of occurrence in a free-swimming untrained animal. The sketches were mounted on index cards and presented in random fashion to each rater separately. A coefficient of concordance (W) of 0.598 was found for agreement between trainers on the ranking of various behaviors; this value is significant at the 0.001 level, indicating a high degree of agreement (Siegl, 1956).

To test the possibility that the trainers were judging complexity rather than novelty in ranking, another questionnaire was prepared requesting ranking according to relative degree of complexity of action. Because some of the original group of 12 trainers were unavailable for retesting, the questionnaire was presented to a group of 49 naive students. The coefficient of concordance (W) for agreement between students was +0.295, significant at the 0.001 level. When the ranking for complexity and frequency were contrasted for each behavior, it was found that some agreement existed between the scores given by the two rating groups, Spearman Rank Correlation (RHO) +0.54, significant at the 0.05 level.

Thus, there seems to be some agreement between complexity and frequency, which should be expected, since complex behaviors require more muscle expenditure than simple ones. Furthermore, analysis was biased by the fact that the experienced group was asked to rate all behaviors serially, and had no way other than complexity to rate the several behaviors that many of them stated they had never seen. However, the agreement between complexity and frequency was not as large between groups as it was within groups; allowing for the fact that the use of two

First, you will notice the absence of *F*-tests or *t*-tests. Indeed, the paper is noticeably lacking in statistical analyses. In lieu of such analysis the writers present a well-documented protocol of the changing behavior of the porpoise. The authors also skillfully utilize cumulative frequency graphs. Pryor et al. provide a daily graph (Figures 19.2 to 19.5) in which the porpoise's behavior is charted using the coordinates of responses (vertical axis) and time (horizontal axis).

It is important for you to study these graphs, as they are used as the primary source of results. In Figure 19.2 there are three distinct components: (1) the number of actions, (2) the time in minutes, and (3) the behavior (inverted swim, porpoise, and corkscrew). Notice the relationship among the three behaviors and how the inverted swim rapidly increased. The same three components are in Figure 19.3, and numerous other behavior characteristics were observed. Trace the development of the flip in Figure 19.3.

rating groups makes it impossible to generalize the rating comparisons in a strict sense, the low frequency assigned to some noncomplex behaviors by the experienced group suggests that complexity and novelty are not necessarily positively correlated.

RESULTS

Sessions 1 to 14

In the first session, Hou was admitted into the experimental tank and, when given no commands, breached. Breaching, or jumping into the air and coming down sideways, is a normal action in a porpoise. This response was reinforced, and the animal began to repeat it on an average of four times a minute for 8 min. Toward the end of the 9-min session it porpoised, or leaped smoothly out of the water and in, once or twice.

Hou began the third session by porpoising; when this behavior was not reinforced, the animal rapidly developed a behavior pattern of porpoising in front of the trainer, entering the water in an inverted position, turning right side up, swimming in a large circle, and returning to porpoise in front of the trainer

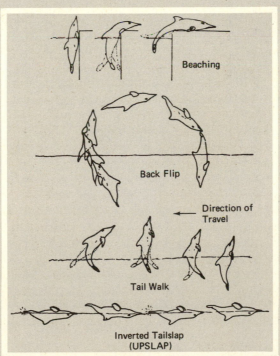

Figure 19.1   Four reinforced novel behaviors, including one shaped behavior—the tail walk.

again. It did this 25 times without interruption over a period of 12.5 min. Finally, it stopped and laid its head against the pool edge at the trainer's feet. This behavior, nicknamed "beaching," was reinforced and repeated (Figure 19.2). Sessions 5, 6, and 7 followed the same pattern.

The trainers decided to shape specific responses in order to interrupt Hou's unvarying repetition of a limited repertoire. Session 8 was devoted to shaping a "tail walk," or the behavior of balancing vertically half out of the water. The tail walk was reinforced in Session 9, and Sessions 10 and 11 were devoted to shaping a "tail wave," the response of lifting the tail from the water. The tail wave was emitted and reinforced in Session 12.

At the end of Session 10, Hou slapped its tail twice, which was reinforced but not repeated. At the end of Session 12, Hou departed from the stereotyped pattern to the extent of inverting, turning right side up, and then inverting again while circling. The experimenters observed and reinforced this underwater revolution from a distance, while leaving the experimental area.

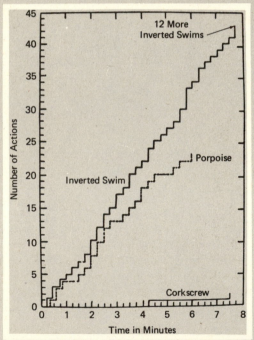

Figure 19.2 Cumulative record of Session 7, a typical early session, in which the porpoise began emitting the previously reinforced response. This response gradually extinguished when another response was formed.

Although a weekend then intervened, Hou began Session 13 by swimming in the inverted position, then right side up, then inverted again. This behavior, dubbed a "corkscrew," was reinforced, and by means of an increasing variable ratio, was extended to five complete revolutions per reinforcement. In Session 14, the experimenters rotated their positions, and reinforced any descent by the animal toward the bottom of the tank, in a further effort not only to expand Hou's repertoire but also to interrupt the persistent circling behavior.

Sessions 15 and 16

The next morning, as the experimenters set up their equipment, Hou was unusually active in the holding tank. It slapped its tail twice, and this was so unusual that the trainer reinforced the response in the holding tank. When Session 15 began, Hou emitted the response reinforced in the previous session, of swimming near the bottom, and then the response previous to that of the corkscrew, and then fell into the habitual circling and porpoising, with, however, the addition of a tailslap on reentering the water. This slap was reinforced, and the animal then combined slapping with breaching, and then began slapping disassociated from jumping; for the first time it emitted responses in all parts of the tank, rather than right in front of the trainer. The 10-min session ended when 17 tailslaps had been reinforced, and other non-reinforced responses had dropped out.

Session 16 began after a 10-min break. Hou became extremely active when the trainer appeared and immediately offered twisting breaches, landing on its belly and its back. It also began somersaulting on its long axis in midair.

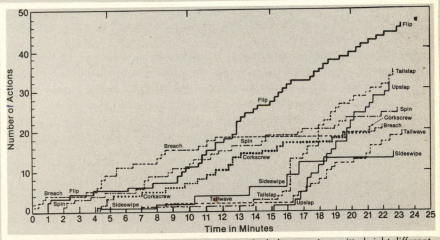

Figure 19.3   Cumulative record of Session 16, in which the porpoise emitted eight different types of responses, four of which were novel (flip, spin, sideswipe, and upslap).

The flip occurred 44 times, intermingled with some of the previously rein-forced responses and with three other responses that had not been seen before: an upside-down tailslap, a sideswipe with the tail, and an aerial spin on the short axis of the body (Figure 19.3).

This session also differed from previous ones in that once the flip had become established, the other behaviors did not tend to drop out. After 24 min, the varied activity—tailslaps, breaches, sideswipes with the tail, and the new behavior of spinning in the air—occurred more rather than less frequently, until the session was brought to a close by the trainer. The previous maximum number of responses in a given session was 110 (in Session 9, a 31-min session). In Session 16, Hou emitted 192 responses in a 23-min session, an average of 8.3 responses per min compared to a previous maximum average of 3.6 responses per min.

By Session 16, the experimenters had apparently been successful in establish-ing a class of responses characterized by the description "only new kinds of responses will be reinforced," and consequently the porpoise was emitting an extensive variety of new responses. The differences between Session 16 and previous sessions may be seen by comparing the cumulative record for Session 16 (Figure 19.3) with that of Session 7, a typical earlier session (Figure 19.2).

Sessions 17 to 27

In Sessions 17 to 27, the new types of responses emitted in Session 16 were selected, one by one, for reinforcement, and some old responses were rein-forced again so that they could be photographed. Other new responses, such as unclassifiable twisting jumps, and sinking head downwards, occurred sporadi-cally. The average rate of response and the numbers of types of responses per session remained more than twice as high as pre–Session 16 levels.

Hou's general activity changed in two other ways after Session 16. First, if no reinforcement occurred in a period of seven minutes, the rate and level of activity declined but the animal did not necessarily resume a stereotyped behavior pattern. Secondly, the animal's activity now included much behav-ior typically associated in cetaceans with situations producing frustration or aggressiveness, such as slapping the water with head, tail, pectoral fin, or whole body (Burgess, 1968).

Sessions 28 to 33

In all of the final sessions, the criterion that the behavior must be a new one was enforced. A new behavior that had been seen but not reinforced previ-ously, the inverted tailslap, had been reinforced in Session 27. Session 28 began with a variety of responses, including another that had been seen but not reinforced before, a sideswipe at water surface with the tail, which was reinforced. In Session 29, Hou's activity included an inverted leap that fulfilled the criterion (Figure 19.4). In Session 30, Hou offered 60 responses over a period of 15 min, none of which were considered new and were not therefore reinforced.

In Sessions 31, 32, and 33, held the next day, Hou's behavior was more completely controlled by the criteria that only new types of responses were reinforced and that only one type of response was reinforced per session. In Session 31, Hou entered the tank and, after a preliminary jump, stood on its tail and clapped its jaws at the trainer, who, taken by surprise, failed to reinforce the maneuver. Hou then emitted a brief series of leaps and then executed a backwards aerial flip that was reinforced and immediately repeated 14 times without intervening responses of other types. In Session 32, after one porpoise and one flip, Hou executed an upside-down porpoise, and, after it was reinforced, repeated this new response 10 times, again without other responses (Figure 19.5).

In the third session of the day, Hou did not initially emit a response judged new by the observers. After 10 min and 72 responses of variable types, the rate of response declined to 1 per min and then gradually rose again to seven responses per minute after 19 min. No reinforcements occurred during this

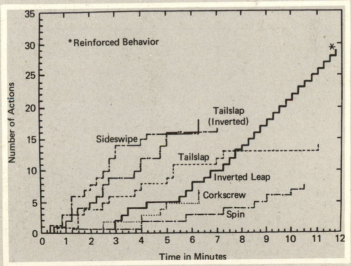

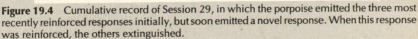

Figure 19.4 Cumulative record of Session 29, in which the porpoise emitted the three most recently reinforced responses initially, but soon emitted a novel response. When this response was reinforced, the others extinguished.

period. At the end of 19 min, Hou stood on its tail and clapped its jaws, spitting water towards the trainer; this time the action was reinforced, and was repeated five times.

Hou had now produced a new behavior in six out of seven consecutive sessions. In Sessions 31 and 32, Hou furthermore began each session with a new response and emitted no unreinforceable responses once reinforcement was presented. This establishment of a series of new types of responses was considered to be the conclusion of the experiment.

DISCUSSION

Over a period of 4 yr since Sea Life Park and the neighboring Oceanic Institute were opened, the training staff has observed and trained over 50 cetaceans of seven different species. Of the 16 behaviors reinforced in this experiment, five (breaching, porpoising, inverted swimming, tail slap, sideswipe) have been observed to occur spontaneously in every species; four (breaching, tail walk, inverted tailslap, spitting) have been developed by shaping in various animals but very rarely occur spontaneously in any; three (spinning, back porpoise, forward flip) occur spontaneously only in one species of *stenella* and have never been observed at Sea Life Park in other species; and four (corkscrew, back flip, tailwave, inverted leap) have never been observed to occur spontaneously. While this does not imply that these behaviors do not sometimes occur spontaneously, whatever the species, it does serve to indicate that a

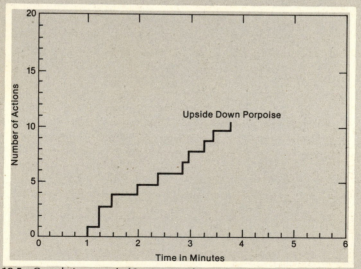

Figure 19.5   Cumulative record of Session 32. The porpoise emitted only a novel response in this session.

single animal, in emitting these 16 types of responses, would be engaging in behavior well outside the species norm.

A technique of reinforcing a series of different, normally occurring actions, in a series of training sessions, did therefore serve, in the case of Hou, as with Malia, to establish in the animal a highly increased probability that new types of behavior would be emitted.

j This ability to emit an unusual response need not be regarded as an example of cleverness peculiar to the porpoise. It is possible that the same technique could be used to achieve a similar result with pigeons. If a different, normally occurring action in a pigeon is reinforced each day for a series of days, until the normal repertoire (turning, pecking, flapping wings, etc.) is exhausted, the pigeon may come to emit novel responses difficult to produce even by shaping.

k A similar process may be involved in one traditional system of the training of five-gaited show horses, which perform at three natural gaits, the walk, trot, and canter, and two artificial gaits, the slow gait and the rack. The trainer first reinforces the performance of the natural gaits and brings this performance under stimulus control. The discriminative stimuli, which control not only the gait, but also speed, direction, and position of the horse while executing the gait, consist of pressure and release from the rider's legs, pressures on the reins and consequently the bit, shifting of weight in the saddle, and sometimes signals with whip and voice. To elicit the artificial gait, the trainer next presents the animal with a new group of stimuli, shaking the bit back and forth in the horse's mouth and vibrating the legs against the horse's sides, while preventing the animal from terminating the stimuli (negative reinforcement)

DISCUSSION

The implication of training creative behavior in the porpoise is discussed in relation to achieving similar results with the pigeon (**j**) and the horse (**k**), and possible common features of the present study and previous work by Maltzman are suggested (**l**).

QUESTIONS

1. In this study identify the following:

conditioned reinforcer	cumulative frequency graph
discriminative stimulus	shaping
operant	creativity (operational definition)
drive	response-reinforcement latency

by means of the previously reinforced responses of walking, trotting, or cantering. The animal will emit a variety of responses that eventually may include the pattern of stepping, novel to the horse though familiar to the trainer, called the rack (Hildebrand, 1965). The pattern, however brief, is reinforced, and once established is extended in duration and brought under stimulus control. (The slow gait is derived from the rack by shaping.)

Comparison may be made here between this work and that of Maltzman (1960). Working in the formidably rich matrix of human subjects and verbal behavior, Maltzman described a successful procedure for eliciting original responses, consisting of reinforcing different responses to the same stimuli, essentially the same procedure followed with Hou and Malia. It is interesting to note that behavior considered by the authors to indicate anger in the porpoise was observed under similar circumstances in human subjects by Maltzman: "An impression gained from observing Ss in the experimental situation is that repeated evocation of different responses to the same stimuli becomes quite frustrating; Ss are disturbed by what quickly becomes a surprisingly difficult task. This disturbed behavior indicates that the procedure may not be trivial and does approximate a nonlaboratory situation involving originality or inventiveness, with its frequent concomitant frustration."

Maltzman also found that eliciting and reinforcing original behavior in one set of circumstances increased the tendency for original responses in other kinds of situations, which seems likewise to be true for Hou and Malia. Hou continues to exhibit a marked increase in general level of activity. Hou has learned to leap tank partitions to gain access to other porpoises, a skill very seldom developed by a captive porpoise. When a trainer was occupied at an adjoining porpoise tank Malia jumped from the water, skidded across 6 ft of wet pavement, and tapped the trainer on the ankle with its rostrum, or snout, a truly bizarre act for an entirely aquatic animal.

2. Can you suggest another technique for measuring the responses?
3. Take another species—for example, dog, goldfish, squirrel—and write a reinforcement schedule for the training of novel behavior. Operationally define your terms.
4. Why did the researchers avoid extensive statistical analysis in this paper?
5. Write a brief essay describing the usefulness of naturalistic observations. What are their strengths? Their weaknesses?
6. Write a lesson plan for 6-year-old children in which creative reactions are shaped and reinforced. What limitations to your lessons do you see?
7. From this research, have you changed your definition of "creativity"? If so, how?
8. List the control measures used in this experiment.

Individual differences in the ability to create unorthodox responses no doubt exist; Malia's novel responses, judged *in toto*, are more spectacular and "imaginative" than Hou's. However, by using the technique of training for novelty described herein, it should be possible to induce a tendency toward spontaneity and creative or unorthodox response in most individuals of a broad range of species.

REFERENCES

Burgess, K. (1968). The behavior and training of a Killer Whale at San Diego Sea World. *International Zoo Yearbook, 8,* 202–205.

Hildebrand, M. (1965). Symmetrical gaits of horses. *Science, 150,* 701–708.

Maltzman, I. (1960). On the training of originality. *Psychological Review, 67,* 229–242.

Norris, K. S. (1965). Open ocean diving test with a trained porpoise *(Steno bredanensis). Deep Sea Research, 12,* 505–509.

Siegel, S. (1956). *Nonparametric statistics for the behavioral sciences.* New York: McGraw-Hill.

Skinner, B. F. (1962). Two synthetic social relations. *Journal of the Experimental Analysis of Behavior, 5,* 531–533.

NOTE: Contribution No. 35, the Oceanic Institute, Makapuu Oceanic Center, Waimanalo, Hawaii. Carried out under Naval Ordinance Testing Station Contract # N60530-12292, NOTS, China Lake, California. A detailed account of this experiment, including the cumulative records for each session, has been published as NOTS Technical Publication # 4270 and may be obtained from the Clearing House for Federal Scientific and Technical Information, U.S. Department of Commerce, Washington, D.C. A 16-mm film, "Dolphin Learning Studies," based on this experiment, has been prepared by the U.S. Navy. Persons wishing to view this film may inquire of the Motion Picture Production Branch, Naval Undersea Warfare Center, 201 Rosecrans Street, San Diego, California 92132. The authors wish to thank Gregory Bateson of the Oceanic Institute, Dr. William Wesit of Reed College, Portland, Oregon, and Dr. Leonard Diamond of the University of Hawaii for their extensive and valuable assistance; also Dr. William McLean, Technical Director, Naval Undersea Research and Development Center, San Diego, California, for his interest and support.

Chapter 20

Humor

INTRODUCTION

In our daily interactions with other people, we are constantly interpreting people's actions to try to ascertain the true feelings or the intentions of those people. Someone may compliment you because he or she genuinely likes what you are doing or because he or she wants to "butter you up." Politicians may publicly endorse a particular program because they actually believe in the program or because they want the votes that they think such an endorsement will bring. We know that many times the words and actions of people do not accurately reflect their feelings. Consequently, without our necessarily being aware of it, we use a series of tests to determine whether or not a person's actions reflect his or her actual beliefs.

SPECIAL ISSUES

Attribution Theory

Social psychologists are beginning to analyze the tests that people use to determine the motives of others. Many of these tests seem to fall quite nicely into a framework that is generally called *attribution theory*. This theory sets down some of the principles of attributing causes to the behavior of others. Many experiments have tested hypotheses derived from attribution theory. This article by Suls and Miller is an interesting application of the theory to humor. In this experiment the subjects read about a person who tells an anti–women's liberation joke to an audience that was either conservative or liberal and that reacts with either laughter or a glare. The subjects were then asked to rate how chauvinistic the joke teller was. Thus, we have an interesting case where the subjects are trying to ascertain the true feelings of the joke teller based on the characteristics and reactions of the audience.

HUMOR AS AN ATTRIBUTIONAL INDEX

JERRY M. SULS
STATE UNIVERSITY OF NEW YORK AT ALBANY

RICHARD L. MILLER
GEORGETOWN UNIVERSITY

The present study examined the circumstances under which observers make attributional judgments about an actor on the basis of his humor. Subjects read a description of a group discussion in which a man told an anti-women's liberation joke to a group of women. The variables manipulated were the audience's anticipated reaction (approval or disapproval) and the audience's actual reaction (approval or disapproval). The results indicated that the actor was seen as least chauvinistic when disapproval was anticipated but approval was received. The actor was seen as most chauvinistic when disapproval was both anticipated and received.

a Although a considerable amount of attention has been paid to the cognitive and motivational aspects of humor (Goldstein & McGhee, 1972; Levine, 1969; Suls, 1972), little is known about the social importance of humor. The present paper reports an exploratory study concerned with what kinds of social perceptions are made about persons who tell jokes under various social circumstances.

b Based on theories of attribution (Jones & Davis, 1965; Kelley, 1971) it would seem that the degree to which someone's joke is presumed to reflect his true beliefs would depend on the situational constraints present when the joke is told. Two attributional principles may provide a useful framework for examining the social consequences of joke telling. The first, Kelley's (1971) augmentation principle, states that a facilitative cause for a behavior is seen as more effective if the behavior occurred despite the opposing effect of an inhibitory cause. The second, the discounting principle, claims that the role of a given cause in producing a given effect is discounted if other plausible (facilitative)

SOURCE: Reprinted by permission from *Personality and Social Psychology Bulletin*, 1976, 2, 256–259.

ANALYSIS

REVIEW OF LITERATURE

The first section of this paper points out the purpose of this research as well as the theory behind it. Paragraph **(a)** outlines the general problem to be explored,

causes are also present. According to the augmentation principle, if an actor behaves contrary to situational pressures (expected disapproval), his personal belief should more likely be seen as causing his behavior than if the behavior were consistent with situational pressures (expected approval). The discounting principle suggests that when disapproval is expected, the actor's belief is the only plausible (facilitative) cause but when approval is expected, both the actor's belief and social approval are plausible causes and thus render the role of the actor's belief and social approval more ambiguous. Thus, both principles may lead observers to attribute more sincerity to an actor when negative consequences are the expected outcome of his behavior (Jones, Davis, & Gergen, 1961; Mills & Jellison, 1967). **On this basis we would expect that a**

c   **joke told to a potentially hostile audience would be perceived as more indicative of the joke teller's attitude than would a joke told to a potentially friendly audience.**

Although the actual reaction to any behavior may be distinguished from the anticipated reaction, little attention has been given to the effects of actual reactions on attributions. For example, Jones and Davis' theory stresses the subjective anticipation of decision-outcomes rather than their actual occurrence. There is, however, good reason to consider audience reaction to humor. If an audience reacts with disapproval or hostility, it may indicate that they took the joke seriously (i.e., that they perceived the joke teller as

d   expressing his opinion which is contrary to theirs). If, on the other hand, the audience reacts favorably, this may reflect either their agreement with the joke's content or their perception that the joke teller is being playful. Whether the audience reacts with approval or disapproval may be an important cue for the observer's judgment of the joke as indicative of the actor's true beliefs. The present exploratory study attempted to determine the effects of anticipated and actual reactions to humor on the perception of an actor's true beliefs.

while paragraph (b) explains the theory behind the research, which in turn leads to a logical derivation of the first hypothesis (c).

In the next paragraph (d) the authors discuss another problem they will explore in the experiment—the audience's reaction to humor. The authors offer no hypothesis for this variable; rather, they point out that audience reaction may be an important cue in judging an actor's true beliefs.

METHOD

This type of research is fairly easy to execute. The equipment costs are minimal, as the experimental manipulation consists of different booklets of mimeographed material that are given to different subjects. This is sometimes called "pen-

METHOD

e

Subjects were 86 students enrolled in introductory psychology classes. There were approximately an equal number of males and females. The subjects were asked to participate in a study on social perception and were given one-page descriptions of a group interaction to read.

f

The description indicated that a group of four people—three women and one man—participated in a discussion of several social issues: welfare, busing, and so on. All four discussants were described as knowing one another socially through their local civic association and as being reasonably well acquainted thus giving the actor presumably a basis upon which to form expectations about the consequences of his behavior. To manipulate the anticipated audience reaction, one-half of the subjects read information indicating that the three women were politically and socially very conservative; the other half of the subjects received information indicating the women were politically and socially "very liberated." During the course of the conversation the man in the group, John, told a joke which "put down" certain aspects of the women's liberation movement. At this point in

cil-and-paper" research, as this is the only equipment needed. In these kinds of experiments the subjects are usually tested in large groups, which allows the data to be collected quickly and at little cost. The availability of equipment, the cost of the research, the availability of subjects, and the time involved are all important concerns in doing research, and this type of research appears to solve most of these problems.

In (e) we learn that the subjects were from introductory psychology classes, with a roughly equal number of males and females. The experiment was described to them as one on "social perception"; the stimulus material was a single-page description of a group interaction.

The group interaction is explained in (f). Note that there are two independent variables: (1) audience attitudes and (2) audience reaction. There are also two levels of each of the independent variables. For the first variable, the audience was either very conservative or very liberal. For the second independent variable, the audience reaction was either laughter or a glare. Thus, we have a 2 × 2 factorial design similar to that explained in Chapter 3. In a 2 × 2 design there are four treatment groups (2 × 2 = 4), which are diagrammed below for this experiment:

		AUDIENCE REACTION	
		LAUGHTER	GLARE
AUDIENCE ATTITUDE	Conservative	1	2
	Liberal	3	4

In the description John tells an anti–women's liberation joke. In experimental group 1 the audience is conservative and responds with laughter; in experimental group 2 the audience is conservative and responds with a glare; and so on.

the conversation the women either "laughed heartily with John" or "glared silently at him." This served as the manipulation of actual audience reaction. One-half of the subjects read the first reaction; the other half read the second. After more information about the group interaction, i.e., other topics covered, subjects were asked a number of questions about the discussion including an estimate of John's opinion about the women's liberation movement. Subjects were also asked to rate how likable they found John and to indicate their own opinion of women's liberation. The subjects made their responses on 7-point Likert-type scales with smaller numbers representing greater "chauvinism," "likableness," and "unfavorableness."

g

Table 20.1   THE EFFECTS OF ANTICIPATED AND ACTUAL AUDIENCE REACTIONS ON ATTRIBUTIONS OF BELIEF AND LIKING

AUDIENCE	REACTION	ATTRIBUTIONS OF BELIEF	LIKING
Liberated	Laugh ($n = 22$)	4.46	2.13
	Glare ($n = 20$)	2.76	3.48
Conservative	Laugh ($n = 21$)	3.53	3.26
	Glare ($n = 21$)	4.09	2.86

After the subjects had read the description, they indicated on Likert-type scales **(g)** how chauvinistic and how likable they thought John was, as well as answering other questions. A typical 7-point Likert scale follows. In this example, subjects would be asked to check the space that best represents how chauvinistic they think John is.

_____ extremely chauvinistic
_____ somewhat chauvinistic
_____ slightly chauvinistic
_____ neutral or undecided
_____ slightly liberated
_____ somewhat liberated
_____ extremely liberated

Subjects would be given a numerical score from 1 to 7 based on the space they checked. Low scores indicate greater chauvinism.

RESULTS

Table 20.1 presents the mean liking score and the mean chauvinism score for each of the four treatment groups.

Since the study deals with women's liberation, it is reasonable to assume that the women may respond differently than would men to the various experimental treatments. To test this assumption, the means of the male subjects across the four treatments were compared to those of the female subjects to see if the means dif-

RESULTS

h
Since a preliminary analysis of the data indicated no sex differences across conditions, the data were collapsed across sex. **A 2 (anticipated reaction) × 2 (actual reaction) analysis of variance was performed on the subjects' estimates of John's attitude toward women's liberation.** The analysis of variance indicated a significant main effect of actual audience reaction ($F = 4.72$, $df = 1/82$, $p < .03$) and a significant anticipated reaction × actual reaction interaction ($F = 15.72$, $df = 1/82$, $p < .001$). Post-hoc comparisons indicated that the actor was seen as most chauvinistic when he addressed a potentially hostile audience and received disapproval as a result of telling his joke. The actor was seen as least chauvinistic when he addressed a potentially hostile audience that reacted favorably. In contrast, when the actor addressed a potentially friendly audience, attributions were less extreme and did not differ across conditions of actual reaction.

The liking ratings were also subjected to a 2 × 2 analysis of variance, which indicated a main effect of actual reaction ($F = 4.18$, $df = 1/82$, $p < .04$) and a significant two-way interaction ($F = 11.96$, $df = 1/82$, $p < .01$). Post-hoc comparisons indicated that the actor was seen as most likable when disapproval was anticipated but approval was received; the actor was seen as least likable when disapproval was anticipated and received.

fered (it becomes a 2 × 2 × 2 design when we include the male-female factor). In (**h**) we learn that there were no such sex differences.

A 2 × 2 analysis of variance of John's attitude toward women's liberation indicated a significant main effect of actual audience reaction and a significant interaction effect. The authors describe this in (**i**). However, a graphic representation would be helpful here. We have included one in Figure 20.1, where the two types of audiences are represented on the horizontal axis, and the two types of audience reactions are represented by separate lines. The analysis of variance (**i**) indicated a significant main effect of audience reaction, which indicates that, in general, John was seen to have a more liberated attitude when people laughed at his joke. (You should reread the section on factorial designs in Chapter 3 for a discussion of main and interaction effects in the analysis of variance.)

However, the interaction effect described in (**i**) qualifies the main effect. The post-hoc comparisons (i.e., statistical tests done on all *pairs* of means) indicate that there was no statistically significant difference between the means when the audience was conservative (potentially friendly), shown by X in Figure 20.1. However, there was a significant difference when the audience was liberal (potentially hostile), shown by Y in Figure 20.1. The authors base their interpretation of the results on this analysis.

In (**j**) the authors report the analysis for the liking ratings. Since they are similar to the ratings for chauvinism, the authors also explain why this is so.

An examination of the attribution means and the liking means shows that when the actor was seen as least chauvinistic he was seen as most likable and when most chauvinistic least likable. This pattern is explicable since the subjects' mean opinion of women's liberation was 4.75 indicating that they were favorable toward women's liberation. Not surprisingly, subjects liked those who they believed held similar beliefs (i.e., others who were in favor of women's liberation).

DISCUSSION

The results indicated that anticipated and actual reaction to the joke interacted such that the most extreme attributions were made where the anticipated reaction was negative. For the condition where disapproval was both anticipated and received, the observer was reasonably certain that the actor's joke reflected his beliefs. This was probably true since (a) there was no apparent external cause for telling the joke and (b) the audience took it seriously. Conversely, where it was anticipated that a joke would receive a hostile reaction but instead was greeted with approval, observers attributed the least belief-humor consistency. It is reasonable to assume that a group would only

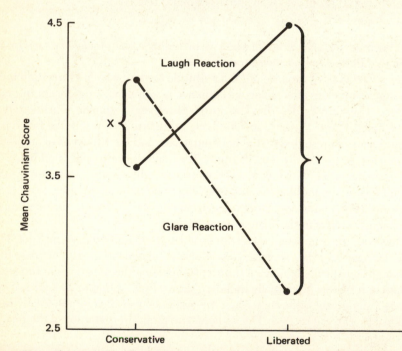

Figure 20.1   The interaction effect of the anti–women's liberation when the audience was conservative (X) and when the audience was liberal (Y).

laugh at a joke that was hostile to their views if the group knew that the joke teller was "teasing" and did not really agree with the content of the joke at all. In fact in order to laugh the audience must have assumed that the actor was in agreement with them. This means that observers could see the actor's beliefs as directly opposite from those expressed in the joke.

The results indicated that less extreme attributions occurred when anticipated audience reaction was favorable. In the case where approval was anticipated and received, the actor may have told the joke simply to get approval or the audience didn't take the joke seriously, or they agreed with the joke. Due to the many possible causes for the actor's behavior, this is an ambiguous situation, so it is reasonable that the observers made less certain inferences about the actor's beliefs. Also, when approval was expected but not received a problematic situation for the observer was created. Why would a group not laugh at a joke which was consistent with their beliefs? The answer could be that either the joke was perceived to be in bad taste or the audience thought the joke teller was being facetious by ridiculing their views. For either reason observers did not rate the actor as holding the view expressed by the joke.

k

To summarize, the results suggest that anticipated and actual audience reaction interact such that the actor was seen as *most* in agreement with the beliefs expressed in his joke when disapproval was anticipated but approval was received. Less extreme attributions occurred when approval was anticipated regardless of the actual reaction since the possible external cause (expected approval) seemed to render ambiguous the actor's real belief.

The present results suggest that under certain social circumstances humor is used as an attributional index. In addition the findings suggest that the resulting attributions may differ depending on whether actual and anticipated social reactions are consistent or inconsistent, positive or negative. It remains to be determined what other situational elements increase or decrease the information value of humor and whether the actual × anticipated reaction interaction holds for nonhumorous attitudinal statements.

DISCUSSION

The discussion section is fairly brief but still interprets the results. Here the authors attempt to explain the pattern of results. Some of this explanation is necessarily speculation and you may be able to think of other causes of the results. The authors end with a summary of the results (k).

REFERENCES

Goldstein, J. H., & McGhee, P. E. (Eds.). (1972). *The psychology of humor.* New York: Academic Press.

Jones, E. E., & Davis, K. E. (1965). From acts to dispositions: The attribution process in person perception. In L. Berkowitz (Ed.), *Advances in experimental social psychology* (Vol. 2). New York: Academic Press.

Jones, E. E., Davis, K. E., & Gergen, K. E. (1961). Role playing variations and their informational value for person perception. *Journal of Abnormal and Social Psychology, 63,* 302–310.

Kelley, H. (1971). *Attribution in social interaction.* Morristown, N.J.: General Learning Press.

Levine, J. (Ed.). (1969). *Motivation in humor.* New York: Atherton.

Mills, J., & Jellison, J. (1967). Effect on opinion change of how desirable the communication is to the audience addressed. *Journal of Personality and Social Psychology, 6,* 98–101.

Suls, J. (1972). A two-stage model for the appreciation of jokes and cartoons. In J. H. Goldstein & P. E. McGhee (Eds.), *The psychology of humor.* New York: Academic Press.

QUESTIONS

1. Explain how the first hypothesis (c) is a logical derivation of the theory in (b).
2. Could some hypotheses about the effect of audience reaction (d) be derived from the theory in (b)? If so, what would they be?
3. Draw a figure illustrating the results of the liking ratings and interpret the results.
4. How would you assign subjects to treatments in this experiment?
5. An obvious advantage of this type of "pencil-and-paper" research is that complex social situations can be studied under controlled conditions rather easily. What are the disadvantages of this type of research?
6. Design an experiment to test the effect of audience attitudes and audience reaction on attitude attribution in a situation that does not involve humor.

Chapter 21

Therapy for Anger

INTRODUCTION

In our contemporary society stress and anger are often apparent. The normal stressors that we encounter in daily living cause irritations, which frequently are manifest in anger. In spite of the frequency of such reactions and a need for greater understanding of how to treat them, it is somewhat surprising that no one has studied the phenomenon more thoroughly. In this article Raymond Novaco studies "anger disorders" and their treatment.

Several important problems encountered in research in clinical psychology are introduced in this article. Be aware of what those problems might be.

SPECIAL ISSUES

Single-Subject Design in Clinical Studies

In this study only one subject was used. In other parts of this book we have discussed small *n* designs (see Chapter 19). Studies using a single subject present special problems. In these experiments measurements are taken before and after the introduction of the independent variable, meaning the contrasts are made during different periods of time. During the initial period, careful measures are made to establish a base rate (or baseline) of the behavioral characteristics that exist before the introduction of the independent variable. These behavioral characteristics may be measured by multidimensional scales that measure several behavioral traits. If a researcher is interested in the emotional changes that occur after the introduction of an independent variable—say, the injection of a mood-altering drug—the baseline might include a description of several characteristics, such as expression of happiness, spontaneity, and sociability. After the introduction of the independent variable, in this case the mood-altering drug, subsequent evaluations on the same multidimensional characteristics are made. These measures are contrasted with baseline data and constitute the main effect of the study.

Because individual differences play a critical role in single-subject designs, a thorough description of the subject is necessary. In research involving psychotherapy, this description includes a "case history," which we can see in this study by Novaco. The author reviews the clinical background of the subject, which includes a history of the subject's work situation, symptoms, hospitalization, reaction to stressful situations, and previous treatment. Identifying these characteristics is particularly important in clinical studies as they provide the reader with a background upon which to base further generalization of the treatment.

Stress Inoculation: A Cognitive Therapy for Anger and Its Application to a Case of Depression

RAYMOND W. NOVACO
University of California, Irvine

Clinical interventions for anger disorders have been scarcely addressed in both theory and research in psychotherapy. The continued development of a cognitive behavior therapy approach to anger management is presented along with the results of its application to a hospitalized depressive with severe anger problems. The treatment approach follows a procedure called "stress inoculation," which consists of three basic stages: cognitive preparation, skill acquisition and rehearsal, and application practice. The relationship between anger and depression is discussed.

a

The treatment of anger-based disorders has escaped the attention of both psychotherapy theory and research. This is a puzzling state of affairs when one considers the abundance of laboratory research on aggression. An approach to the treatment of chronic anger and its experimental analysis has been presented by Novaco (1975). Although scattered reports of circumscribed interventions with anger exist (Herrell, 1971; Kaufmann & Wagner, 1972; Rimm, deGroot, Boord, Reiman, & Dillow, 1971), there is a distinct need for concerted work in this area. In an effort to promote a more detailed conception of anger problems and of therapeutic interventions for anger, the present article describes the further development of a cognitive behavior therapy and presents the results of its application to a hospitalized depressive with severe anger problems.

SOURCE: Article is abridged and reprinted by permission from *Journal of Clinical Psychology*, 1977, *45*, 600–608, and from the author.

ANALYSIS

REVIEW OF LITERATURE

The first paragraph (a) is an excellent example of an introductory statement. The author shows the need for the experiment, hints as to the treatment, and suggests how such results may be applied. Note that a formal hypothesis is not stated. This represents a departure from most of the previous studies analyzed in this book, but the practice is not erroneous. The author has provided sufficient detail for the reader to infer the hypothesis without formally stating it.

Persons having serious problems with anger control have been successfully treated by a cognitive behavior therapy approach (Novaco, 1975, 1976b). In the present article, this treatment approach has been further developed to follow a procedure called *stress inoculation* that has been applied to problems of anxiety (Meichenbaum, 1975; Meichenbaum & Cameron, Note 1) and pain (Turk, Note 2). The stress inoculation therapy consists of developing the client's cognitive, affective, and behavioral-coping skills and then providing for the practice of these skills with exposure to regulated doses of stressors that arouse but do not overwhelm the client's defenses. The stress inoculation approach involves three basic steps or phases: (a) cognitive preparation, (b) skill acquisition and rehearsal, and (c) application practice. This approach to anger problems was first developed by me with regard to the training of police officers (Novaco, in press). The present article describes the further development of this approach as a clinical procedure and its application to the anger problems of a depressed patient in an acute psychiatric ward.

METHOD

Treatment Procedure

b ### Cognitive Preparation

The initial phase educates clients about the functions of anger and about their personal anger patterns, provides for a shared language system between client and therapist, and introduces the rationale of the treatment approach. This process is facilitated through an instructional manual for clients.[1] The text of the manual describes the nature and functions of anger, when anger becomes a problem, what causes anger, and how anger can be regulated.

The components of the cognitive preparation phase are (a) identifying the persons and situations that trigger anger; (b) fostering recognition of the difference between anger and aggression; (c) understanding the cognitive, somatic, and behavioral determinants of anger, with particular emphasis on anger-instigating self-statements; (d) discriminating justified from unnecessary anger; (e) recognizing the signs of tension and arousal early in a provocation sequence; and (f) introducing the anger management concepts as coping strategies.

c ### Skill Acquisition and Rehearsal

The process of skill acquisition involves a familiarization with the three sets of coping techniques, modeling of the techniques by the therapist, and then rehearsal

---

[1] The client instructional manual and the therapist treatment procedure manual may be obtained from the author.

by the client. At the cognitive level, the client is taught how to alternatively view situations of provocation by changing personal constructs (Kelly, 1955) and by modifying the exaggerated importance often attached to events (Ellis, 1973).

Two central cognitive devices for anger regulation are a task orientation to provocation and coping self-statements. A task orientation involves attending to desired outcomes and implementing a behavioral strategy directed at producing these outcomes. Self-instructions are used to modify the appraisal of provocation and to guide coping behavior. Pragmatically, the self-instructions are designed to apply to various stages of a provocation sequence: (a) preparing for a provocation, (b) impact and confrontation, (c) coping with arousal, and (d) subsequent reflection in which the conflict is either unresolved or resolved. Stages (c) and (d) provide for the possible failure of self-regulation and for the prolongation or reinstigation of arousal by ruminations.

At the affective level, the client is taught relaxation skills and is also encouraged to maintain a sense of humor as a response that competes with anger.

Relaxation training was conducted via the Jacobsen method of tensing and relaxing sequential sets of muscle groups. Mental as well as physical relaxation is emphasized, as relaxation imagery is used to induce a light somnambulistic state. Deep breathing exercises supplement the muscle tension procedures. Previous research (Novaco, 1975) has found that the cognitive-coping component is comparatively more effective than the relaxation-training component. However, relaxation training enables the client to become aware of the tension and agitation that lead to anger and provides an important sense of mastery over troublesome internal states.

The behavioral goals of treatment are to promote the effective communication of feelings, assertive behavior, and the implementation of task-oriented, problem-solving action. Generally speaking, the client is induced to use anger in a way that maximizes its adaptive functions and minimizes its maladaptive functions (Novaco, 1976a). Among the positive functions of anger are its energizing effects, the sense of control that it potentiates, and its value as a cue to cope with the problem situation.

The therapeutic procedure enables clients to recognize anger and its source in the environment and to then communicate that anger in a nonhostile form. This controls the accumulation of anger, prevents an aggressive overreaction, and provides a basis for changing the situation that caused the anger.

Inducing a task-oriented response set facilitates the occurrence of problem-solving behavior. It is emphasized that anger is an emotional reaction to stress or conflict that is due to external events being at variance with one's liking. Poor behavioral

Method

In the introduction, the author defines the "stress inoculation" approach and in sections (b), (c), and (d) describes the stages of this approach. Of special interest is his attempt to operationally define the concepts, thus allowing subsequent inves-

adjustment is linked to the inability to provide problem-solving responses that are effective in achieving desired goals (Platt & Spivack, 1972; Shure & Spivack, 1972). Anger management involves a strategic confrontation whereby the person learns to focus on issues and objectives. The execution of such behavior is explicitly prompted and progressively directed by the use of coping self-statements.

d   Application Practice

The value of application practice has been emphasized by a variety of skills-training approaches (D'Zurilla & Goldfried, 1971; Meichenbaum, 1975; Richardson, 1973; Suinn & Richardson, 1971). The regulation of anger is a function of one's ability to manage a provocative situation, and in this treatment phase, clients are given an opportunity to test their proficiency. The client is exposed to manageable doses of anger stimuli by means of imaginal and role-playing inductions. The content of the simulations is constructed from a hierarchy of anger situations that the client is likely to encounter in real life. By progressively working on each hierarchy scene, first imaginally and then by role playing, the client is enabled to sharpen anger management skills that had been rehearsed with the therapist.

Case History

The client was a 38-year-old male who had been admitted to the psychiatric ward of a community hospital with the diagnosis of depressive neurosis. Upon admission he was judged to be grossly depressed, having suicidal ruminations and progressive beliefs of worthlessness and inadequacy. He was a credit manager for a national business firm and had been under considerable job pressure. Quite routinely, he developed headaches by midafternoon at work. He had recurrent left anterior chest pain that was diagnosed by a treadmill stress test procedure as due to muscle tension.

Their client had been hospitalized for 3 weeks when the attending psychiatrist referred him to me for the treatment of anger problems. At that time I had initiated a staff training program for the treatment of anger. The principal behavior setting in which problems with anger control emerged were at work, at home,

tigators to replicate the study. Notice the components of cognitive preparation in (a) through (f) in section **(b)** and the various stages of a provocation sequence in (a) through (d) in section **(c)**.

Case History

Single-subject studies, especially of a clinical nature, typically involve a comprehensive case history in which the salient psychological characteristics of the subject are documented **(e)**. The case history is designed to reveal the basic descriptive facts of the subject but not to interpret these facts. The writing style is lucid and without embellishment.

and at church. The client was md and had six children, one of which was hyperactive.

Circumstances at work had progressively generated anger and hostility for this man's superiors, colleagues, and supervisees. His anger at work was typically overcontrolled. He would actively suppress his anger and would then periodically explode with a verbal barrage of epithets, curses, and castigations when a conflict arose. At home, he was more impulsively aggressive. The accumulated tensions and frustrations at work resulted in his being highly prone to provocation at home, particularly in response to the disruptive behavior of the children. Noise, disorders, and the frequent fights among the children were high anger elicitors. Unlike his behavior at work, he would quickly express his anger in verbal and physical outbursts. Although not an abusive parent, he would readily resort to physical means and threats of force (e.g., "I'll knock your goddamn head off") as a way to control the behavior of his children. The children's unruly behavior often became a problem during church services. The client's former training in a seminary disposed him to value family attendance at church, but serious conflict was often the result. In an incident just prior to hospitalization, the client abruptly removed two of his boys from church for creating a disturbance and threatened them to the extent that one ran away. At this point he had begun to realize that he was reacting "out of proportion," but he felt helpless about instituting the desired changes in behavior.

During hospitalization, treatment sessions were conducted three times per week for 3½ weeks. Following discharge, follow-up sessions were conducted biweekly for a 2-month period. During these sessions, anger diary incidents were discussed, and there was continued modeling, rehearsal, and practice of coping procedures.

Dependent Measures

f Proneness to provocation was first assessed by means of an inventory of 80 situation descriptions for which the respondent rates anger on a 5-point scale. This is a revised version of the instrument reported in Novaco (1975). The current scale has been found to have a high degree of internal consistency across various subject populations.

g **Behavior ratings were obtained by a clinical psychologist with 9 years of postdoctoral experience.** Dichotomous ratings were obtained on 14 behavior descriptions indexing the expression of anger (e.g., impatient with others, expresses resentment toward others, threatens to harm someone, quick to react with antagonism) and 8 behavioral descriptions indexing constructive coping and improvement (e.g., hu-

Dependent Measures

Measurement of a person's response in a clinical study frequently poses a problem of objectivity. What indices of reduced anger could be used that both would reflect a valid measure of emotional change and yet could be used by subsequent researchers in a replication of the study? The author is very specific in identifying the special tests (**f**) and behavioral ratings (**g** and **i**). Furthermore, in (**h**), the

morous and good natured, optimistic about future, tolerant and considerate of others, deals with conflict constructively, is positively assertive). Scaled ratings were obtained on three behavioral dimensions: (a) "The person is relaxed and at ease," (b) "the person demonstrates hostility or anger (expressed anger)," and (c) "the person is restraining anger and resentment (suppressed anger)." These items were rated on 8-point scales ranging from "not at all" to "exactly so" verbally anchored at each interval point.

The reliability of the behavior ratings for the dichotomous items was examined for behavior in group therapy, occupational therapy, and recreation for a total of nine observation periods with an average of eight patients observed in each setting. Four psychiatric nurses with an average of 6 years of experience served as the raters along with the clinical psychologist. The mean percentage of agreement was 84.6% for each pair of raters on the 22 items. For the 3 scaled items, the ratings of the psychologist were correlated with those of three psychiatric nurses who,

h

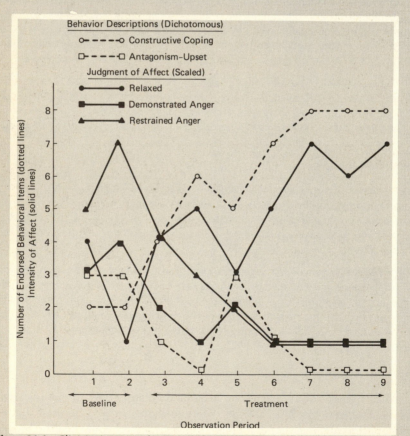

Figure 21.1 Clinician's ratings of patient's behavior for anger and anger management during hospitalization.

respectively, observed the patient in three group therapy sessions. The average correlations were .7581 for Item 1, .8593 for Item 2, and .6621 for Item 3. Thus, greatest consistency was obtained for ratings of "expressed anger" and least consistency for ratings of "suppressed anger." The psychologist's ratings were also correlated with the patient's ratings of himself for behavior in group therapy, which resulted in correlations of .9242, .6770, and .7521 for the respective items. The judgments of this observer were concluded to be sufficiently reliable for the present case analysis.

Self-monitoring of anger reactions by the patient is integral to the treatment procedure, as in other self-control therapies (Mahoney & Thoresen, 1974). The patient was asked to maintain a diary of anger experiences for which two ratings on a 7-point scale were obtained—(a) the degree of anger arousal experienced and (b) the degree of anger management achieved. Although the reduction in frequency and intensity of anger is a central goal of treatment, it is assumed that in certain instances, the arousal of anger is justified and appropriate. Therefore, it is also important to assess the degree to which anger regulation resulted in a constructive response to provocation. This patient's anger experiences were concentrated at home and at work. Only rarely was he provoked while in the hospital. Hence, the self-ratings were obtained for weekend home visits while on pass from the hospital and then on a daily basis following discharge and return to work.

RESULTS

Pretreatment assessment of the patient's proneness to provocation by means of the anger inventory resulted in a total inventory score of 301. At discharge,

reliability of the ratings is assessed. The patient also monitored his or her own anger response (**j**).

One may argue that the dependent measures in this experiment lack the objectivity of measures made in some nonclinical experiments. For example, the dependent measure of a rat pressing a bar in a Skinner box leaves little room for ambiguity. But judgments about emotional states in clinical research can also achieve a high degree of objectivity; one just needs to specify, as completely as possible, the measurement instruments and their validity.

RESULTS

In this section a comparison is made between the inventory scale given during pretreatment and the scale given at discharge (**k**). The results are also presented in Figure 21.1 [see (**1**)]. In this figure the researcher has shown five different measures of the patient's behavior for anger and anger management. Study the relationship between "relaxed" behavior and "restrained anger" in this curve.

k | the inventory was readministered and a total score of 258 was obtained. This decrease of 43 points represents a decrease of greater than 1 standard deviation based on the recent random college sample and on the chronic anger clients of previous research. It also is comparable to the change obtained on this instrument for subjects in a controlled treatment design (Novaco, 1975).

l | Continuous measures of change were obtained from the behavior rating scales and from the client's self-ratings. Behavior ratings of anger were obtained during group therapy. Since the patient had already been in the hospital for 3 weeks prior to the referral for treatment, only a 1-week baseline could be obtained. **Results for the behavior description measures and for the scaled ratings of affect are contained in Figure 21.1.** It can be seen that over the course of treatment, the patient underwent improvement on each index. The number of positive behavior descriptions endorsed and the scaled judgments of relaxed appearance increased over time, whereas antagonistic behavior and judgments of demonstrated and suppressed anger decreased.

m | **After the thirdweek of treatment, the attending psychiatrist, the staff, and the patient concurred in the decision for discharge.**

n | Convergent evidence of the patient's improvement was obtained from the self-monitored ratings of anger and anger management. Data were recorded on a weekend pass during the hospitalization period and on a continuous daily basis after discharge. **Results for the 60 days of recorded self-monitoring for frequency, degree of anger, and degree of control are contained in Figure 21.2.** The first 15 days of observation represent 5 weekends during the hospitalization period. As previously indicated, little anger was experienced with regard to events in the hospital environment, and these data are not plotted.

In the absence of a more satisfactory baseline period (which could not be extended due to the patient's length of stay in an acute facility prior to referral for treatment), changes in behavior must be examined as treatment progresses. On the weekend prior to treatment, the mean frequency was 4.0 per day, and by the patient's report this was quite standard. It can be seen that the mean frequency of anger incidents decreased from 1.55 per day during the first 20-day period to 1.10 per day for the second 20-day interval, and to .40 per day for the third interval. The mean anger magnitude ratings for the intervals were 4.30, 4.62, and 4.25, whereas the mean ratings of anger manage-

In section (**m**) we learn that the psychiatrist, the staff, and the patient decided the patient should be discharged.

The long-term effects of the treatment are shown in (**n**) and in Figure 21.2.

In section (**o**) the author gives a very personal insight into the success of the treatment and the severity of the original problem.

ment were 4.30, 4.97, and 4.50, respectively. This indicates that when he did become angry, the degree of regulation was commensurate with or greater than the degree of arousal. It should also be noted that incidents of high achievement in self-control were occasionally not recorded by the client, who, like many depressives, was reluctant to "make too much" of his accomplishments. Hence, he periodically neglected to record an incident of high control and low anger. Although ratings were unfortunately not obtained from his wife, she consistently remarked that he was showing progressive improvement, in marked contrast to his previous behavior.

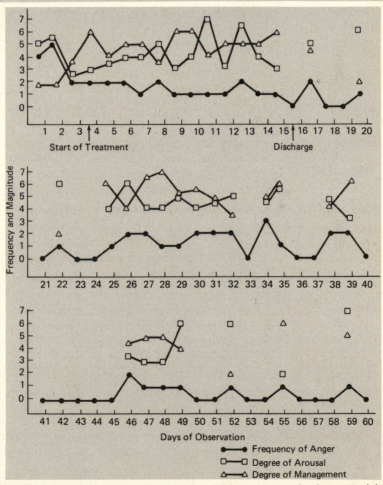

Figure 21.2 Self-monitored ratings of frequency of anger, magnitude of anger, and degree of anger management.

o

Among the most apparent and important behavioral changes achieved was the reduction of impulsive aggression as a punitive response to disruptive behavior by the children. During the treatment period numerous incidents of *physical* fighting among the boys were recorded in the diary as having aroused anger. There were 10 recorded incidents in the home, 5 in the car, and 2 in church. Considering the latter two settings, one can appreciate the magnitude of the disturbance and its potential impact. The incidents described as having occurred in the car, for example, were indeed unnerving. Fist fights between the boys resulted in bleeding mouths, black eyes, and lumps on the head. In one instance, a weight-lifting bar was used as a weapon. Despite the magnitude of these disturbances, the client only struck his children three times during the 3-month treatment period. Prior to treatment it had been quite routine for him to hit the boys in response to fighting or prephysical quarrels.

DISCUSSION

The self-regulation of anger is an important facet of the ability to cope with stress. The present article has reported the continued development of a cognitive behavior therapy approach to the treatment of anger problems and the extension of these procedures to a hospitalized depressed patient. **The stress inoculation model, first developed by Meichenbaum (1975), has been extended to the domain of anger and has been shown to be effective with someone having a severe psychological disturbance.**

p

The severity of the clinical problem is a nontrivial factor. That this factor is not to be overlooked is demonstrated by comparing the length of treatment in the present case with that of previous controlled research. In the clinical sample investigated by Novaco (1975), extensive pretreatment assessment eliminated all but those subjects having serious anger problems. Novaco (1975), however, used an outpatient sample, and significant results were achieved in 6 treatment sessions. Yet in the present case, comparable results required 11 sessions in the hospital and 4 as an outpatient. In addition, the clinical procedures had improved, an instructional manual had been incorporated, and relaxation therapy was available on a daily basis by tape. In the present case the severity of the anger problem was interlocked and compounded with the psychological deficits associated with depression.

REFERENCE NOTES

1. Meichenbaum, D., & Cameron, R. (1973). *Stress inoculation: A skills training approach to anxiety management.* Unpublished manuscript, University of Waterloo, Ontario, Canada.

2. Turk, D. (1974). *Cognitive control of pain: A skills-training approach.* Unpublished manuscript, University of Waterloo, Ontario, Canada.

REFERENCES

D'Zurilla, T., & Goldfried, M. (1971). Problem solving and behavior modification. *Journal of Abnormal Psychology, 78,* 107–126.

Ellis, A. (1973). *Humanistic psychology: The rational-emotive approach.* New York: Julian Press.

Feshbach, N. D. (1975). Empathy in children: Some theoretical and empirical considerations. *Counseling Psychologist, 5,* 25–30.

Herrell, J. M. (1971). Use of systematic desensitization to eliminate inappropriate anger. *Proceedings of the 79th Annual Convention of the American Psychological Association, 6,* 431–432. (Summary)

Kaufmann, L., & Wagner, B. (1973). Barb: A systematic treatment technology for temper control disorders. *Behavior Therapy, 3,* 84–90.

Kelly, G. A. (1955). *The psychology of personal constructs* (Vol. 2). New York: Norton.

Meichenbaum, D. A. (1975). Self-instructional approach to stress management: A proposal for stress inoculation training. In C. Spielberger & I. Sarason (Eds.), *Stress and anxiety* (Vol. 2). New York: Wiley.

Novaco, R. W. (1974). The treatment of anger through cognitive and relaxation controls. *Journal of Consulting and Clinical Psychology, 44,* 681.

Novaco, R. W. (1975). *Anger control: The development and evaluation of an experimental treatment.* Lexington, Mass.: Lexington Books.

Novaco, R. W. (1976). The functions and regulation of the arousal of anger. *American Journal of Psychiatry, 133,* 1124–1128.

Novaco, R. W. (In press). A stress inoculation approach to anger management in the training of law enforcement officers. *American Journal of Community Psychology.*

Platt, J. J., & Spivack, G. (1972). Problem-solving thinking of psychiatric patients. *Journal of Consulting and Clinical Psychology, 39,* 148–151.

Richardson, F. (1973). A self-study manual for students on coping with test-taking anxiety. (Tech. Rep. No. 25) Austin: University of Texas Press.

Rimm, D. C., deGroot, J. C., Boord, P., Reiman, J., and Dillow, P. V. (1971). Systemic desensitization of an anger response. *Behavior Research and Therapy, 9,* 273–280.

Shure, M. B., & Spivack, G. (1972). Means-ends thinking, adjustment, and social class among elementary-school-aged children. *Journal of Consulting and Clinical Psychology, 38,* 348–353.

Suinn, R., & Richardson, F. (1971). Anxiety management training: A nonspecific behavior therapy program for anxiety control. *Behavior Therapy, 2,* 498–510.

Weissman, M. M., Klerman, G. L., & Paykel, E. S. (1971). Clinical evaluation of hostility in depression. *American Journal of Psychiatry, 128,* 261–266.

Requests for reprints should be sent to Raymond W. Novaco, Program in Social Ecology, University of California, Irvine, California 92717.

DISCUSSION

In keeping with the format outlined in this book, the author discusses his research in the context of a theoretical issue raised in the introduction **(p)**.

QUESTIONS

1. Discuss the special problems in developing objective dependent measures in clinical research.
2. The results are not discussed in terms of statistical probability. Why not?
3. Why were pretreatment measures made?
4. Why did the investigator choose to use multidimensional measures of behavior?
5. What precautions were undertaken to assure the objectivity of behavioral ratings?
6. Why is a detailed case history part of this study?
7. What special ethical issues are raised by this study?

Chapter 22

Prosocial Behavior

INTRODUCTION

One of the most hotly debated subjects in recent years has been the effect of television viewing on children's behavior. The average American child spends more hours per week watching television than attending school, and several educators have suggested that television may have replaced school as the primary "socializing" influence outside of the family.

The effects of television on children's behavior has been the subject of numerous psychological experiments, governmental studies, and congressional investigations. Most of this effort has centered on whether television violence leads to violence in society. Less often studied is whether television can lead to *prosocial behaviors,* which are behaviors that are believed to benefit others, such as helping people in distress.

This experiment is an experimental study of the effect of television viewing on prosocial behavior. We have left this case for your analysis. In your analysis we suggest that you follow this outline:

1. What is the general topic of the research?
2. What pertinent literature is reviewed?
3. What is the research problem?
4. What are the authors testing that has not been done in previous research; that is, what is unique about this experiment?
5. What is the hypothesis?
6. Name the independent variable(s).

7. Name the dependent variable(s).
8. How were subjects assigned to treatments?
9. What controls were used to prevent extraneous variables from confounding the results?
10. How were the data collected?
11. How were the data analyzed?
12. Summarize the results. Draw a bar graph illustrating the results.
13. Summarize the discussion.

EFFECTS OF A PROSOCIAL TELEVISED EXAMPLE ON CHILDREN'S HELPING

JOYCE N. SPRAFKIN, ROBERT M. LIEBERT, AND RITA WICKS POULOS
STATE UNIVERSITY OF NEW YORK AT STONY BROOK

The possibility that regularly broadcast entertainment television programs can facilitate prosocial behavior in children was investigated. Thirty first-grade children, 15 boys and 15 girls, were individually exposed to one of three half-hour television programs: a program from the *Lassie* series which included a dramatic example of a boy helping a dog, a program from the *Lassie* series devoid of such an example, or a program from the family situation comedy series *The Brady Bunch.* The effects of the programming were assessed by presenting each child with a situation that required him to choose between continuing to play a game for self-gain and helping puppies in distress. Children exposed to the *Lassie* program with the helping scene helped for significantly more time than those exposed to either of the other programs.

Since its spectacular rise as an entertainment medium in the 1950s, television's possible effects on the behavior of the young has captured the interest of many investigators. Throughout the years the most compelling question has remained the possible detrimental effects of violent programming, resulting in considerable knowledge of the influence of televised aggression (e.g., Liebert, Neale, & Davidson, 1973). Remarkably little is known, on the other hand, about the degree to which the medium might also inculcate prosocial attitudes and behaviors. The present report is concerned with this latter issue.

That certain commercially broadcast programs may serve as positive socializers is suggested by numerous investigations employing specially prepared programming presented in a television format. Such studies have shown that modeling cues provided by a television set can increase the generosity of young observers (Bryan, 1970), foster adherence to rules (Stein & Bryan, 1972; Wolf & Cheyne, 1972), augment delay of gratification (Yates, 1974), and facilitate positive interpersonal behaviors (O'Connor, 1969). Only one study, so far as we know, endeavored to determine the potential of *broadcast* televised materials for instigating prosocial forms of behavior in children. Over the course of 4 weeks, Stein and Friedrich (1972) exposed 3- to 5-year-olds to one of three television diets: aggressive (Batman and Superman cartoons), prosocial (episodes taken from "Mister Rogers' Neighborhood"), or

SOURCE: Reprinted by permission from *Journal of Experimental Child Psychology,* 1975, *20,* 119–126.

neutral (such scenes as children working on a farm). Generally confirming earlier work, certain types of prosocial behavior were found to increase after exposure to the "Mister Rogers" program, relative to the other two types of input. In this important demonstration, as the investigators themselves indicated (Friedrich & Stein, 1973), it was not feasible to control for factors such as the format of the programs, the characters displayed, and the relative balance of entertainment vs. instructional components in each diet. Equally important, "Mister Rogers" is not a commercially produced network program and its direct, exhortative format also might serve to limit the size of its child audience.

In the experiment reported, we wished to determine whether the presence of a specific act of helping that appeared in a highly successful commercial series would induce similar behavior in young viewers when they later found themselves in a comparable situation. To control for various broad characteristics of the program, children in the two principal groups viewed one of two programs from the series *Lassie*, complete with commercial messages. The *Lassie* programs both featured the same canine and human main characters, took place in the same setting, apparently were written for the same audience, and presumably were motivated by the same entertainment and commercial interests. Whereas the programs necessarily differed somewhat in story line, the critical difference was that one of them had woven into it the particular prosocial example while the other did not. This combination of similarities and differences, it was reasoned, would optimize detecting the potential effect of specific modeling cues from entertainment television upon youngsters' later behavior. As a further comparison, a third group was also included, in which a generally pleasant situation comedy program, *The Brady Bunch,* was viewed.

METHOD

Design and Participants

A 3 × 2 factorial design was used, employing the television experience (observation of prosocial *Lassie*, neutral *Lassie*, or *Brady Bunch* program) and the sex differentiation of the subjects. The participants were 30 children, 15 boys and 15 girls, from four first-grade classes in middle-class, predominantly white suburban neighborhood. All prospective subjects were given forms to be signed by their parents authorizing their participation. The letter described the nature of the project, but requested parents to withhold this information from their children. Subjects were selected randomly from the pool of youngsters whose parents returned the consent forms; approximately 80% did so. Within each class and sex, the children were assigned randomly to the treatment conditions.

One experimenter escorted the child from the classroom to the television viewing room, conducted the screening of the videotape, and then brought the child to the second experimenter in an adjacent room. The second experimenter, who was blind to the treatment condition, administered the dependent measure. The former role was assumed by four graduate students over the course of the study; the latter role was always filled by the same 24-yr-old white, female graduate student.

Apparatus and Materials

The television viewing room contained a black and white television monitor with a 9-in. screen, programmed by a hidden Sony video recorder. The experimental room, in which assessment was conducted, contained two pieces of apparatus: a Point Game and a Help button, separated by a distance of 5 ft and each placed on a child-size desk.

The Point Game consisted of a response key and a display box. The response key was a 1½ by ¼ in. rectangular blue button mounted on a 4 by 6½ in. inclined platform which served as a handrest. The display box was a 6 by 8 in. gray metal box on which a blue 15-W light bulb and a Cramer digital timer were mounted. The 10's and 1's columns of the timer were covered with black tape. When the button was pressed, the bulb lit and the timer was activated.

The Help button was a small, green, circular button mounted on a 7 by 5 in. gray metal box with the word "Help" printed in black letters directly below it. The Help button activated a second Cramer digital timer which was located outside the experimental room.

Earphones were located next to the game on the desk. They were connected via an extension cable to a tape recorder located outside the experimental room. The tape contained 30 sec of silence followed by 120 sec of dogs barking.

Procedure

Introduction to the Treatment Condition

The first experimenter escorted the child from the classroom to the television viewing room, which was a small classroom containing desks, chairs, and the videotape monitor. She told the child that he was going to play a game shortly, but that Miss [*the second experimenter*] was not ready for him yet. She then turned on the video recorder surreptitiously, turned on the monitor openly, and suggested that the child watch while she did some work in the back of the classroom. All children appeared quite interested in the half-hour programs and the experimenter worked busily to discourage the child from initiating conversation.

Experimental and Control Conditions

Children in the experimental condition viewed the prosocial *Lassie* program. The episode dealt largely with Lassie's efforts to hide her runt puppy so that it would not be given away. The story's climax occurred when the puppy slipped into a mining shaft and fell onto the ledge below. Unable to help, Lassie brought her master, Jeff, and Jeff's grandfather to the scene. Jeff then risked his life by hanging over the edge of the shaft to save the puppy. The selection of the program was based on the inclusion of this dramatic helping scene.

In one control condition, the neutral *Lassie* program was shown. Featuring the same major characters as the prosocial program, it dramatized Jeff's attempts to avoid taking violin lessons. It was devoid of any example of a human helping a dog, but necessarily featured the animal in a positive light.

In the second control condition, the youngsters viewed *The Brady Bunch,* a nonaggressive, nonanimated children's program which did not feature a dog. The program depicted the youngest Brady children's efforts to be important by trying to set a record for time spent on a seesaw. It provided a measure of children's willingness to help after exposure to a popular program that showed positive interpersonal family encounters but no cues pertinent to either human or canine heroism.

Assessment of Willingness to Help Puppies in Distress

Assessment consisted of placing the child in a situation that required him to choose between continuing to play a game for self-gain and trying to get help for puppies in distress.

1. *Introduction of the game.* After the television program was viewed, the first experimenter told the child that Miss [*the second experimenter*] was ready for him, escorted the child to the experimental room, performed appropriate introductions, and departed. The child was invited to play the Point Game in which he could earn points by pressing a button that lit a bulb. The number of points obtained was displayed on a timer visible to him. The experimenter showed the child a display of prizes, situated on two desks, that varied in size and attractiveness—from small, unappealing erasers to large, colorful board games. It was noted that "some prizes are bigger and better than others," and that "the more points you get in the game, the better your prize will be." The experimenter then told the child that she was going to let him play the game alone so that she could finish some work down the hall.

2. *Introduction of distress situation.* The experimenter started to leave, then returned and exclaimed:

Oh, there's something else I'd like you to do while I'm gone. You see, I'm in charge of a dog kennel a few miles from here—that's a place that we keep puppies until a home can be found for them. I had to leave the puppies alone so that I could come here today, but I know whether the dogs are safe or not by listening through these earphones [indicating] which are connected to the kennel by wires. When I don't hear any noises through the earphones, I know the dogs are OK—they're either playing quietly or sleeping. But if I hear them barking, I know something is wrong. I know that they are in trouble and need help. If I hear barking, I press this button [indicating Help button], which signals my helper who lives near the kennel, and when he hears the signal, he goes over to the kennel to make sure nothing bad happened to the dogs.

While I'm gone, I'd like you to wear the earphones for me while you play your game. If you don't hear any barking, everything is OK at the kennel; but if you hear barking, you can help the puppies if you want by pressing the Help button. You might have to press the button for a long time before my helper hears the signal, and there's a better chance he'll hear it if you press it a lot. You can tell when he has heard the signal when you hear the dogs stop barking.

Remember, this game is over as soon as I come back—you won't be able to get any points after that. Try to get as many points as you can because the more points you get, the better your prize will be. You know, if the puppies start barking, you'll have to choose between helping the puppies by pressing the Help button and getting more points for yourself by pressing the blue button. It's up to you.

After being assured that the child understood the situation, the experimenter placed the earphones on the child and left the room. When outside the experimental room, she immediately turned on the tape recorder and a stopwatch. The child first heard 30 sec of silence followed by 120 sec of increasingly frantic barking. The experimenter recorded the latency of helping, reentered the experimental room 60 sec after the cessation of barking, and awarded a prize that was loosely based on the number of points earned. After the child was escorted to his classroom, the total seconds of helping was recorded from the timer located outside the experimental room.

RESULTS AND DISCUSSION

The primary dependent measure was the number of seconds the Help button was pressed. The mean helping scores for all cells of the design, and the standard deviations for each, are shown in Table 22.1. A 3 (Treatment) × 2 (Sex) analysis of variance of these scores yielded only one significant effect, for treatment conditions ($F \times 4.97$, $p < .025$); the main effect for Sex was not significant ($F = 2.18$). Dunnett's t tests revealed that subjects who saw the prosocial *Lassie* program helped significantly more than those in the neutral *Lassie* ($t = 2.08$, $p < .05$) or the *Brady Bunch* condition ($t = 2.79$, $p < .01$). The latter groups did not differ from one another ($t = .66$).[1]

[1] A parallel 3 × 2 analysis of variance of latency to help scores revealed no significant differences, but the tendency was consistent with amount of helping: 24.6 for the prosocial *Lassie,* 36.1 for the neutral *Lassie,* and 55.4 for the family situation comedy.

There was also a marginal tendency for sex and treatment condition to inter-act ($F = 2.80$, $p < .10$). As is apparent descriptively in Table 22.1, the effect appeared because the impact of the prosocial *Lassie* program, relative to the neutral one, tended to be greater for girls than for boys; moreover, the girls tended to respond differently to the neutral *Lassie* and *Brady Bunch* programs whereas the boys did not. Both of these tendencies may be due, in part, to a possible ceiling effect for the boys; in the two control conditions they tended to help more than their female counterparts. Further, nine out of the ten boys in the two control conditions rendered some help, whereas only five out of ten girls in the control conditions did so. (Four of the five boys who saw the prosocial *Lassie* program, and all five of the girls, rendered some help to the distressed puppies.)

Despite these different tendencies, the overall results of this experiment dis-close clearly the validity of the basic demonstration we sought to produce: At least under some circumstances, a televised example can increase a child's willingness to engage in helping behavior. This finding extends the earlier work with simulated television materials and the experimental field study reported by Stein and Friedrich (1972) by demonstrating the influence of commercial broadcast programming designed primarily as entertainment. The present design also permitted us to isolate the particular aspect of the program, a prosocial example by the protagonist, as the central ingredient necessary for such an effect to occur; the alternative *Lassie* show and the generally warm *Brady Bunch* program produce significantly less helping. Inasmuch as the former control program featured the same major characters, setting, and general dramatic style as the prosocial *Lassie* show, it is unlikely that factors other than the specific modeled example influenced helping. This conclusion is further bolstered by the comparison of helping responses be-tween children who saw either the neutral *Lassie* or *Brady Bunch* shows; the mere presence of a canine hero in the former did not itself significantly facili-tate helping of puppies in distress. Theoretically, then, the present results

Table 22.1   MEANS [AND *SD*] OF HELPING SCORES IN SECONDS FOR ALL GROUPS

	SEX OF SUBJECT		
	FEMALE	MALE	COMBINED
Prosocial *Lassie*	105.48	79.80	92.64
	[20.37]	[47.38]	[36.94]
Neutral *Lassie*	34.80	68.20	51.50
	[46.30]	[53.36]	[50.28]
Brady Bunch	8.50	66.50	37.50
	[14.18]	[45.55]	[44.13]

support the social learning view of Bandura (1969, 1971) and Liebert (1972, 1973; Liebert, Neale, & Davidson, 1973) that the effects of television on behavior are mediated by specific modeling cues and the interpretation of these cues by the child, rather than general format and other global considerations. What is more, the dependent measure involved helping the experimenter as well as helping the puppies; to the extent that the children responded to this aspect of the situation, the results represent even more generalization from the specific altruistic behavior modeled in the prosocial show.

The practical implications of the present demonstration are also clear: It is possible to produce television programming that features action and adventure, appeals to child and family audiences, and still has a salutary rather than negative social influence on observers. A detailed understanding of the modeling processes involved in such influences may contribute to the production of other socially desirable programs in the future.

REFERENCES

Bandura, A. (1969). *Principles of behavior modification.* New York: Holt, Rinehart & Winston.

Bandura, A. (1971). Analysis of modeling processes. In A. Bandura (Ed.), *Psychological modeling.* New York: Aldine-Atherton.

Bryan, J. H. (1970). Children's reactions to helpers: Their money isn't where their mouths are. In J. Macauley & L. Berkowitz (Eds.), *Altruism and helping behavior.* New York: Academic Press.

Friedrich, L. K., & Stein, A. H. (1973). Aggressive and prosocial television programs and the natural behavior of preschool children. *Society for Research in Child Development, 38* (whole Monogr.).

Liebert, R. M. (1970). Television and social learning: Some relationships between viewing violence and behaving aggressively (overview). In J. P. Murray, E. A. Rubinstein, & G. A. Comstock (Eds.), *Television and social behavior. Vol. II: Television and social learning.* Washington, D.C.: U.S. Government Printing Office.

Liebert, R. M. (1973). Observational learning: Some social applications. In P. J. Elich (Ed.), *Fourth western symposium on learning.* Bellingham, Wash.: Western Washington State College.

Liebert, R. M., Neale, J. M., & Davidson, E. S. (1973). *The early window.* New York: Pergamon Press.

O'Connor, R. D. (1969). Modification of social withdrawal through symbolic modeling. *Journal of Applied Behavior Analysis, 2,* 15–22.

Stein, A. H., & Friedrich, L. K. (1973). Television content and young children's behavior. In J. P. Murray, E. A. Rubinstein, & G. A. Comstock (Eds.), *Television and social behavior. Vol. II: Television and social learning.* Washington, D.C.: U.S. Government Printing Office.

Stein, G. M., & Bryan, J. H. (1972). The effect of a television model upon rule adoption behavior of children. *Child Development, 43*, 268–273.

Wolf, T., & Cheyne, J. (1972). Persistence of effects of live behavioral, televised behavioral, and live verbal models on resistance to deviation. *Child Development, 43,* 1429–1436.

Yates, G. C. P. (1974). Influence of televised modeling and verbalisation on children's delay of gratification. *Journal of Experimental Child Psychology, 18,* 333–339.

This study was supported, in part, by grants from General Foods Corporation and General Mills, Inc. to Eli A. Rubinstein, John M. Neale, and the second and third authors for the investigation of television's prosocial influence. Grateful acknowledgment is due the Wrather Corporation for providing the *Lassie* programs, and to the principal, Mrs. Ann Littlefield, and staff of Smithtown Elementary School, Smithtown, New York, for valuable collaboration in this work. Special thanks also are due Elaine Brimer, Ann Covitz, Francine Hay, David Morgenstern, and Steven Schuetz for their capable assistance as experimenters.

ADDITIONAL QUESTIONS

1. What design principles are illustrated in this experiment? Is the sample potentially biased by this constraint? Explain.
2. Think of another experiment that would test some of the hypotheses or issues raised by this experiment.
3. Why did the experimenter get parental consent? When do you need the consent of parents or the subjects themselves before an experiment?
4. Do you think that the children believed that the situation was real or contrived? Explain.
5. The experiment involved deception in that the children were led to believe they were monitoring a dog kennel. Discuss the issue of deception in psychological research. Under what conditions is it permissible?
6. The experimenters apparently did not debrief the subjects by telling them the true nature of the experiment after it was over. When is debriefing necessary or desirable, and when can it be ignored?
7. Compare your answers to questions 3, 5, and 6 with the ethical principles in Chapter 8.

Chapter 23

Birdsong Learning

INTRODUCTION

Growing up in the rolling hills of Nebraska, I fondly remember the first time I learned about identifying birds through their songs. "A bobwhite can be easily recognized by its song because he says his name," my father told me before whistling a tune to imitate the sound of the bird, "bob-bob-white." I was convinced that this perfect mnemonic device was but one example of the divine harmony between verbal labels and nature's phenomena—in this case, the name of a bird and its song. But my early enthusiasm was quickly dashed when I tried to apply the same technique to the sound of a robin or a wren. And the complex melody of the meadowlark was nothing like the word *meadowlark* (even though the "caw" of the blackbird did seem to fit that animal).

It is somewhat gratifying to know that researchers are also concerned with the problems of labeling birdsongs, and in this article the authors look at the identification of birdsongs by means of a clever technique. Richard Walk and Michael Schwartz study birdsong identification by using visual stimuli as mnemonic devices. You will find that the experiments deal with an interesting theoretical problem as well as illustrating a unique design problem.

This article was selected because it contains several important design features. As you study this experiment, pay particular attention to these elements: (1) This is a multiple-experiment study. The researchers present two separate experiments that address a common problem but deal with separate aspects of the problem. (2) The experimenters pretested subjects to establish a base rate of birdsong identification. (3) The design is primarily a random group design that uses a visual code group and a no-code group, but it also has elements of a repeated measure design in that two measures were taken on both groups. (4) The experimenters report the loss of detailed records for some subjects. Notice how they handle this problem.

Special Issues

Multiple Experiments

In the past several years psychological research has become increasingly complex. Because of the expanded nature of the subject, researchers often undertake studies that examine several aspects of a problem. Some APA journals encourage multiple-experiment manuscripts, a fact that even a casual reading of the current literature in psychology will confirm.

Multiple-experiment articles have several strengths. In these articles, it is possible to approach a psychological issue from various experimental viewpoints, with each single experiment providing a portion of the answer. Several experiments may reveal a more complete picture of the issue. Another aspect of multiple-experiment studies is that "programmatic" research can be reported in a single experiment. Many contemporary researchers develop a long-range, progressive research scheme. This scheme may involve a series of experiments in which later experiments are predicated on the outcome of earlier experiments. This type of research is particularly amenable to multiple-experiment articles.

Birdsong Learning and Intersensory Processing

RICHARD D. WALK
George Washington University, Washington, D.C.

MICHAEL L. SCHWARTZ
Yale University School of Medicine, New Haven, Connecticut

Two experiments were performed in which subjects learned to attach names to birdsongs. In the first experiment, subjects who were instructed to generate their own visual codes were much superior to those not given any instructions except those of learning the birdsongs. In the second experiment, both those given a model code for half of the birdsongs and those who made their own visual codes were superior to controls without visual codes. The experiments show the way in which learning in the auditory modality can benefit from visual symbols.

Eleanor Gibson, in her influential book *Perceptual Learning and Development,* reports that W. H. Thorpe used visual spectrograms to help his students learn to identify birdsongs (Gibson, 1969, p. 211). This is an example of the use of a visual model or crutch to help to learn a complex auditory pattern. **It has implications for intersensory processing, the way in which one modality can influence learning in another one.** Birdsongs are readily available in our environment and on commercial phonograph records, yet few of us can identify with certainty any bird sounds except the most obvious, such as the caw of the crow or the cock-a-doodle-doo of the rooster. But these common bird sounds are not songs. Byron once characterized, with poetic insight, a birdsong as "the sweetest song ear ever heard."

a

The present series of experiments concerned the learning of birdsongs. The first experiment investigated the effect on such learning of having the subjects

SOURCE: Reprinted from *Bulletin of the Psychonomic Society,* 1982, *19,* 101–104, by permission of the Psychonomic Society and the authors. The original article has been abridged.

ANALYSIS
REVIEW OF LITERATURE

The review of literature is unusually brief and proselike. (The authors even quote a poet!) However, the essential qualities are represented: the basic theory that motivates the study (**a**), the problem (**b**), and the hypothesis (**c**), the latter two in the form of a question.

construct their own visual code of the birdsong as compared with those given instructions just to learn the songs. The second experiment had an additional group that had model visual patterns of half of the birdsongs.

b
c   **Do the visual codes help learning? What are the theoretical implications of such auditory-visual studies?**

GENERAL METHOD

Stimulus Materials

Birdsongs

A sample of 10 common birdsongs was selected from a record that contained a large number of birdsongs. The songs chosen were: Baltimore oriole, cardinal, common goldfinch, mockingbird, purple finch, robin, song sparrow, towhee, tufted titmouse, and winter wren.

d   **The songs were placed in random order on a cassette tape.** A total of six random orders, one for each phase, was constructed as follows: Phase 1 (pretest)—birdsongs followed by 8-sec pause; Phases 2–5 (training)—birdsongs identified at end of 8-sec pause; Phase 6 (posttest)—birdsongs followed by 8-sec pause. Each phase played all 10 birdsongs, with the trial identified for a phase so it could be matched with the answer sheet, for example: *"Trial one.* Birdsong [e.g., cardinal]. *Trial two.* Birdsong [e.g., winter wren] . . . ,"* and so on. With six phases and 10 trials for each phase, the experiment had a total of 60 trials. In the training section, as described, the name of the song came at the end of the 8 sec, whereas no feedback was given for pretest and posttest. The tape lasted about 25 min because of the taped instructions, the birdsongs, pauses for writing the answers, and 30-sec pauses between

e   phases.[1]

GENERAL METHOD

The general method section describes the stimulus material and the procedure used in both experiments.

Then the birdsongs used in the experiment are identified, and the authors tell us that the songs were presented in six random orders (**d**). Presumably this was done to guard against any systematic errors due to presentation order. (We know that the first and last items of a series are learned better than the middle items; that effect may interact with the independent variable.) The exact makeup of the stimulus material is then disclosed. The authors describe in a reference note (**e**) the specific phonograph record used in obtaining the stimulus material.

The procedure used in both experiments is described in (**f**).

Procedure

f

All subjects were told that they would hear a tape with birdsongs on it and that they would be asked to match the songs they heard with the names of the birds listed on the response sheet. The taped instructions reiterated this. On the pretest, the taped instructions said that no identification would be given of a song, but it asked that the subjects guess on every trial. The training phases (Phases 2–5) told the subjects to guess before the song was identified at the end of 8 sec. The posttest instructions told the subjects the birdsongs would not be identified and asked them to answer every trial.

EXPERIMENT 1

In this experiment, some of the subjects were asked to construct their own visual code to help identify the birdsongs.

Method

Subjects

The subjects were 87 students in a laboratory course in psychology at George Washington University. They were from five different laboratory sections.

Procedure

Two laboratory sections were given no additional instructions. The other three sections were told that a way to assist learning would be to take notes on the characteristics of each birdsong. For example, high and low marks might illustrate high and low notes in the songs.

Analysis

g

Not all subjects requested to use notes actually did so. Those requested to use notes, but who used none at all, compose a second "no-code" group, referred to in the results as "No-Code 2," meaning a second group of subjects without a visual code.

EXPERIMENT 1

After the general method the experimenters turn to the specific method used in Experiment 1. This section is easy to understand and requires no comment except for the analysis section (**g**). This section indicates that not all subjects used techniques suggested by the experimenters. Nevertheless, data generated by these subjects are of interest and Walk and Schwartz tell how these data were analyzed.

RESULTS

The results in terms of accuracy by group are shown in Table 23.1. An analysis of variance showed that the groups did not differ on the pretest; all were at the chance level of about 10% correct. On the posttest, however, all groups were more accurate. Pretest-posttest difference scores showed that both the first no-code group ($p < .01$) and the second no-code group ($p < .05$) had significantly improved scores and the visual code group was markedly improved ($p < .001$). An analysis of variance of posttest scores of the three groups found a significant group effect [$F(2,84) = 17.13$, $p < .01$], and the student's t range statistic demonstrated that the visual code group was significantly superior ($p < .01$) to the other two groups.

The performance of the three groups on each individual birdsong is shown in Table 23.2. The use of a visual code helped performance on all songs, except that the superiority of the visual code group was slight for the common goldfinch. The use of a visual code particularly helped in discriminating the songs of the mockingbird, robin, towhee, and tufted titmouse. The high performance of the visual code group on the Baltimore oriole, cardinal, towhee, and winter wren should also be noted. Of course, contextual factors, such as the place of a song on the list and the confusability of competing songs, is a factor in the task.

Table 23.1   AVERAGE PERCENT CORRECT IDENTIFICATIONS FOR THE THREE GROUPS IN EXPERIMENT 1

	GROUP		
TIME	NO-CODE 1	NO-CODE 2	VISUAL CODE
Pretest	9	11	10
Posttest	24	25	53
N	38	17	32

RESULTS

The results of the first experiment are clearly presented. The authors use a standard statistical test (an analysis of variance) and report the level of significance. The data are also shown in two tables: The first contains the average correct identifications for the three groups and the second contains the percentage of correct identifications for each birdsong. The authors comment on the degree of improvement for some species (**h**). The section concludes with some brief comments, but the major discussion is postponed until the results of Experiment 2 are presented.

To conclude Experiment 1, the use of a visual code markedly improved performance on this auditory task. But the visual code must be used. Those given instructions to use the code who ignored the instructions performed no better than the control group that did not receive such instructions. Further discussion of these results is deferred until after Experiment 2 is presented.

EXPERIMENT 2

Method

Subjects

The main subjects were 37 students from three laboratory sections. Detailed records were lost for an additional 34 subjects from two laboratory sections.

Procedure

Two of the laboratory sections had the same procedure as Experiment 1. The third laboratory section had a model code pattern made for five of the birdsongs (see Figure 23.1). These birdsongs were chosen to have equal performance on Experiment 1 by those who used a visual code. The songs chosen,

Table 23.2 AVERAGE PERCENT CORRECT IDENTIFICATIONS ON THE POSTTEST FOR EACH BIRDSONG, BY GROUP, IN EXPERIMENT 1

| | GROUP | | |
| | NO-
CODE 1 | NO-
CODE 2 | VISUAL
CODE |
BIRDSONG			
Baltimore Oriole	37	29	63
Cardinal	47	47	81
Common Goldfinch	18	6	25
Mockingbird	11	12	47
Purple Finch	13	12	34
Robin	24	0	47
Song Sparrow	24	47	56
Towhee	21	35	72
Tufted Titmouse	11	24	44
Winter Wren	32	35	66

EXPERIMENT 2

In this article two experiments are presented—a practice that is becoming increasingly common in the psychological literature.

The method section does not repeat the information in the first experiment but adds information that is specific to the new procedure. Because the authors could not describe the model given to the subjects, they present a sample of the model in Figure 23.1.

mockingbird, robin, song sparrow, towhee, and tufted titmouse, were correctly identified 52.8% of the time on the posttest by the visual code group of Experiment 1. The other birdsongs had 53.4% correct identifications on the posttest by the visual code group.

The modeled birdsong group was given a dittoed sheet with models of the songs (Figure 23.1). These subjects were not allowed to turn the sheet over until after the pretest phase of the experiment.

Analysis

One laboratory section ($n = 18$) was a modeled code group; the other ($n = 16$) was a group that constructed its own visual codes. All subjects requested to use a visual code actually did so.

RESULTS

The main results are the three sections for which complete data were available. The other two sections will serve to fill in and help give perspective to the experiment.

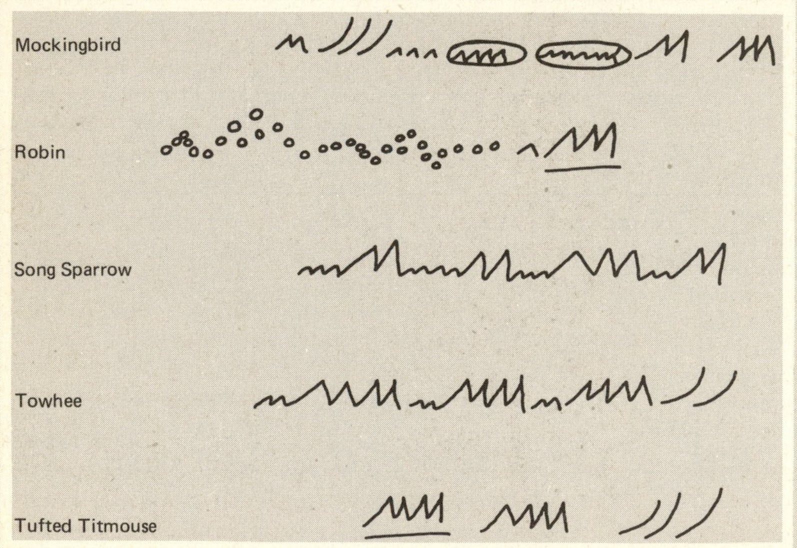

Figure 23.1 The model or pattern furnished subjects in the model code group of Experiment 2.

Experiment 2 raises several issues. The first is the loss of detailed records for part of the subjects. One may argue that a professional article should reflect the purest techniques, so if detailed records were lost then another group should have been used to obtain a set of complete records. But consider how the experimenter

The results are shown in Table 23.3. They are very comparable to those of the first experiment. Pretest scores were around the chance level of 10%; the no-code group had 24% on the posttest and the visual code group had over 50%. The subjects furnished model patterns of the birdsongs were similar in their performance to those who made their own visual code.

Statistical analysis confirmed this. The groups did not differ on the pretest. A posttest showed a significant difference [$F(2,34) = 8.61$, $p < .01$] among groups, and the two code groups were significantly superior on the student's t range statistic ($p < .01$) to the group not instructed to use a visual code.

Two further questions are relevant to this analysis. Are those with a pattern model superior on the modeled birdsongs to those who construct their own visual code? Does the use of a patterned model for half of the songs generalize to improve performance on the other songs?

This is also shown in Table 23.3. Both the modeled group and the own-code group were superior to the no-code group on the five modeled songs [$F(2,34) = 7.13$, $p < .01$]. The two code groups were significantly different from the no-code group but not from each other. The answer to the first question is negative. The second question is more complex. On songs without a model, the analysis of variance among groups was again significant [$F(2,34) = 6.05$, $p < .01$], but the significant difference in the means was for the own-code group compared with the no-code group, and the difference among means

Table 23.3 AVERAGE PERCENT CORRECT IDENTIFICATIONS FOR EACH GROUP IN EXPERIMENT 2, SHOWING POSTTEST PERFORMANCE ON MODELED AND NONMODELED SONGS

			POSTTEST		
GROUP	N	PRETEST	OVERALL	MODEL SONGS°	NO MODEL
No Code	14	10	24	21	26
Own Code	11	7	59	55	64
Model Code	12	11	50	57	43

°Only the "model code" group actually had a model or pattern of the five birdsongs.

used the data generated by the group whose records were lost (**i**). These data were not considered to be the primary findings but were presented as supplementary information. Although the authors went ahead and used the data cautiously, we suggest that your research be restricted to groups on which complete records are maintained. There may be some exceptions to this rule, as in the case of field-based experiments where a nonrepeatable situation occurs (e.g., a death of a national figure, a riot, an accident). In these situations the capricious nature of the event and the subject sample may mean that some data and/or characteristics of the sample may be unavailable or lost.

almost showed the own-code group (64%) significantly superior to the patterned model group (43%). The modeled song group seemed somewhat superior on nonmodeled songs to the no-code group (43% vs. 26% accuracy), but the difference was not significant and the trend may have been due to an artifact. The artifact is that the song models raise the chance level on the nonmodeled songs. Thus, to answer the second question, we have no evidence that the use of models generalizes (perhaps by increasing attention to identifying features) to improve performance on songs without a model.

i Results from those for whom statistical analysis could not be performed were similar. The own-code subjects had 9% correct on the pretest and 47% on the posttest, with 40% correct on the other group's modeled songs and 54% on the nonmodeled songs. The modeled group had 11% correct on the pretest and 46% on the posttest, with 54% accuracy on the songs with a model and 37% on the songs without one. Compared with the controls, with 26% accuracy, the 37% correct on the songs without a model for the modeled group would seem to show that the modeled experience does not generalize to help performance on the nonmodeled songs. The slight superiority could be an artifact.

Thus, those for whom no analysis could be performed reinforced the conclusions from the other subjects. The modeled songs are a definite aid to learning, but the use of the models does not seem to generalize beyond them.

DISCUSSION

j These experiments show that visual codes, whether generated by the subjects themselves or from models provided by the experimenters, help subjects to learn auditory songs. The results seem to be fairly specific: Those subjects instructed to use a code who used none did not benefit, and learning of the nonmodeled songs in Experiment 2 was not helped by the models provided for some of the songs. Explanation of the effect in terms of the rehearsal of, participation in, or attention to the task provided by the suggestion of or presence of the codes does not seem to be appropriate because of the specificity. We seem to have specific intersensory facilitation based on vision.

DISCUSSION

In section (j) the authors repeat the major findings of their two experiments and provide an abbreviated interpretation of their data. The remainder of the discussion section is devoted to a description of their results and the research of others. They also discuss the theoretical implication of their experiments. The discussion section ends with several questions that remain unanswered (k) and may serve as the springboard for further experimental work.

Birdsongs are hard to describe in words, but complex visual patterns can help in their identification. Neither the subjects' own codes or the models shown in Figure 23.1 are obvious sound spectrograms, but they do contain inter-modal elements that aid learning. Some of the self-generated visual codes from Experiment 1 are reproduced in Walk (1978). They have few obvious intersubject similarities, and yet they did aid learning.

The broader theoretical question concerns high-level similarities and differences in the coding of different modalities. One thinks of Gibson's (1969) concept of "higher order invariances" or similarities, such as sharpness or smoothness, that interconnect modalities. Even 6- to 8-month-old infants can identify as similar or different patterns in the visual and auditory modalities (Allen, Walker, Symonds, & Marcell, 1977).

Birdsongs have no obvious matching codes like those the culture has provided for reading and music. One could teach oneself musical notation, of course. Sound spectrograms may help, but our subjects did not need them.

We need more research on how the modalities are related to each other. Where should the present research lead? Is it best to ask what "distinctive features" best interconnect the two modalities and how we can characterize the intersensory principles of perceptual organization? Or will any symbol help? We doubt that. If we named the birdsongs (the "squeaky one," "the busy one," etc.), we might get just as much facilitation as we do with visual models, but we could not randomly connect the names (or the models) to the songs.

QUESTIONS

1. How were subject responses gathered? Suggest an alternative way that may be more naturalistic.
2. Suggest and design an experiment that uses verbal labels, rather than visual models, as a mnemonic device in birdsong learning.
3. Why did the authors divide the subjects who were instructed to use the notes (Experiment 1) into two groups: those who actually used the notes and those who did not? Why did they use these two groups in their statistical analysis?
4. In Experiment 2, the authors compare some models that were given to subjects as a means of encoding the birdsongs. What are the implications for these subjectively drawn models, especially with regard to developing comparative experiments?
5. What purpose did the pretest serve (Experiment 1)? Was it necessary? Why or why not?
6. Comment on the loss of subjects' data.

REFERENCE NOTE

1. Birdsongs were taken from the phonograph record "A field guide to bird songs of Eastern and Central North America" (revised) (HM-4671). Boston: Houghton Mifflin, 1971. A copy of the tape used in this experiment can be obtained from the first author at cost.

REFERENCES

Allen, T. W., Walker, K., Symonds, L., & Marcell, M. (1977). Intrasensory and intersensory perception of temporal sequences during infancy. *Developmental Psychology, 13*, 225–229.

Gibson, E. J. (1969). *Principles of perceptual learning and development.* New York: Appleton-Century-Crofts.

Walk, R. D. (1978). Perceptual learning. In E. C. Carterette & M. P. Friedman (Eds.), *Handbook of perception (Vol. 9). Perceptual processing.* New York: Academic Press.

Reprints may be obtained from Richard D. Walk, Department of Psychology, George Washington University, Washington, D.C. 20052.

Chapter 24

Note Taking

INTRODUCTION

The second most common activity of college students is probably note taking. In courses such as physics, anatomy, philosophy, sociology, history, biology, English, football, and anthropology, students and notebooks are ubiquitously entwined. And woe to the unfortunate student who misplaces his or her treasured notes before an examination. But does note taking help recall? This very practical question has not escaped the eye of educational psychologists, who often have investigated the question. This experiment by Judith Fisher and Mary Harris is an example of how a very practical educational problem can be brought under experimental control. Fisher and Harris ask: "Is note taking beneficial for the student, and if so, does the process of note taking itself aid the student in recalling the material or does having an external record of the lecture to refer to afterward aid the recall?" Of particular interest in this experiment are the conflicting results regarding the effect of note taking on recall (**a**), with some experiments reporting a significantly lower recall for note takers than for students who did not take notes.

ATTRIBUTES OF FIELD EXPERIMENTS

Field experiments generally lack the control of laboratory experiments, but they are not devoid of control. In many experiments (for example, this article) the researcher can specify the nature of the subjects, the independent variable, the dependent variable, and many situational stimuli that may influence the results. Before considering some specific issues raised by this article, a few of the positive attributes of the field techniques is that they can be conducted in the subject's natural setting, with little disturbance of his or her normal behavior. This tends to enhance the credibility of the study. In some forms of field research, the subject is unaware that he or she is a subject. For example, a social psychologist may be interested in comparing the ratio of Perot to Clinton bumper stickers on a college campus and in a church parking lot. The subject is not even present—except by proxy.

Special Issues

Subject Variables

Fisher and Harris are mindful of the possible influence of gender on their results and test its influence **(k).** Although they found a gender influence (males did more poorly than females), gender did not significantly interact with the experimental conditions. These results are in themselves interesting, and, as the researchers are quick to point out in their discussion **(p),** further investigation into the effect is warranted. These results contradict other experiments that show that subjects do better on learning-memory tasks when the experimenter is of the opposite sex. The gender effect in this experiment may be only slightly affected by the lecturer being a woman and more greatly affected by the sample of male students who were enrolled in a course in human growth and development.

Effect of Note Taking and Review on Recall

JUDITH L. FISHER AND MARY B. HARRIS
UNIVERSITY OF NEW MEXICO

Male and female college students were randomly assigned to five treatment groups combining different note-taking and review combinations. Recall was measured immediately and three weeks later. The results showed that a combination of taking notes and reviewing one's own notes produced the most recall, while not taking notes and reviewing the lecture "mentally" produced the least recall. Females recalled significantly more data than males, but opinions concerning note taking and efficiency of notes were not related to recall outcome. Quality of notes was positively correlated with free-recall score for two of the three note-taking groups and with short-term objective test score for two of the three note-taking groups.

Although the practice of taking notes and subsequent review of those notes for study purposes is widespread, the experimental evidence with respect to the efficacy of note taking is not clear. DiVesta and Gray (1972) have shown that note taking leads to an increase in the number of ideas recalled from a prose passage. Crawford (1925b) found that note-takers had significantly higher immediate and long-term recall scores than non–note-takers, and Howe (1970b) showed that the probability of recalling an item that occurred in the subject's notes was about seven times that of items not in his notes. **Other studies, however, have shown no significant differences between different**
a   **note-taking strategies and not taking notes on a measure of recall** (Howe, 1970a; McClendon, 1958) **and even significantly lower recall scores for note-takers than non–note-takers during a film presentation** (Ash & Carlton, 1953).

SOURCE: Reprinted with permission from *Journal of Educational Psychology*, 1973, 65, 321–325.

ANALYSIS
REVIEW OF LITERATURE

In the review of literature **(a)** the authors point out conflicting results and the problem.

METHOD

This study tries to discover if note taking and review of notes aid students in recall, investigate the nature of note taking (encoding versus an external memory

The experimental effects of review on recall are more definitive. Howe (1970a) found that subjects who were allowed to review notes that they had taken had significantly higher mean recall scores than subjects who attended to an interfering task. DiVesta and Gray (1972) found that rehearsal immediately after listening increased the number of words recalled.

The functions of note taking have been discussed in various publications. Notes have been described as serving as an "external memory device" (Miller, Galanter, & Pribram, 1960) for storing data for later retrieval and study. Note taking has also been described as a kind of "encoding" process, in which the learner reorganizes the input data, and by putting it in his own words, in a sense, transforms the data and "makes it his own" (DiVesta & Gray, 1972; Howe, 1970b).

DiVesta and Gray (1972) as well as Howe (1970a) see the second or "encoding" function of note taking as the more important of the two functions in terms of learning. DiVesta and Gray in fact suggest that the external memory function of note taking may possibly lead the learner to feel that having a "good set of notes" is the equivalent of studying and influence him to bypass any form of transformational coding. Howe (1970a) also discusses the importance of the second function of note taking and points out that if the external memory function were the only reason for taking notes, "It would be simpler to provide mimeographed outlines leaving the student free to attend and react to other things. . . . (p. 100)."

b **The present study sought to discover if note taking and review of notes do aid students in short-term and long-term recall of material presented in a typical lecture setting.** It further investigated whether note taking serves as an encoding device, an external memory device, or a combination of the two, by manipulating the various review conditions of the study. It also attempted to assess whether there was a relationship between students' opinions about taking notes (liking, disliking, etc.) and their subsequent recall of material and if there were any significant correlations between quality of notes taken and subsequent recall, as well as efficiency of notes taken and subsequent recall.

c **It was predicted that subjects who took notes, regardless of review condition, would perform better than subjects who did not take notes (the "encoding" function) and that those who had notes to review (either their own**

function), and assess the relationship between a student's attitude toward note taking and recall. In (b) the experimental plan is clearly identified. In most of the experiments reviewed in this book the authors wisely add a sentence or two in the introduction that sharply delineate the experimental plan. That technique helps the reader understand the experiment.

In (c) the hypothesis is stated in the form of a prediction—an acceptable

or the lecturer's notes) would perform better than those with no notes to
review (the "external memory device" function). In addition, it was predicted that those who took notes but did not review them would perform

d   better than those who took no notes but reviewed the lecturer's notes, if the
encoding process is really the more important determinant of learning, as
DiVesta and Gray (1972) and Howe (1970b) postulate.

METHOD

Design

e   The experiment consisted of **three phases: a 40-minute period during which the
subjects heard a lecture, a 10-minute review period, and a 30-minute testing
period.** A 15-minute posttest was also given approximately three weeks later. During the lecture phase, a random 60% of the students were instructed to take notes
on the lecture being presented, while the other 40% were instructed not to take any
notes. Each of these two groups was further subdivided during the 10-minute review period, when subjects were instructed either to review the lecture "mentally," to review the notes they had taken, or to review a set of the lecturer's notes,
which was provided. This design resulted in five groups or treatment conditions:

f   (a) took no notes—reviewed the lecturer's notes (NN-RLN), (b) took no notes—
mental review (NN-MR), (c) took notes—mental review (N-MR), (d) took notes—
reviewed own notes (N-RON), and (e) took notes—reviewed the lecturer's notes
(N-RLN).

Subjects

g   The subjects were **83 female and 29 male college students,** 19 to 51 years of age,
in two Human Growth and Development classes. Students in each classroom were
randomly assigned to the five conditions. **The lecturer was a female graduate
student in education.**

Materials

h   A 40-minute lecture on personality testing containing 1750 words was delivered to
each class by the experimenter. **The topic was chosen for its relevance to the class**

alternative to a formal statement of the hypothesis. A second hypothesis (**d**) is made
regarding the effect on recall of note taking without a review of the notes. In this
experiment the authors hypothesized that the subjects assigned to the note-taking
conditions would perform better on recall tasks than the subjects who did not take
notes and, further, that the subjects allowed to review their notes would do better
than subjects who did not review. The experimenters also predicted that, if encod-

curriculum. Every effort was made to keep experimental conditions as close to actual classroom conditions as possible.

There were five types of packets compiled according to experimental treatment condition. Each packet consisted of a cover sheet, which contained a questionnaire concerning note-taking habits, and instructions to either take or not take notes. Paper was provided for those in the three note-taking groups. Later pages instructed the subject about review procedures, providing subjects in the NN-RLN and N-RLN conditions with a copy of the lecturer's notes. The last three pages of each packet consisted of a free-recall page where the subject was instructed to put down all that he could remember of the lecture and a two-page objective test consisting of 15 multiple-choice and four short-answer items. These packets were put in random order before being distributed to the classes. The posttest consisted of nine multiple-choice items, which had been asked on the short-term objective test, four new multiple-choice items, and one three-part short-answer question; assignment of questions to the posttest and short-term objective test was done randomly.

ing was indeed important, those who took notes but could not review them would do better than those who did not take notes but who were allowed to review the lecturer's notes. The experimental design can be conceptualized as follows:

	REVIEW CONDITION		
Note-takers	Review lecturer's notes	Mental review	Review own notes
Non–note-takers	Review lecturer's notes	Mental review	—

In (e) the time sequence of the test session is specified. Subjects were assigned to five groups (f), which are shown below:

Note Condition	Review Condition	Subject Assignment			
	Review lecturer's notes	S_{A1}	S_{A2}	$S_{A3} \ldots$	S_{An}
Non–note-takers	Mental review	S_{B1}	S_{B2}	$S_{B3} \ldots$	S_{Bn}
	Review lecturer's notes	S_{C1}	S_{C2}	$S_{C3} \ldots$	S_{Cn}
Note-takers	Mental review	S_{D1}	S_{D2}	$S_{D3} \ldots$	S_{Dn}
	Review own notes	S_{E1}	S_{E2}	$S_{E3} \ldots$	S_{En}

In (g) Fisher and Harris tell us that the lecturer was a female graduate student and that female subjects outnumbered male subjects by nearly 3 to 1; we will discuss this more later.

In the description of materials (h) the authors state that "every effort was made to keep experimental conditions as close to actual classroom conditions as

Procedure

After the subjects assumed their regular seats, the packets were distributed. Subjects were instructed to fill out the questionnaire and read the instructions at the bottom of the page. These instructed the subject either to take notes on the following blank pages or to listen to the lecture without taking notes or turning the first page of the packet. The lecture was then delivered. At the conclusion of the lecture, the subjects were told to turn to page A and follow the instructions. The instructions on this page varied with experimental treatment, so that subjects in the N-MR and NN-MR groups were instructed to review mentally what they heard in the lecture. Subjects in the N-RON group were instructed to review the notes they had taken, while those in the N-RLN and NN-RLN groups were instructed to review the set of the lecturer's notes, which was provided in the packet. At the end of 10 minutes, the subjects were told to turn to page B, where they were instructed to write down as much of the lecture as they could remember. They were also instructed to continue to the last two pages when they had completed the free-recall section. These last two pages contained multiple-choice and short-answer questions. At the end of 30 minutes all papers which had not been turned in were collected. No mention of a posttest was made. Three weeks after the lecture was given, the experimenter brought the posttest of long-term recall to class and gave it to those subjects who were present and who had heard the original lecture. Since this was the last week of classes of the regular year, only 71 of the original subjects were present to take the posttest.

possible." Thus, the experimenters hoped to minimize the possible artificial influence of a laboratory setting.

Each subject received a packet that contained everything needed in the initial part of the experiment. It is especially important to note that the dependent variable in this experiment was evaluated by several tests: a free-recall test, a multiple-choice test, and a short-answer test. The posttest conditions also contained several types of tests. These multiple measures are analyzed separately in the results section. In an effort to specify direct cause-and-effect relationships between experimental conditions and results, it is often necessary to make complex measures of the subjects' responses. Such was the case in this experiment. Note taking may influence test results, but Fisher and Harris go beyond this to find out exactly what type of test results are influenced most.

The experimenters were careful to set up an objective scoring system so that a reliable score could be assigned to the ambiguous data.

You should also note that the free-recall scores and the objective-test scores were highly correlated, which would suggest that both tests are a measure of similar information or abilities.

The results were measured in three dependent measures (i). Prior to the main statistical analysis, the authors ran a test for homogeneity of data (j). Had the data not been homogeneous, it would have required special treatment.

RESULTS

i **The three basic dependent measures were those of short-term free-recall scores short-term objective test scores, and long-term objective posttest scores.** Since t tests revealed no significant differences between the two classes on any of the measures and since such differences would not have been of theoretical interest anyway, data from the two classes were pooled in all further analyses. Hartley's F_{max} tests revealed that variances of scores on

j both the free-recall test ($F_{max} = 3.29$, $df = 5/25$) and the short-term objective test ($F_{max} = 1.09$, $df = 5/25$) were homogeneous. The 5×2 unweighted means analyses of variance of treatment and sex **revealed significant treatment effects and significant sex differences, indicating that while women in all treatment conditions showed a higher mean score on the free-recall test ($\bar{X}_F =$**

k **18.58, $\bar{X}_M = 12.17$; $F = 7.295$, $df = 1/102$, $p < .01$) and short-term objective test ($\bar{X}_F = 15.19$, $\bar{X}_M = 11.56$; $F = 9.850$, $df = 1/102$, $p < .01$) than did men, no sex by treatment interactions approaching statistical significance were found.** Women did not score significantly higher than men on the posttest ($F = 3.267$, $df = 1/61$). Therefore, in order to increase the n in each group, data from men and women were pooled in all further analyses.

Scores on the short-term free-recall and objective tests were significantly positively correlated ($r = .70$, $p < .01$), as were the free-recall and posttest scores ($r = .66$, $p < .01$) and the short-term objective test and posttest scores ($r = .75$, $p < .01$).

The free-recall tests were scored by giving 1 point for each idea recalled from the lecture, according to a master list of the 101 points contained in the lecture, 91 of which were also contained in the lecturer's notes given to the subjects for review. All tests were scored by one judge who was unaware of the subject's experimental condition. A second judge was then trained in the scoring procedure using 20 protocols, and then both raters scored a second 20 randomly selected protocols, producing an interscorer reliability coeffi-

l cient of .98. The mean scores for the five experimental groups on the free-recall test are presented in Table 24.1. **A one-way analysis of variance revealed a significant treatment effect ($F = 3.06$, $df = 4/107$, $p < .05$), and a post**

m hoc comparison using **Scheffé's method indicated that the groups could be ordered in the following fashion:** N-RON (weight $= +3$) > NN-RLN ($+2$) > N-RLN ($+1$) > N-MR (-1) > NN-MR (-5) ($F = 11.95$, $p < .05$).

The mean scores for the short-term objective test are also presented in Table

n 24.1. A one-way analysis of variance revealed a significant treatment effect

In (**k**) the authors are mindful of gender differences.

A principal finding is in (**l**), and then the data are further processed by means of Scheffé's test (**m**), which specifically orders the significant groups from each other. The authors use this procedure again for the short-term objective test (**n**).

(F = 2.89, df = 4/107, p < .05), and a post hoc comparison using Scheffé's method indicated that the groups could be ranked in the following order: N-RON (weight = +3) > NN-RLN (+2) > N-RLN (−1) = N-MR (−1) > NN-MR (−3) (F = 11.18, p < .05).

The mean scores for the posttest are also presented in Table 24.1. Although they are in approximately the same order as the other two measures, and of course not independent of them, an analysis of variance showed no significant differences among them (F = 1.84, df = 4/64).

To determine if there was a relationship between subjects' opinions about note taking and subsequent recall, the answers to the question "How do you feel about taking notes?" on the first page of the packet were analyzed. The three "no comment" remarks received were discarded, and the remainder were classified as either negative remarks or positive (wholly or conditionally) remarks. The t tests between these two groups yielded no significant differences on any of the three dependent measures, the free-recall test ($\bar{X}_N$ = 14.39, n = 31, $\bar{X}_P$ = 17.87, n = 78; t = 1.707, df = 107), the objective test ($\bar{X}_N$ = 12.71, n = 31, $\bar{X}_P$ = 14.78, n = 78; t = 1.54, df = 107), or the posttest measure ($\bar{X}_N$ = 10.05, n = 20, $\bar{X}_P$ = 10.47, n = 49; t = .24, df = 67).

The quality of the notes taken was defined as the number of ideas from the lecture that were included in the notes. A scorer, blind as to the subject's experimental condition, scored all the notes, and a randomly selected 20 sets were scored by a second blind judge. The correlation between the two sets of

Table 24.1   MEAN SCORES OF EXPERIMENTAL GROUPS

DEPENDENT MEASURE	EXPERIMENTAL GROUP				
	N-RON	NN-RLN	N-RLN	N-MR	NN-MR
Free recall	21.8	19.2	17.8	15.3	12.4
Short-term objective	17.5	15.9	13.7	13.7	11.5
Posttest	13.6	13.9	10.7	9.1	8.1

Note: Abbreviations: N-RON = took notes—reviewed own notes; NN-RLN = took no notes—reviewed the lecturer's notes; N-RLN = took notes—reviewed the lecturer's notes; N-MR = took notes—mental review; and NN-MR = took no notes—mental review.

The discussion section leads off with a succinct statement of what the experimenters found (**o**). Throughout the remainder of the discussion, the authors place that finding into a large conceptual framework and compare it to other empirical findings (e.g., **p**). The final conclusion (**q**) is direct and to the point.

ratings was .98. The efficiency of the notes was defined, as Howe (1970b) suggested, by dividing the quality of the notes (i.e., number of ideas) by the total number of words. The correlations of the quality and efficiency of notes with the objective and free-recall tests are presented in Table 24.2. Although all six correlations between quality of the notes taken and the other measures are positive, only the correlations with free recall for the N-RON and N-RLN groups and with objective scores for the N-RLN and N-MR groups are statistically significant. None of the correlations involving efficiency of note taking were significant.

Discussion

The results of the present study appear to indicate that **note taking serves both an encoding function and an external memory function, with the latter function being the more important.** The group showing the best performance, the N-RON group, was able to utilize both functions, whereas the group with the lowest recall scores, the NN-MR group, could neither encode the data nor use notes as an external storage mechanism. Subjects in the second highest group, the NN-RLN group, were able to use notes as a memory aid but were unable to encode the data directly; those in the second lowest group, the N-MR group, were able to take advantage of the encoding but not external storage function of note taking. It thus appears that of the two functions, the one of serving as an external memory device has the greater facilitating effect on recall.

The fact that subjects in the NN-RLN group performed better than those in the N-RLN group was not predicted, since it was thought that those in the latter group would benefit from both functions of note taking, whereas those in the former group would use notes only as an external memory storage device. A

Table 24.2 CORRELATIONS OF QUALITY AND EFFICIENCY OF NOTES WITH SHORT-TERM OBJECTIVE TEST AND FREE-RECALL SCORES

		QUALITY		EFFICIENCY	
GROUP	n	OBJECTIVE	FREE RECALL	OBJECTIVE	FREE RECALL
N-RON	20	.28	.53°	−.23	−.39
N-RLN	26	.42°	.51°°	.15	−.03
N-MR	25	.44°	.37	.14	.07

Note: Abbreviations: N-RON = took notes—reviewed own notes; N-RLN = took no notes—reviewed the lecturer's notes; and N-MR = took notes—mental review.
  °$p > .05$.
  °°$p > .01$.

possible explanation for this result is that the substitution of the lecturer's notes for the subject's own notes in some way interfered with the consolidation process, thus decreasing recall.

The analyses of the posttest data were not significant, although the means were in approximately the same direction as the short-term free-recall and objective test means, which suggests that the same processes were operating here. An experimental mortality rate of 38% might have affected significance, as might the fact that subjects did not expect to be retested on the material.

Since a significant relationship was not found between opinions about note taking and subsequent recall, it would appear from these data that liking, disliking, or other opinions connected with note taking do not have an effect on recall.

The correlations between quality or efficiency of notes and objective or free-recall data for the three note-taking groups indicated that subjects with notes of good quality generally recalled more data than those with notes of poor quality, but that efficiency of notes did not correlate significantly with other measures. Crawford too found that there was a significant positive correlation between the number of points recorded in his subjects' lecture notes and the number recalled at the time of a quiz he gave (Crawford, 1925a). The present data would appear to confirm this finding, but are in opposition to Howe's report of a significant positive correlation between efficiency of notes and a measure of free recall (Howe, 1970b).

A further finding of the present study was that females had significantly higher mean recall scores than males for all treatment conditions on both the short-term free-recall and the objective tests. This result is in line with the finding of Todd and Kessler (1971), who found that women recalled significantly more words than did men. **Further investigation into the reason for this effect is needed, in order to determine whether the subject matter, sex of the instructor, or a true sex difference in memory is responsible for the results.**

P

QUESTIONS

1. The authors note that the failure to find statistically significant differences between the experimental treatments in the three-week follow-up test may have been due to the 38% subject attrition rate. Why would this make a difference? Can you think of a way in which the authors could have eliminated this attrition?
2. Search the current literature for a subsequent article on note taking and recall and prepare a report on it.

q | In conclusion, the findings of this study indicate that in a typical college lecture setting, the taking of notes of good quality and the subsequent review of these notes are associated with more efficient recall.

REFERENCES

Ash, P., & Carlton, B. J. (1953). The value of note-taking during film learning. *British Journal of Educational Psychology, 23,* 121–125.

Crawford, C. C. (1925a). The correlation between lecture notes and quiz papers. *Journal of Educational Research, 12,* 282–291.

Crawford, C. C. (1925b). Some experimental studies on the results of college note-taking. *Journal of Educational Research, 12,* 379–386.

Deese, J. (1958). *The psychology of learning.* New York: McGraw-Hill.

DiVesta, F. J., & Gray, G. S. (1972). Listening and note taking. *Journal of Educational Psychology, 63,* 8–14.

Howe, M. J. A. (1970a). Note-taking strategy, review, and long term retention of verbal information. *Journal of Educational Research, 63,* 100.

Howe, M. J. A. (1970b). Using students' notes to examine the role of the individual learner in acquiring meaningful subject matter. *Journal of Educational Research, 64,* 61–63.

McClendon, P. (1958). An experimental study of the relationship between the note-taking practices and listening comprehension of college freshmen during expository lectures. *Speech Monographs, 25,* 222–228.

Miller, G. A., Galanter, E., & Pribram, K. (1960). *Plans and the structure of behavior.* New York: Holt, Rinehart and Winston.

Todd, W. B., & Kessler, C. C. (1971). Influence of response mode, sex, reading ability, and level of difficulty on four measures of recall of meaningful written material. *Journal of Educational Psychology, 62,* 229–234.

3. Design a factorial experiment on note taking and recall in which the sex of the subjects and lecturers are integral components.

4. Draw a graph using the data presented in Table 24.1.

5. Based on this research, what advice would you give a college student about his or her note taking?

6. The students who served as subjects in this experiment were part of a captive audience—apparently they had to participate in the study. Also, the study was apparently taken during class time. Do these procedures meet the ethical standards presented in Chapter 8?

7. What control did the experimenters use to be certain that the subjects followed the procedures of the condition to which they were assigned? Why are such controls necessary?

8. The experimenter distributed the entire packet of materials before the study started. What potential problems does this present? What controls are necessary? Why?

Chapter 25

Weight Loss

INTRODUCTION

Obesity is a major concern in America. Millions of Americans are overweight. The causes of this condition have been defined as either *endogenous*—originating from some abnormal condition within the body—or *exogenous*—stemming from the excessive ingestion of food. Most obese people fall with the latter category. People may eat too much for a variety of reasons: habit, social pressure, boredom, tension, fear, happiness, loneliness, . . . the list is seemingly endless.

One "cure" for obesity is simple: Eat less. Its execution, however, is nearly impossible for some people. Throughout our country, countless millions pursue the shapely, idealized form of a slender Hollywood starlet or a svelte athlete through fad diets, self-help courses, group therapy, fat farms, diet books, and a warehouse full of gymnastic paraphernalia, all promising the faithful user a trimmer physique. Fat is big business.

Psychologists have been interested in obesity for a long time. Experimental psychologists have studied obesity's origins and the factors that cause people to ingest more food than their body can metabolize. Practicing clinical psychologists have been more interested in therapeutic procedures that effectively reduce body weight and facilitating the client's capacity to cope with the causes of his or her problem through psychotherapy. The effectiveness of psychotherapeutic techniques is always difficult to evaluate because, in experimental terms, the dependent variable is often poorly defined. However, in research on the efficacy of weight-loss techniques, the obvious dependent variable is the number of pounds lost as an immediate result of therapy and as a long-term result of therapy.

Of all the therapeutic techniques developed, one has great intuitive appeal. It is the self-help strategy, which involves self-reward, self-punishment, or a combination of both. In spite of the wide use of these techniques in popular obesity clinics and clinical settings, the empirical evaluation of their efficacy for both the immediate term and the long term has been woefully lacking. This experiment by Mahoney, Moura, and Wade was selected because it effectively demonstrates how a topic in clinical psychology can be brought under experimental control. Note that the problems this experiment raises—for example, definitions of dependent and independent variables, and the subject sample—are significantly more difficult that those in other examples in this book. You should read carefully how the researchers attended to these factors and model your own experiments after this one.

SPECIAL ISSUES

Research in Clinical Psychology

Experimental research in psychology is flexible enough to study many different forms of behavior. Yet no matter how diverse the behavior being studied, most forms of experimentation follow a fundamental model: Subjects are treated to experimental variables, and the consequence of the treatment is evaluated. In brief, that is the method used by Mahoney, Moura, and Wade in their article on weight loss and weight-loss techniques. In this experiment the subjects were people who responded to a newspaper article. They were treated to experimental variables (weight-loss techniques), and the consequence (weight loss) was evaluated.

Although this experimental model is the same as that of most research reported in the psychological literature, this paper was selected because it represents a type of clinical research in which subject characteristics are an integral part of the design. The clinical nature of the design is also unique. Clinical research does not differ in form from other research in psychology, but it frequently differs in the nature of the subject, the treatment, and the dependent variable. Let's consider each of these topics in this experiment.

Subject Characteristics in Clinical Research. In some clinical research the subjects are, of course, selected from a markedly abnormal population. For example, an investigator may direct his or her efforts toward paranoid subjects, homicidal schizophrenic subjects, or manic-depressive subjects. No matter what the clinical symptoms of the subjects, however, care must be taken to define unambiguously their characteristics. And that very issue raises a substantial problem in clinical research. By definition, "abnormal" subjects differ from "normal" subjects. By using a diagnostic description of subjects, researchers introduce an additional variable to the experiment that may affect the reliability of the results—the definition of subjects. However, the researcher need not be thwarted because of definitional problems. He or she can reduce the variability introduced by subject diagnosis by carefully delineating the precise characteristics of the experimental subjects. It is of no use to define abnormal subjects as "schizophrenics" or "obese" or "phobic." A more precise definition is necessary. In the method section of this paper you will see an example of how *obesity* was carefully defined (percentage of body weight overweight, height/weight ratio). In addition, the researchers apply an eligibility criterion that further delineated the subject sample.

The Independent Variable in Clinical Research. A second characteristic of clinical research is the nature of the treatment, or independent variable. Sometimes this variable is similar to the independent variables in other research discussed in this book, such as when an experimenter studies the effect of a specific drug on behavior. Frequently, however, the independent variable in clinical research involves the introduction of a psychotherapeutic technique. Here again the researcher needs to worry about the ambiguity he or she may be introducing. To simply say that the subjects underwent psychotherapy not only belies the complexity of the psychotherapeutic process but also is a poor research technique. Therefore, great care also must be exercised in defining the type of psychotherapy. In this experiment, the researchers spell out in detail the psychotherapeutic methods applied to their subjects.

The Dependent Variable in Clinical Research. The final factor in our skeleton of design principles in clinical research is the dependent variable, or the subjects' response to the treatment. Clinical research frequently studies complex behavioral changes. When this is the case, the changes under investigation need to be clearly identified and evaluated. In this experiment, the dependent variable is weight loss, which is an uncomplicated factor. However, many clinical studies yield complex data that require sophisticated analysis. (See Chapter 21.)

Treatment research is obviously valuable in clinical psychology and in psychological theory. But it frequently raises definitional problems that must be solved for the research to be reliable. We selected this article because it represents an example of clinical research and its unique problems. The researchers demonstrate how the definitional problems in clinical research can be controlled by accurate and detailed delineation of the variables.

RELATIVE EFFICACY OF SELF-REWARD, SELF-PUNISHMENT, AND SELF-MONITORING TECHNIQUES FOR WEIGHT LOSS

MICHAEL J. MAHONEY, NANCY G. M. MOURA, AND TERRY C. WADE
STANFORD UNIVERSITY

Obese adults ($n = 53$) were randomly assigned to five groups: *(a)* self-reward, *(b)* self-punishment, *(c)* self-reward and self-punishment, *(d)* self-monitoring, and *(e)* information control. All *Ss* were given information on effective stimulus control techniques for weight loss. This constituted the sole treatment for control *Ss*. Self-monitoring *Ss* were asked to weigh in twice per week for four weeks and to record their daily weight and eating habits. Self-reward and self-punishment *Ss*, in addition to receiving self-monitoring instructions, were asked to reward or fine themselves a portion of their own deposit contingent on changes in their weight and eating habits. After four weeks of treatment, self-reward *Ss* lost significantly more weight than either self-monitoring or control *Ss*. At four-month follow-up, those *Ss* lost more weight. These findings are interpreted as providing a preliminary indication that self-reward strategies are superior to self-punitive and self-recording strategies in the modification of at least some habit patterns.

a | Despite the fact that self-control strategies have become increasingly popular in clinical and applied settings (Kanfer & Phillips, 1970), there has been an **appalling lack of research on the processes and parameters of these techniques** (Mahoney, 1974). In particular, there have been neither comparisons among the various techniques nor attempts to isolate their active components. **The present study addressed itself to an evaluative comparison of three of the more popular self-control techniques, self-reward, self-punishment, and self-monitoring.** Since the latter is also a component of the former, a partial component analysis was also provided.

SOURCE: Reprinted with permission from *Journal of Consulting and Clinical Psychology,* 1973, *40,* 404–407.

ANALYSIS
REVIEW OF LITERATURE

In the first part of the experiment **(a)** Mahoney et al. point out the "appalling lack of research" on weight loss techniques that use self-control strategies. This experiment compared three popular techniques. Section **(a)** contains a very abbreviated statement of the problem and research plan, which is augmented in the

To provide a stringent test of the efficacy of the above techniques, they were each applied in the modification of a chronic and resistant habit pattern. The pattern chosen was overeating since, in addition to providing observable indices of treatment effectiveness, obesity has been one of the most resistant of maladaptive habit patterns (Stunkard, 1958). Moreover, previous studies have demonstrated that a core of behavioral techniques can—when consistently applied—dramatically improve human weight control (Harris, 1969; Penick, Filion, Fox, & Stunkard, 1971; Stuart, 1967, 1971; Wollersheim, 1970). **In general, behavior modification approaches to obesity have emphasized the alteration of everyday eating habits by manipulation of eating-related environmental cues (i.e., stimulus control). To date, these techniques have been combined with a variety of auxiliary maintenance strategies such as group support, covert sensitization, and therapist approval.** The present study focused on the effectiveness of various self-control strategies in motivating and maintaining individual applications of cue-altering weight control techniques. Specifically, **Ss were instructed to employ self-reward, self-punishment, or self-monitoring techniques in their modification of both their body weight and their eating habits.** A control group was provided with identical information on stimulus control techniques, but they did not receive self-regulatory instructions.

b

c

METHOD

Subjects

The Ss were invited by a newspaper ad to participate in a program for self-managed weight control. Eligibility criteria included **(a) a minimum age of 17 years, (b) nonpregnant status, (c) physician's consent, and (d) a minimum of 10% overweight.** Applicants who were concurrently enrolled in a reducing club and/or undergoing other treatments for weight loss were ineligible. Also, individuals reporting recent dramatic changes in body weight were excluded. A total of 53 Ss (48 females, 5 males) enrolled and completed the program. Their average age was 37.9 years. Individual body weights varied from 107 to 270 with a mean of 166.3 pounds.

d

next paragraph and in the methods section. Despite the obvious redundancy, this style sets the stage for the rest of the article. Good scientific writing conveys a message in the most concise form possible without being so lean that the reader is required to reread each passage for its content. The examples of good experimental design in this book embody the balance between redundancy and parsimony.

Previous studies that attempted to determine the relative efficacy of these techniques combined them with other supportive techniques, such as group sup-

Prior to group assignment, the degree of obesity for each *S* was computed. This was done by dividing an individual's current weight by his ideal weight (Stillman & Baker, 1967), thereby controlling for height factors. The degree of obesity for all *S*s in the study ranged from 13% with a mean of 48.6%. **The *S*s were ranked according to degree of obesity and randomly assigned to experimental groups.**

Procedures

All *S*s were required to place a refundable deposit of $10 with the *E*s for the duration of treatment (four weeks). A commercially available bathroom scale was employed. At their initial weigh in, all *S*s were given a small booklet describing stimulus control approaches to weight loss. Treatment procedures varied as follows:

Self-Reward: Group 1 (*n* = 12)

The *S*s in this group were asked to deposit an additional $11 with *E* for purposes of self-reward. They were asked to weigh in biweekly for 7 weigh-ins and to keep a daily graph of their weight at home. The *S*s were provided with a weight chart and a behavioral diary in which they were to record the daily frequency of *(a)* "fat thoughts" (discouraging self-verbalizations), *(b)* "thin thoughts" (encouraging self-verbalizations), *(c)* instances of indulgence (eating a fattening food or excessive

port, therapist approval, and covert sensitization, thus failing to produce clear-cut results. In **(b)** the purpose of this study is stated, and in **(c)** a brief version of the research plan, including a control group, is stated. The independent variable—the various treatment groups—is also stated.

METHOD

The method section identifies the subjects, including the criteria for their selection **(d)**, and describes the procedures of each treatment in great enough detail to allow replication. We learn that the subjects, for example, were required to be over 17 years of age, to be nonpregnant, to be at least 10% overweight, and to be cleared by a physician to participate. We also learn that the experimenters were not allowed to express approval or disapproval about weight change during the weigh-in sessions and that the experimenters' contact with the subjects was kept to a minimum of 5 to 10 minutes per weigh-in. A postquestionnaire was used to assess a subject's use of any method not specified in his or her treatment group.

In research in which the principal effect is susceptible to wide individual differences, it is essential that variation caused by those differences be tightly controlled. In this experiment the assignment of subjects to treatment groups **(e)** was based on a matched-subject design, in which, for example, one very obese subject in the self-reward group would have a counterpart in the other groups. After ranking

quantity), and *(d)* instances of restraint (refusing a fattening food or reducing food intake). Self-reward was to take place at weigh-ins. Each *S*'s $21 was transformed into 21 shares whose value increased when other *S*s forfeited deposits or self-punished. A type of bank account was set up for each individual. The *S*s began with an empty account and could self-reward by requesting that a deposit be made to their account. Each *S* had access to three shares per weigh-in. It was recommended that *S*s self-reward two shares for a weight loss of one pound or more since their last weigh-in. The remaining share was to be self-awarded if adaptive behaviors (thin thoughts and restraint) had outnumbered nonadaptive behaviors since the last weigh-in. Beyond these recommendations, no external constraints were placed on *S*'s standards or execution of self-reward. When an *S* chose not to self-reward at a weigh-in, his/her shares were placed in a community pool and divided among all other shareholders (thereby increasing the value of a share). The *S*s were allowed three absences but were thereafter fined three shares for each absence. At their final weigh-in, they received the amount they had self-rewarded ($1/share) and were later mailed a dividend check covering increases in share value.

Self-Punishment: Group 2 (*n* = 12)

Procedures in this group were exactly parallel to those of Group 1 except that *S*s began with a full 21-share account and were instructed to fine themselves shares for lack of weight loss and/or lack of behavior improvement. Self-punished shares were placed in the community pool and divided among all remaining shares. When *S*s did not self-punish at a weigh-in, their shares remained in their account and were refunded to them at the final weigh-in. The share amounts and goals were identical to those in Group 1.

Self-Reward and Self-Punishment: Group 3 (*n* = 8)

Conditions in this group combined those of Groups 1 and 2. The *S*s began with an empty account and could either deposit to it (self-reward) or fine themselves (self-punish) up to three shares per weigh-in.

the subjects, the authors chose to randomly assign them to experimental and control groups. This procedure is based on the assumption that random assignment will homogeneously distribute subjects throughout the treatment groups. It might have proved embarrassing to the researchers if, by some quirk of probability, all of the very obese subjects had fallen into one group and all the mildly obese had fallen into another. But if you look at the results section, you will notice that Mahoney et al. tested for both the pretreatment degree of obesity for each group (**g**) and the pretreatment number of pounds overweight (**h**) and found the differences among the groups to yield a very low *F* ratio, which suggested that subjects in one group did not significantly differ in the pretreatment weigh-in.

Self-Monitoring: Group 4 (*n* = 5)

The *S*s in this group were asked to weigh in biweekly and to record their weight and adaptive and nonadaptive eating habits. The standard weight loss and behavior improvement goals used in Groups 1–3 were also suggested for these *S*s. In short, conditions in this group duplicated those in the first three groups with the exception that no additional deposit was required and the self-reward and self-punishment strategies were not discussed.

Information Control: Group 5 (*n* = 16)

The *S*s in this group received stimulus control booklets but did not participate in a second weigh-in until the four-week treatment period had ended. No self-monitoring materials were provided, and it was recommended that *S*s refrain from any weigh-ins at home during that time.

f

Efforts were made to avoid any *E* approval or disapproval for weight change, and to minimize *E* contact (weigh-ins took 5–10 minutes). **After four weeks all *S*s were weighed and instructed to continue self-application of their respective techniques. A postquestionnaire inquired about the use of any extraneous methods. Four months after their initial appointment, a follow-up weigh-in was conducted.**

Results

g

h

i

Data analyses on the pretreatment degree of obesity for each group revealed no significant differences ($F = .31$, $df = 4/48$). Likewise, the groups did not initially differ on number of pounds overweight ($F = .44$). **A posttreatment analysis of number of pounds lost yielded an overall *F* of 4.49 ($p < .005$).** Newman-Keuls comparisons of treatment means showed that the self-reward *S*s had lost significantly more pounds than either the self-monitoring ($p < .025$) or the control group ($p < .025$). The self-punishment group did not differ significantly from any other. A difference approaching significance at the .05

After the authors describe the treatment procedures, they note the principal dependent variables (**f**). It is important to notice that after the conclusion of the formal part of the experiment, a follow-up weigh-in was conducted. Many times weight-loss programs are temporarily effective, sometimes even spectacularly, but the long-range effects often are not effective. The follow-up procedure is an important one that is too often lacking in psychological research.

Results

The authors present the results in terse, almost blunt, language (**i**): "the self-reward *S*s had lost significantly more pounds than either the self-monitoring . . . or

j

level was obtained when Group 3 Ss (self-reward and self-punishment) were compared with those in the self-monitoring group. **Average number of pounds lost per individual was 6.4, 3.7, 5.2, .8, and 1.4 for the five groups, respectively.**

k

An analysis of percentage of body weight lost also revealed significant group differences ($F = 3.44$, $p < .025$). Newman-Kuels comparisons indicated that the self-reward group lost significantly more in percentage of body weight than either the self-monitoring ($p < .05$) or the control group ($p < .05$).

An analysis of "follow-through" was performed for Groups 1–3 in order to assess any differential tendencies toward self-reward or self-punishment. Since the weigh-ins constituted the most accurately measured and observable behavior of Ss, analyses were done on the consistency with which Ss self-administered rewards or fines for weight change. Using the recommended standard (loss of one pound per weigh-in), Ss were scored on whether they appropriately transacted two shares for success or failure at the above criterion. At first glance the data indicate that follow-through was higher for self-reward. There were no instances where an individual made his weight criterion and failed to self-reward, whereas there were numerous instances where individuals failed to make the criterion but failed to self-punish. However, a further analysis revealed quite a few instances where individuals self-rewarded even though they had not met the weight criterion. (It should be recalled that if a self-reward S had not met the criterion and appropriately chose not to self-reward, he was inadvertently *punished* by the automatic transfer of his shares to the community pool.) A follow-through analysis using both halves of the dichotomy revealed no significant intergroup differences ($F = .53$, $df = 3/33$). The rate of follow-through was 58.1% for self-reward, 57.6% for self-punishment, and 67.1% for the combined group.

In order to control for nonspecific factors associated with weigh-ins, Groups 1–4 Ss who began the study but completed fewer than four (out of seven)

the control group. . . ." The average weight loss per individual by group is given in **(j)**. A graphic representation comparing immediate loss and long-term loss would have been very useful. We have included this in Figure 25.1.

Another way of analyzing the data is by measuring the percentage of weight lost, which is given in **(k)**.

Throughout the remainder of the section, the authors demonstrate careful experimental analysis by examining all suspected relevant factors. In the four-month follow-up, note that some subjects were lost, for a variety of reasons. So many were lost from one treatment group **(l)** that the group had to be dropped from subsequent analysis. Subject loss poses a difficult experimental question. On one hand, the data from the follow-up are very important—some may say critical—to the experiment; on the other hand, it may be that the remaining subjects are not

weigh-ins were excluded from all data analyses. This restriction affected Group 4 (self-monitoring) more than the others. Only five (out of nine original) Group 4 Ss met the attendance criterion. In the first three groups, attendance was enhanced by levying fines after three absences. An analysis of attendance variations among Ss who did meet the criterion revealed no significant differences ($F = .86$).

1 Thirty-one Ss appeared for the four-month follow-up weigh-in. Of these, seven had to be excluded because of intervening or attendance factors (e.g., hormone shots, health farms, etc.). Because data were available from only

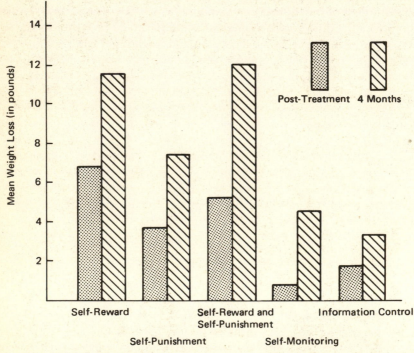

Figure 25.1 Weight loss by treatment groups.

representative of the entire treatment group. For example, if some subjects in one treatment group ate savagely for four months, they may have been so ashamed that they would not submit to a follow-up. Note that the authors struggle with this technical issue (**1**) and resolve it as best they can. There is no easy answer to this dilemma, and data gathered from treatments in which subject loss is a factor should be interpreted with great caution.

two self-monitoring Ss, this group had to be excluded from follow-up analyses. An analysis of the four-month data revealed that the self-reward group and the combined group (self-reward and self-punishment) had both lost greater percentages of their body weight than information control Ss ($p < .05$). Self-punishment Ss did not differ significantly from controls. Because follow-up Ss differed significantly in initial weight, analysis of number of pounds lost had to be converted to a proportion (number of pounds lost divided by number of pounds overweight). After an overall F of 5.03 ($p < .02$), individual Newman-Keuls comparisons showed that both the self-reward and the self-reward and self-punishment group had lost significantly more weight than either the self-punishment ($p < .05$) or the control group ($p < .05$). No other group comparisons were significant. Over the entire four months, the average number of pounds lost per individual was 11.5 for self-reward, 7.3 for self-punishment, 12.0 for self-reward and self-punishment, 4.5 for self-monitoring, and 3.2 for controls.

DISCUSSION

m

The foregoing results provide some preliminary information on the relative effectiveness of several self-control strategies. In general, **it would appear that self-reward strategies may provide an effective incentive component in weight loss attempts.** Only those groups given self-reward opportunities differed significantly from any others in successful weight loss. Data from information control Ss indicate that the simple provision of relevant information on stimulus control techniques causes only minimal change in body weight. Moreover, variables associated with self-recording, frequent weigh-ins, and (minimal) E contact failed to produce substantial weight loss. This finding is in contrast with recent evidence on the reactive effects of self-monitoring

DISCUSSION

In the final section, Mahoney et al. cautiously state their main finding **(m)**. This is a preliminary answer to the question that initiated this study: Which self-control techniques are most efficient in controlling overeating? The experimenters comment on the problem of subject attrition **(n)** and suggest a deposit scheme similar to the one mentioned in the results section to counteract this attrition rate. The authors then state the implications of this study **(o)**: (1) self-reward strategies seem to be superior to self-punishment and/or self-monitoring techniques and (2) the parameters and components of successful self-reward strategies need further investigation in order to determine their applicability to problems other than obesity, to determine the effects of magnitude and scheduling of rewards, and so on.

(e.g., Broden, Hall, & Mitts, 1971; McFall, 1970; McFall & Hammen, 1971). However, subsequent research (Mahoney, 1974) has suggested that specific self-monitoring applications may vary considerably in their reactivity and in the permanence of their effects.

It should be noted that empirical comparison of self-reward and self-punishment strategies is complicated by such factors as control for frequency of application (which is, in turn, altered by their relative effectiveness). Moreover, **it would be desirable for subsequent studies to motivate all Ss' appearance at weigh-ins and follow-up by some form of deposit. Although there were no group differences in dropout rate, equal deposits are also advisable to insure against pretreatment motivational variations.** Finally, care should be taken to avoid E punishment of appropriate self-regulation. In the present study, follow-through rates for self-reward Ss may have been decreased by the fact that these Ss were adventitiously fined for not self-rewarding (irrespective of goal attainment). A methodologically improved study (Mahoney, 1974) has revealed average self-reward follow-through rates in excess of 90%.

The implications of the present findings are twofold. First, it would appear that self-reward strategies may be superior to self-punishment and/or self-recording techniques in the modification of at least some habit patterns. Second, the parameters and components of successful self-reward strategies need to be investigated. For example, how important are the factors of magnitude, scheduling, and focus in self-reward systems? Can the pattern of the present findings be generalized to behavior problems other than obesity? These and other issues must await clarification by further research in self-control.

QUESTIONS

1. What techniques would you suggest to prevent subject dropout?
2. Suggest another dependent variable that might be useful in obesity research.
3. What practical application could be made of these results?
4. Why were subjects required to deposit $10 at the beginning of the experiment?
5. What effects, if any, would the sex of the experimenter have on the results? The age of the experimenter?
6. Would magnitude and type of reward have an effect on the results? If so, design an experiment to test these potential effects.
7. Identify at least three other weight-loss strategies that could be experimentally tested.

REFERENCES

Broden, M., Hall, R. V., & Mitts, B. (1971). The effect of self-recording on the classroom behavior of two eighth-grade students. *Journal of Applied Behavior Analysis, 4,* 191–199.

Harris, M. B. (1969). Self-directed program for weight control: A pilot study. *Journal of Abnormal Psychology, 74,* 263–270.

Kanfer, F. H., & Phillips, J. S. (1970). *Learning foundations of behavior therapy.* New York: Wiley.

Mahoney, M. J. (1974). Self-reward and self-monitoring techniques for weight loss. *Behavior Therapy, 5,* 48–57.

McFall, R. M. (1970). The effects of self-monitoring on normal smoking behavior. *Journal of Consulting and Clinical Psychology, 35,* 135–142.

McFall, R. M., & Hammen, C. L. (1971). Motivation, structure, and self-monitoring: The role of nonspecific factors in smoking reduction. *Journal of Consulting and Clinical Psychology, 37,* 80–86.

Penick, S. B., Filion, R., Fox, S., & Stunkard, A. J. (1971). Behavior modification in the treatment of obesity. *Psychosomatic Medicine, 33,* 49–55.

Stillman, I. M., & Baker, S. S. (1967). *The doctor's quick weight loss diet.* New York: Dell.

Stuart, R. B. (1967). Behavioral control over eating. *Behavior Research and Therapy, 5,* 357–365.

Stuart, R. B. (1981). A three-dimensional program for the treatment of obesity. *Behavior Research and Therapy, 9,* 177–186.

Stunkard, A. J. (1958). The management of obesity. *New York State Journal of Medicine, 58,* 79–87.

Wollersheim, J. P. (1970). The effectiveness of group therapy based upon learning principles in the treatment of overweight women. *Journal of Abnormal Psychology, 76,* 462–474.

The authors would like to thank Albert Bandura for his assistance and suggestions.

Chapter 26

Parents' Views

INTRODUCTION

Differences between the sexes have been both a topic of interest to experimental psychologists asnd a lively social issue. Some assert that the behavior of males differs from that of females because of biological differences, while others point to environmental factors. Both groups can cite an impressive series of empirical results to support their assertion. This has led to many psychologists recognizing that human behavior is the result of both inherited biological factors and environmental factors.

An important factor in the determination of behavior is the role of social expectation. Traditionally, boys are expected to be engineers, doctors, lawyers, aviators, and mechanics, while girls are expected to be nurses, homemakers, secretaries, garment workers, and models—to name but a few of the stereotypic occupations. Although these role types are being challenged in today's world, the influence of subtle forms of expectancy on behavior in still with us.

Of particular interest to Rubin, Provenzano, and Luria is the question of how early in an infant's life behavioral expectancies are formed. In order to answer this question, the researchers evaluated parents' views of the characteristics of newborn children.

We think this article contains some surprising findings, but it also demonstrates some important design issues. The article is reprinted from the American Journal of Orthopsychiatry, which is a publication of the American Orthopsychiatric Association, but frequently publishes articles of interest to psychologists. The style is very similar to APA style, but there are some minor changes. Since psychologists occasionally read and publish their results in publications in journals not published by the APA, it is important to be acquainted with these sources and their style.

You should analyze this article. As you read the article, consider some of the following questions:

1. Notice how the researchers develop their argument in the review of literature section. What differences do you see in this article compared to previous articles?
2. What is the general conclusion reached in this paper?
3. What measures were made?
4. How did the authors treat their data?

THE EYE OF THE BEHOLDER: PARENTS' VIEWS ON SEX OF NEWBORNS

JEFFREY Z. RUBIN, FRANK J. PROVENZANO, AND ZELLA LURIA

Thirty pairs of primiparous parents, fifteen with sons and fifteen with daughters, were interviewed within the first 24 hours postpartum. Although male and female infants did not differ in birth length, weight, or Apgar scores, daughters were significantly more likely than sons to be described as little, beautiful, pretty, and cute, and as resembling their mothers. Fathers made more extreme and stereotyped rating judgments of their newborns than did mothers. Findings suggest that *sex-typing* and *sex-role socialization* have already begun at birth.

As Schaffer (1971) observed, the infant at birth is essentially an asocial, largely undifferentiated creature. It appears to be little more than a tiny ball of hair, fingers, toes, cries, gasps, and gurgles. However, while it may seem that "if you've seen one, you've seen them all," babies are *not* all alike—a fact that is of special importance to their parents, who want, and appear to need, to view their newborn child as a creature that is special. Hence, much of early parental interaction with the infant may be focused on a search for distinctive features. Once the fact that the baby is normal has been established, questions such as, "Who does the baby look like?" and "How much does it weigh?" are asked.

Of all the questions parents ask themselves and each other about their infant, one seems to have priority: "Is it a boy or a girl?" The reasons for and consequences of posing this simple question are by no means trivial. The answer, "boy" or "girl," may result in the parents' organizing their perception of the infant with respect to a wide variety of attributes—ranging from its size to its activity, attractiveness, even its future potential. It is the purpose of the present study to examine the kind of verbal picture parents form of the newborn infant, as a function both of their own and their infant's gender.

The study reported here is addressed to parental perceptions of their infants at the point when these infants first emerge into the world. If it can be demonstrated that parental sex-typing has already begun its course at this earliest of moments in the life of the child, it may be possible to understand better one of the important antecedents of the complex process by which the growing child comes to view itself as boy-ish or girl-ish.

Based on our review of the literature, two forms of parental sex-typing may be expected to occur at the time of the infant's birth. First, it appears likely that parents will view and label their newborn child differentially, as a simple function of the infant's gender. Aberle and Naegele (1952) and Tasch (1952), using only fathers as subjects, found that they had different expectations for sons and daughters: sons were expected to be aggressive and athletic, daughters were expected to be pretty, sweet, fragile, and delicate. Rebelsky and Hanks (1971) found that fathers spent more time talking to their daughters than their sons during the first three months of life. While the sample size was too small for the finding to be significant, they suggest that the role of father-of-daughter may be perceived as requiring greater nurturance. Similarly, Pedersen and Robson (1969) reported that the fathers of infant daughters exhibited more behavior labeled (by the authors) as "apprehension over well-being" than did the fathers of sons.

The second form of parental sex-typing we expect to occur at birth is a function both of the infant's gender *and* the parent's own gender. Goodenough (1957) interviewed the parents of nursery school children, and found that mothers were less concerned with sex-typing their child's behavior than were fathers. More recently, Meyer and Sobieszek (1972) presented adults with videotapes of two seventeen-month-old children (each of whom was sometimes described as a boy and sometimes as a girl), and asked their subjects to describe and interpret the children's behavior. They found that male subjects as well as those having little contact with small children, were more likely (although not always significantly so) to rate the children in sex-stereotypic fashion—attributing "male qualities" such as independence, aggressiveness, activity, and alertness to the child presented as a boy, and qualities such as cuddliness, passivity, and delicacy of the "girl." We expect, therefore, that sex of infant and sex of parent will interact, such that it is fathers, rather than mothers, who emerge as the greater sex-typers of their newborn.

In order to investigate parental sex-typing of their newborn infants, and in order, more specifically, to test the predictions that sex-typing is a function of the infant's gender, as well as the gender of both infant and parent, parents of newborn boys and girls were studied in the maternity ward of a hospital, within the first 24 hours postpartum, to uncover their perceptions of the characteristics of their newborn infants.

Method

Subjects

The subjects consisted of 30 pairs of primiparous parents, 15 of whom had sons, and 15 of whom had daughters. The subjects were drawn from the available population of expecting parents at a suburban Boston hospital serving local, predominantly lower-middle-class families. Using a list of primiparous expectant mothers obtained from the hospital, the experimenter made contact with families by mail several months prior to delivery, and requested the subjects' assistance in "a study of social relations among parents and their first child." Approximately one week after the initial contact by mail, the experimenter telephoned each family, in order to answer any questions the prospective parents might have about the study, and to obtain their consent. Of the 43 families reached by phone, 11 refused to take part in the study. In addition, one consenting mother subsequently gave birth to a low birth weight infant (a 74-ounce girl), while another delivered an unusually large son (166 ounces). Because these two infants were at the two ends of the distribution of birth weights, and because they might have biased the data in support of our hypotheses, the responses of their parents were eliminated from the sample.

All subjects participated in the study within the first 24 hours postpartum—the fathers almost immediately after delivery, and the mothers (who were often under sedation at the time of delivery) up to but not later than 24 hours later. The mothers typically had spoken with their husbands at least once during this 24-hour period.

There were no reports of medical problems during any of the pregnancies or deliveries, and all infants in the sample were full-term at time of birth. Deliveries were made under general anesthesia, and the fathers were not allowed in the delivery room. The fathers were not permitted to handle their babies during the first 24 hours, but could view them through display windows in the hospital nursery. The mothers, on the other hand, were allowed to hold and feed their infants. The subject participated individually in the study. The fathers were met in a small, quiet waiting room used exclusively by the maternity ward, while the mothers were met in their hospital rooms. Every precaution was taken not to upset the parents or interfere with hospital procedure.

Procedure

After introducing himself to the subjects, and their congratulatory amenities, the experimenter (FJP) asked the parents: "Describe your baby as you would to a close friend or relative." The responses were tape-recorded and subsequently coded.

The experimenter then asked the subjects to take a few minutes to complete a short questionnaire. The instructions for completion of the questionnaire were as follows:

On the following page there are 18 pairs of opposite words. You are asked to rate your baby in relation to these words, placing an "x" or a checkmark in the space that best describes your baby. The more a word describes your baby, the closer your "x" should be to that word.

Example: Imagine you were asked to rate Trees.
Good

⌐⌐⌐ Bad

Strong

⌐⌐⌐Weak

If you cannot decide or your feelings are mixed, place your "x" in the center space. Remember, the more you think a word is a good description of your baby, the closer you should place your "x" to that word. If there are no questions, please begin. Remember, you are rating your baby. Don't spend too much time thinking about your answers. First impressions are usually the best.

Having been presented with these instructions, the subjects then proceeded to rate their baby on each of the following eighteen 11-point, bipolar adjective scales: firm–soft; large featured–fine featured; big–little; relaxed–nervous; cuddly–not cuddly; easygoing–fussy; cheerful–cranky; good eater–poor eater; excitable–calm; active–inactive; beautiful–plain; sociable–unsociable; well coordinated–awkward; noisy–quiet; alert–inattentive; strong–weak; friendly–unfriendly; hardy–delicate.

Upon completion of the questionnaire, the subjects were thanked individually, and when both parents of an infant had completed their participation, the underlying purposes of the study were fully explained.

Hospital Data

In order to acquire a more objective picture of the infants whose characteristics were being judged by the subjects, data were obtained from hospital records concerning each infant's birth weight, birth length, and Apgar scores. Apgar scores are typically assigned at five and ten minutes postpartum, and represent the physician's ratings of the infant's color, muscle tonicity, reflex irritability, and heart and respiratory rates. No significant differences between the male and female infants were found for birth weight, birth length, or Apgar scores at five and ten minutes postpartum.*

RESULTS

In Table 26.1, the subjects' mean ratings of their infant, by condition, for each of the eighteen bipolar adjective scales, are presented. The right-extreme

*Birth weight ($\bar{X}_{Sons} = 114.43$ ounces, $\bar{X}_{Daughters} = 11.00$, $t(28) = 1.04$); Birth length ($\bar{X}_{Sons} = 19.80$ inches, $\bar{X}_{Daughters} = 19.96$, $t(28) = 0.52$); 5 minute Apgar score ($\bar{X}_{Sons} = 9.07$, $\bar{X}_{Daughters} = 9.33$, $t(28) = 0.69$); and 10 minute Apgar score ($\bar{X}_{Sons} = 10.00$, $\bar{X}_{Daughters} = 10.00$).

column of Table 26.1 shows means for each scale, which have been averaged across conditions. Infant stimuli, overall, were characterized closer to the scale anchors of soft, fine featured, little, relaxed, cuddly, easygoing, cheerful, good eater, calm, active, beautiful, sociable, well coordinated, quiet, alert, strong, friendly, and hardy. Our parent-subjects, in other words, appear to have felt on Day 1 of their babies' lives that their newborn infants represented delightful, competent new additions to the world!

Analysis of variance of the subjects' questionnaire responses (1 and 56 degrees of freedom) yielded a number of interesting findings. There were *no* rating differences on the eighteen scales as a simple function of Sex of Parent:

Parents appear to agree with one another, on the average. As a function of Sex of Infant, however, several significant effects emerged: Daughters, in contrast to sons, were rated as significantly softer ($F = 10.67$, $p < .005$), finer featured ($F = 9.27$, $p < .005$), littler ($F = 28.83$, $p < .001$), and more inattentive ($F = 4.44$, $p < .05$). In addition, significant interaction effects emerged for seven of the eighteen scales: firm–soft ($F = 11.22$, $p < .005$), large featured–fine featured ($F = 6.78$, $p < .025$), cuddly–not cuddly ($F = 4.18$, $p < .05$), well coordinated–awkward ($F = 12.52$, $p < .001$), alert–inattentive ($F = 5.10$, $p < .05$), strong–weak ($F = 10.67$, $p < .005$), and hardy–delicate ($F = 5.32$, $p < .025$).

Table 26.1   MEAN RATING ON THE 18 ADJECTIVE SCALES AS A FUNCTION OF SEX OF PARENT *(MOTHER VS. FATHER)* AND SEX OF INFANT *(SON VS. DAUGHTER)*[a]

SCALE	EXPERIMENTAL CONDITION				
(I)–(II)	M–S	M–D	F–S	F–D	$\overline{X}$
Firm–Soft	7.47	7.40	3.60	8.93	6.85
Large featured–Fine featured	7.20	7.53	4.93	9.20	7.22
Big–Little	4.73	8.40	4.13	8.53	6.45
Relaxed–Nervous	3.20	4.07	3.80	4.47	3.88
Cuddly–Not cuddly	1.40	2.20	2.20	1.47	1.82
Easygoing–Fussy	3.20	4.13	3.73	4.60	3.92
Cheerful–Cranky	3.93	3.73	4.27	3.60	3.88
Good eater–Poor eater	3.73	3.80	4.60	4.53	4.16
Excitable–Calm	6.20	6.53	5.47	6.40	6.15
Active–Inactive	2.80	2.73	3.33	4.60	3.36
Beautiful–Plain	2.13	2.93	1.87	2.87	2.45
Sociable–Unsociable	4.80	3.80	3.73	4.07	4.10
Well coordinated–Awkward	3.27	2.27	2.07	4.27	2.97
Noisy–Quiet	6.87	7.00	5.67	7.73	6.82
Alert–Inattentive	2.47	2.40	1.47	3.40	2.44
Strong–Weak	3.13	2.20	1.73	4.20	2.82
Friendly–Unfriendly	3.33	3.40	3.67	3.73	3.53
Hardy–Delicate	5.20	4.67	3.27	6.93	5.02

[a]The larger the mean, the greater the rated presence of the attribute denoted by the second (right-hand) adjective in each pair.

The meaning of these interactions becomes clear in Table 26.1, in which it can be seen that six of these significant interactions display a comparable pattern: fathers were more extreme in their ratings of *both* sons and daughters than were mothers. Thus, sons were rated as firmer, larger featured, better coordinated, more alert, stronger, and hardier—and daughters as softer, fine featured, more awkward, more inattentive, weaker, and more delicate—by their fathers than by their mothers. Finally, with respect to the other significant interaction effect (cuddly–not cuddly), a rather different pattern was found. In this case, mothers rated sons as cuddlier than daughters, while fathers rated daughters as cuddlier than sons—a finding we have dubbed the "oedipal" effect.

Responses to the interview question were coded in terms of adjectives used and references to resemblance. Given the open-ended nature of the question, many adjectives were used—healthy, for example, being a high frequency response cutting across sex of babies and parents. Parental responses were pooled, and recurrent adjectives were analyzed by ψ^2 analysis for sex of child. Sons were described as big more frequently than were daughters ($\psi^2(1) = 4.26$, $p < .05$); daughters were called little more often than were sons ($\psi^2(1) = 4.28$, $p < .05$). The "feminine" cluster—beautiful, pretty, and cute—was used significantly more often to describe daughters than sons ($\psi^2(1) = 5.40$, $p < .05$). Finally, daughters were said to resemble mothers more frequently than were sons ($\psi^2(1) = 3.87$, $p < .05$).

DISCUSSION

The data indicate that parents—especially fathers—differently label their infants, as a function of the infant's gender. These results are particularly striking in light of the fact that our sample of male and female infants did *not* differ in birth length, weight, or Apgar scores. Thus, the results appear to be a pure case of parental labeling—what a colleague has described as "nature's first projective test" (personal communication, Leon Eisenberg). Given the importance parents attach to the birth of their first child, it is not surprising that such ascriptions are made.

The results of the present study are, of course, not unequivocal. Although it was found, as expected, that the sex-typing of infants varied as a function of the infant's gender, as well as the gender of both infant and parent, significant differences did not emerge for all eighteen of the adjective scales employed. Two explanations for this suggest themselves. First, it may simply be that we have overestimated the importance of sex-typing at birth. A second possibility, however, is that sex-typing is more likely to emerge with respect to certain classes of attributes—namely, those which denote physical or constitutional, rather than "internal," dispositional, factors. Of the eight different adjective pairs for which significant main or interaction effects emerged, six (75%) clearly refer to external attributes of the infant. Conversely, of the 10 adjective pairs for which no significant differences were found, only three (30%) clearly denote external attributes. This suggests that it is physical and constitutional factors that especially lend themselves to sex-typing at birth, at least in our culture.

Another finding of interest is the lack of significant effects, as a simple function of sex of parent. Although we predicted no such effects, and were therefore not particularly surprised by the emergence of "non-findings," the implication of these results is by no means trivial. If we had omitted the sex of the infant as a factor in the present study, we might have been led to conclude (on the basis of simply varying the sex of the parent) that *no* differences exist in parental descriptions of newborn infants—a patently erroneous conclusion! It is only when the infant's and the parent's gender are considered together, in interaction, that the lack of differences between overall parental mean ratings can be seen to reflect the true differences between the parents. Mothers rate both sexes closer together on the adjective pairs than do fathers (who are the stronger sex-typers), but *both* parents agree on the direction of sex differences.

The central implication of the study, then, is that sex-typing and sex-role socialization appear to have already begun their course at the time of the infant's birth, when information about the infant is minimal. The Gestalt parents develop, and the labels they ascribe to their newborn infant, may well affect subsequent expectations about the manner in which their infant ought to behave, as well as parental behavior itself. This parental behavior, moreover, when considered in conjunction with the rapid unfolding of the infant's own behavioral repertoire, may well lead to a modification of the very labeling that affected parental behavior in the first place. What began as a one-way street now bears traffic in two directions. In order to understand the full importance and implications of our findings, therefore, research clearly needs to be conducted in which delivery room stereotypes are traced in the family during the first several months after birth, and their impact upon parental behavior is considered. In addition, further research is clearly in order if we are to understand fully the importance of early parental sex-typing in the socialization of sex roles.

REFERENCES

Aberle, D., and Naegele, K. (1952). Middleclass fathers' occupational role and attitudes toward children. *American Journal of Orthopsychiatry, 22*(2), 366–378.

Goodenough, E. (1957). Interest in persons as an aspect of sex differences in the early years. *Genetic Psychology Monograph, 55,* 287–323.

Harari, H., and McDavid, J. Name stereotypes and teachers' expectations. *Journal of Educational Psychology.* (In press).

Meyer, J., and Sobieszek, B. (1972). Effect of a child's sex on adult interpretations of its behavior. *Developmental Psychology, 6,* 42–48.

Pedersen, F. and Robson, K. (1969). Father participation in infancy. *American Journal of Orthopsychiatry, 39,* 466–472.

Rebelsky, F., and Hanks, C. (1971). Fathers' verbal interaction with infants in the first three months of life. *Child Development, 42,* 63–68.

Schaffer, H. (1971). The Growth of Sociability. Baltimore: Penguin Books.

Tasch, R. (1952). The role of the father in the family. *Journal of Experimental Education, 20,* 319–361.

Zajonc, R. (1968). Attitudinal effects of mere exposure. *Journal of Personality and Social Psychology Monograph,* Supplement 9, 1–27.

ADDITIONAL QUESTIONS

1. What is the problem?
2. What is the hypothesis?
3. What control issues are raised in this experiment? How did the authors deal with these issues?
4. Were some of the subjects lost? If so, how did the authors deal with this problem?
5. What type of scale was used in the experiment?
6. How would you classify this research?
7. Present the data from Table 26.1 in graphic form.
8. The results are predicated on the assumption that the infants were essentially identical shortly after birth. Discuss this assumption and suggest another design that would have eliminated the need for this assumption.
9. Briefly summarize the authors' discussion section.
10. Does this research alter your concept of sex-role typing? If so, in what way?

Computational Procedures for Basic Statistics

This appendix discusses the mathematical procedures for a few of the statistical tests mentioned in this book. Only the rudiments are described.

CENTRAL TENDENCY

A score representative of all scores in a distribution is called *central tendency*. One measure is the *mean* ($\overline{X}$), or average, which is calculated by adding all scores and dividing by the number of scores:

$$\overline{X} = \frac{\Sigma X}{N}$$

Example: Five students took an examination with a possible high score of 10 and low score of 1. The results were as follows:

Paul   7

Mary 3

Jill   9

Bob   6 $\overline{X} = \frac{\Sigma X}{N} = \frac{30}{5} = 6$

Curt <u>5</u>

$\Sigma X = 30$

$N = $ 5

MEASURE OF VARIABILITY

The *standard deviation* is the most frequently used index of how much scores vary. The formula for calculating a standard deviation is

$$SD = \sigma = \sqrt{\frac{(\Sigma X^2 - N\bar{X}^2)}{(N-1)}}$$

The calculation of the standard deviation for the examination scores mentioned previously follows:

X		X^2
Paul	7	49
Mary	3	9
Jill	9	81
Bob	6	36
Curt	5	25
$\Sigma X =$	30	$\Sigma X^2 = 200$
$N =$	5	

$$\sigma = \sqrt{\frac{(200 - (5 \times 6^2))}{(5-1)}} = 2.24$$

MEASURE OF ASSOCIATION

The strength of association between two sets of data can be determined with the *correlation coefficient*. Two methods for determining correlation follow.

Spearman Rank-Order Correlation

The first is called the *Spearman rank-order correlation coefficient (rho)* and is used with *ordinal* data, or data that can be ranked.

The formula for *rho* is

$$rho = 1 - \frac{6\Sigma d^2}{(N^3 - 1)}$$

Example: In addition to test scores in the previous example, the students also were given a physical examination and ordered on healthiness with a high score indicating good health. The two sets of rank order scores were as follows:

	TEST SCORES (RANK ORDER) X_t	HEALTH SCORES (RANK ORDER) X_h	d DIFFERENCE $(X_t - X_h)$	d^2
Paul	4	5	−1	1
Mary	1	1	0	0
Jill	5	2	3	9
Bob	3	3	0	0
Curt	2	4	−2	4
$N = 5$				$\Sigma d^2 = 14$

$$rho = 1 - \frac{6 \times 14}{(5^3 - 1)} = 0.32$$

The significance of *rho* can be determined by finding the *critical values* (a probability level set by the experimenter that reaches *statistical significance*) in Table B.1 in Appendix B.

The critical value for *rho* is found by finding N in Table B.1, which represents the number of pairs. In our example $rho = 0.32$ with $N = 5$. By using $p = 0.0500$ we can see that our value of *rho* is far short of significance, which is not surprising given the limited number of observations.

Pearson Product-Moment Correlation

A second correlation coefficient, called the *Pearson product-moment correlation coefficient* (r), is used to measure the relationship between two interval scales, such as temperature and test scores.

The formula for r is

$$r = \frac{(N\Sigma XY - \Sigma X \Sigma Y)}{\sqrt{N\Sigma X^2 - (\Sigma X)^2}\sqrt{N\Sigma Y^2 - (\Sigma Y)^2}}$$

Example: In addition to test scores, the students also were given an intelligence test. The test had a high score of 20 and a low score of 0. The two sets of scores were as follows:

	TEST SCORES		INTELLIGENCE SCORES		
	X_t	X_t^2	Y_i	Y_i^2	$X_t Y_i$
Paul	7	49	18	324	126
Mary	3	9	5	25	15
Jill	9	81	15	225	135
Bob	6	36	12	144	72
Curt	5	25	10	100	50
$\Sigma X = 30$		$\Sigma X^2 = 200$	$\Sigma Y = 60$	$\Sigma Y^2 = 818$	$\Sigma XY = 398$

$N = 5$

$df = N - 2$

$df = 5 - 2 = 3$

$$r = \frac{(5(398) - (30)(60))}{\sqrt{5(200) - 30^2}\sqrt{5(818) - 60^2}}$$

$$= \frac{(1,990 - 1,800)}{\sqrt{1,000 - 900}\sqrt{4,090 - 3,600}}$$

$$= \frac{190}{\sqrt{100}\sqrt{490}}$$

$$= \frac{190}{10 \times 22.14}$$

$$= 0.86$$

To find the level of significance for r, we need the *degrees of freedom*, or *df*. Then we can consult Table B.2 and determine the level of significance. In our example $r = 0.86$ with $df = 3$, which falls between 0.10 and 0.05.

CHI-SQUARE

The chi-square (χ^2) test is used to determine if the observed frequency of scores differs from the expected frequency of scores. The formula for chi-square is

$$\chi^2 = \Sigma\frac{(O - E)^2}{E}$$

where O is the observed frequency and E is the expected frequency.

Example: Suppose an experimenter was interested in whether the genders watch different television programs. A sample of 100 subjects might be asked which of three television programs they preferred to watch (*Dating Game, 60 Minutes*, or *Wheel of Fortune*). The (fictitious) data were the following:

	DG	60	WF	ROW TOTALS
Males (N = 50)	12	20	18	50
Females (N = 50)	22	10	18	50
Column totals	34	30	36	N = 100

We can find the expected frequencies for each of the cells by using the following formula:

$$E = \frac{\text{row total} \times \text{column total}}{N}$$

In the first cell (male–*Dating Game*) the row total is 50, the column total is 34, and $N = 100$. Thus,

$$E = \frac{50 \times 34}{100} = 17$$

In the second cell (male–*60 Minutes*) the row total is 50, the column total is 30, and $N = 100$. Thus,

$$E = \frac{50 \times 30}{100} = 15$$

If we continued to find all expected frequencies and placed them in a table, it would look like the following:

	DG/E	60/E	WF/E	ROW TOTALS
Males (N = 50)	12/17	20/15	18/18	50
Females (N = 50)	22/17	10/15	18/18	50
Totals	34	30	36	N = 100

Then we could calculate $(O - E)^2/E$ for each cell. To get χ^2 we would add the values as follows:

$$\chi^2 = \frac{(12 - 17)^2}{17} + \frac{(20 - 15)^2}{15} + \frac{(18 - 18)^2}{18} +$$

$$\frac{(22 - 17)^2}{17} + \frac{(10 - 15)^2}{15} + \frac{(18 - 18)^2}{18} = 6.27$$

The degrees of freedom (df) is found by the formula

$$df = (R - 1)(C - 1)$$

where R = rows and C = columns. In our example, $R = 2$ and $C = 3$. To find the probability level for the value of 6.27 we would consult Table B.3. The value with 2 df is slightly more than the 0.05 level.

t-TEST

Here we show the computational procedures for the t-test. The t-test is used to determine if the probability of the differences between two independent groups occurs by chance. Another form of the t-test is available for correlated measures. For further details about the t-test and other statistical procedures mentioned here consult a standard statistics text.

The formula for the t-test for independent groups is

$$t = \frac{(\overline{X}_1 - \overline{X}_2)}{\sqrt{\left(\dfrac{SS_1 + SS_2}{(n_1 - 1) + (n_2 - 1)}\right)\left(\dfrac{1}{n_1} + \dfrac{1}{n_2}\right)}}$$

where $\overline{X}_1$ is the mean for the first group, $\overline{X}_2$ is the mean for the second group, n is the number of subjects in each group, and SS can be found with the following formula:

$$SS = \Sigma X^2 - \frac{(\Sigma X)^2}{n}$$

Example: Ten randomly selected subjects are randomly assigned to two equal groups. One group (control) is asked to copy notes from a score of music as originally composed, and a second group is asked to copy notes from the same score, except that the subjects can hear the music as they are copying the notes. The independent variable is the playing of the music; the dependent variable is the time taken to copy the score.

The (hypothetical) results were as follows:

TIME (IN SECONDS) TO COPY MUSIC

X_1 CONTROL	X_1^2	X_2 EXPERIMENTAL	X_2^2
7	49	5	25
12	144	8	64
10	100	7	49
11	121	10	100
14	196	9	81
$\Sigma X_1 = 54$	$\Sigma X_1^2 = 610$	$\Sigma X_2 = 39$	$\Sigma X_2^2 = 319$

Then we could find the components of the formula, substitute them into the equation, and find t.

$$SS_1 = \Sigma X_1^2 - \frac{(\Sigma X_1)^2}{n_1} = 610 - \frac{(54)^2}{5} = 26.8$$

$$SS_2 = \Sigma X_2^2 - \frac{(\Sigma X_2)^2}{n_2} = 319 - \frac{(39)^2}{5} = 14.8$$

$$\bar{X_1} = \frac{\Sigma X_1}{n_1} = \frac{54}{5} = 10.8$$

$$\bar{X_2} = \frac{\Sigma X_2}{n_2} = \frac{39}{5} = 7.8$$

$$t = \frac{(10.8 - 7.8)}{\sqrt{\left(\frac{26.8 + 14.8}{(5-1)+(5-1)}\right)\left(\frac{1}{5} + \frac{1}{5}\right)}}$$

$$= \frac{3}{\sqrt{\left(\frac{41.6}{8}\right)\left(\frac{2}{5}\right)}} = \frac{3}{\sqrt{2.08}} = \frac{3}{1.44} = 2.08$$

$$df = n_1 + n_2 - 2$$

$$df = 5 + 5 - 2 = 8$$

In Table B.4 we can see that a t value of 2.08 with 8 df falls between p of 0.05 and 0.10 and would be considered significant at $p < 0.10$. Most psychology experiments require at least a $p < 0.05$ to be considered statistically significant.

Example: We can also use a t-test for Model 1 of the wine tasting experiment from Chapter 4.

	X_1 CONTROL	X_1^2		X_2 EXPERIMENTAL	X_2^2
S_1	1	1	S_9	5	25
S_2	2	4	S_{10}	4	16
S_3	2	4	S_{11}	3	9
S_4	1	1	S_{12}	5	25
S_5	2	4	S_{13}	5	25
S_6	1	1	S_{14}	4	16
S_7	1	1	S_{15}	4	16
S_8	4	16	S_{16}	5	25
	$\Sigma X_1 = 14$	$\Sigma X_1^2 = 32$		$\Sigma X_2 = 35$	$\Sigma X_2^2 = 157$

$$SS_1 = \Sigma X_1^2 - \frac{(\Sigma X_1)^2}{n_1} = 32 - \frac{(14)^2}{8} = 3.5$$

$$SS_2 = \Sigma X_2^2 - \frac{(\Sigma X_2)^2}{n_2} = 157 - \frac{(35)^2}{8} = 3.875$$

$$\overline{X}_1 = \frac{\Sigma X_1}{n_1} = \frac{14}{8} = 1.75$$

$$\overline{X}_2 = \frac{\Sigma X_2}{n_2} = \frac{35}{8} = 4.375$$

$$t = \frac{(1.75 - 4.375)}{\sqrt{\left(\frac{3.5 + 3.875}{(8-1) + (8-1)}\right)\left(\frac{1}{8} + \frac{1}{8}\right)}}$$

$$= \frac{-2.525}{\sqrt{\left(\frac{7.275}{14}\right)\left(\frac{1}{4}\right)}} = \frac{-2.525}{\sqrt{0.13}} = -7.00$$

$$df = n_1 + n_2 - 2$$

$$df = 8 + 8 - 2 = 14$$

Significance level: $p < .01$

CORRELATED *t*-TEST (DEPENDENT *t*-TEST)

To test for differences between two distributions that have been matched on some dimension, such as IQ, running speed, expertise in wine tasting, and so on, a correlated *t*-test may be used. The formula is

$$t = \frac{\bar{d}}{\sqrt{\frac{SS_d}{(n-1)}\left(\frac{1}{n}\right)}}$$

Example: This computation is based on Model 2 of the wine tasting experiment in Chapter 4 in which subjects were first evaluated in terms of their wine tasting abilities. The data is shown below:

X_1 EXPERIMENTAL		X_2 CONTROL		d DIFFERENCE $(X_2 - X_1)$	d^2
S_{1a}	2	S_{9a}	4	2	4
S_{2b}	3	S_{10b}	5	2	4
S_{3c}	2	S_{11c}	5	3	9
S_{4d}	1	S_{12d}	5	4	16
S_{5e}	1	S_{13e}	4	3	9
S_{6f}	2	S_{14f}	4	2	4
S_{7g}	3	S_{15g}	5	2	4
S_{8h}	1	S_{16h}	4	3	9
				$\Sigma d = 21$	$\Sigma d^2 = 59$

To find *t*, we must first perform the following computations:

$$SS_d = \Sigma d_2 - (\Sigma d)^2/n$$

$$59 - (21)^2/8 = 3.875$$

$$\bar{d} = \Sigma d/n$$
$$21/8 = 2.625$$

Then we can compute *t*:

$$t = \frac{2.625}{\sqrt{\frac{3.875}{(8-1)}\left(\frac{1}{8}\right)}} = 0.685$$

To find the critical value of *t*, we compute *df* = number of pairs − 1 = 8 − 1 = 7 and use Table B.4 and see that p > .10.

ANALYSIS OF VARIANCE

The *analysis of variance* (ANOVA) is one of the most useful and versatile of all statistics used in psychology today. It can be used with between-subject or within-subject designs and with experiments that have several levels of the independent variable. ANOVA can even treat more complex designs.

The main formula for F is

$$F = \frac{MS_{bg}}{MS_{wg}} \qquad\qquad df_{bg} = k - 1$$

$$MS_{bg} = \frac{SS_{bg}}{df_{bg}} \qquad\qquad df_{wg} = N - k$$

$$MS_{wg} = \frac{SS_{wg}}{df_{wg}}$$

$$SS_{bg} = \frac{(\Sigma X_1)^2}{n_1} + \frac{(\Sigma X_2)^2}{n_2} + \frac{(\Sigma X_3)^2}{n_3} - \frac{(\Sigma X_t)^2}{n_t}$$

$$SS_{wg} = SS_{tot} - SS_{bg}$$

$$SS_{tot} = \Sigma X_t^2 - \frac{(\Sigma X_1 + \Sigma X_2 + \Sigma X_3)^2}{N_t}$$

Example: An experimenter was interested in the effect of imagery on learning. She defined *imagery* as how well a word conjures up a picture in someone's head. In the experiment, a group of three types of words was presented to three independent samples of subjects. For example, one group might have heard the words *arrow, salary,* and *context* and been asked to rate them from high to low in imagery. The subjects studied the words and then were asked to recall them. The independent variable was the three levels of imagery. The dependent variable was the number of items recalled.

The results were as follows:

NUMBER OF WORDS RECALLED AS A FUNCTION OF IMAGERY

GROUP 1 LOW IMAGERY		GROUP 2 MODERATE IMAGERY		GROUP 3 HIGH IMAGERY	
X_1	X_1^2	X_2	X_2^2	X_3	X_3^2
3	9	9	81	12	144
5	25	4	16	14	196
2	4	8	64	12	144
4	16	9	81	10	100
6	36	10	100	15	225
$\Sigma X_1 = 20$	$\Sigma X_1^2 = 90$	$\Sigma X_1 = 40$	$\Sigma X_2^2 = 342$	$\Sigma X_3 = 63$	$\Sigma X_3^2 = 839$
$n_1 = 5$		$n_2 = 5$		$n_3 = 5$	

The calculation of F is as follows:

$$\Sigma X_t^2 = 90 + 342 + 839 = 1{,}271$$

$$\Sigma X_t = 20 + 40 + 63 = 123$$

$$SS_{tot} = 1{,}271 - \frac{(123)^2}{15} = 262.4$$

$$SS_{bg} = \frac{(20)^2}{5} + \frac{(40)^2}{5} + \frac{(63)^2}{5} - \frac{(123)^2}{15} = 185.2$$

$$SS_{wg} = 262.4 - 185.2 = 77.2$$

$$df_{bg} = k - 1 = 3 - 1 = 2$$

$$df_{wg} = N - k = 15 - 3 = 12$$

$$MS_{bg} = \frac{185.2}{2} = 92.60$$

$$MS_{wg} = \frac{77.2}{12} = 6.43$$

$$F = \frac{92.60}{6.43} = 14.40$$

To determine if this F value is statistically significant, consult Table B.5. To find the appropriate F number, two degrees of freedom must be used: a df for the numerator (the df associated with the between-group data) and the df for the denominator (the df associated with the within-group data). These are found by using the following formulas:

$$df_{bg} = k - 1$$

$$df_{wg} = N - k$$

In these equations k is the number of experimental conditions. In this example, $k = 3$. N is the total number of observations. In our example $N = 15$. Thus, df_{bg} is 2 and df_{wg} is 12. At the intersection of 2 and 12 in Table B.5, we find that our value of 14.40 is greater than the value for the 0.01 level of significance. This is expressed as: $F(2,12) = 14.40, p < 0.01$.

Statistical Tables

Table B.1 CRITICAL VALUES OF *rho* (SPEARMAN RANK-ORDER CORRELATION COEFFICIENT)

N	$p = 0.0500$	$p = 0.0100$
5	1.000	—
6	0.886	1.000
7	.786	0.929
8	.738	.881
9	.683	.833
10	.648	.794
12	.591	.777
14	.544	.715
16	.506	.665
18	.475	.625
20	.450	.591
22	.428	.562
24	.409	.537
26	.392	.515
28	.377	.496
30	.364	.478

Computed from Olds, E. G., Distribution of the sum of squares of rank differences for small numbers of individuals, *Annals of Mathematical Statistics*, 1938, IX, 133–148, and the 5% significance levels for sums of squares of rank differences and a correction, *Annals of Mathematical Statistics*, 1949, XX, 117–118, by permission of the Institute of Mathematical Statistics.

Table B.2 CRITICAL VALUES OF *r*
(PEARSON PRODUCT-MOMENT
CORRELATION COEFFICIENT)

df	LEVEL OF SIGNIFICANCE FOR TWO-TAILED TEST		
	0.10	0.05	0.01
1	0.988	0.997	0.9999
2	.900	.950	.990
3	.805	.878	.959
4	.729	.811	.917
5	.669	.754	.874
6	.622	.707	.834
7	.582	.666	.798
8	.549	.632	.765
9	.521	.602	.735
10	.497	.576	.708
11	.476	.553	.684
12	.458	.532	.661
13	.441	.514	.641
14	.426	.497	.623
15	.412	.482	.606
16	.400	.468	.590
17	.389	.456	.575
18	.378	.444	.561
19	.369	.433	.549
20	.360	.423	.537
25	.323	.381	.487
30	.296	.349	.449
35	.275	.325	.418
40	.257	.304	.393
45	.243	.288	.372
50	.231	.273	.354
60	.211	.250	.325
70	.195	.232	.303
80	.183	.217	.283
90	.173	.205	.267
100	.164	.195	.254

Adapted from R. A. Fisher, *Statistical Methods for Research Workers.* 14th Edition. Copyright 1973. Hafner Press.

Table B.3 CRITICAL VALUES OF CHI-SQUARE

df	p = 0.05	p = 0.01
1	3.84	6.64
2	5.99	9.21
3	7.82	11.34
4	9.49	13.28
5	11.07	15.09
6	12.59	16.81
7	14.07	18.48
8	15.51	20.09
9	16.92	21.67
10	18.31	23.21
11	19.68	24.72
12	21.03	26.22
13	22.36	27.69
14	23.68	29.14
15	25.00	30.58
16	26.30	32.00
17	27.59	33.41
18	28.87	34.80
19	30.14	36.19
20	31.41	37.57
21	32.67	38.93
22	33.92	40.29
23	35.17	41.64
24	36.42	42.98
25	37.65	44.31
26	38.88	45.64
27	40.11	46.96
28	41.34	48.28
29	42.56	49.59
30	43.77	50.89

Table B.3 is taken from Table 4 of Fisher & Yates, *Statistical Tables for Biological, Agricultural and Medical Research,* published by Longman Group Ltd., London (previously published by Oliver and Boyd Ltd., Edinburgh). By permission of the authors and publishers.

Table B.4 CRITICAL VALUES OF *t*

df	p = 0.10	p = 0.05	p = 0.02	p = 0.01
1	6.314	12.706	31.821	63.657
2	2.920	4.303	6.965	9.925
3	2.353	3.182	4.541	5.841
4	2.132	2.776	3.747	4.604
5	2.015	2.571	3.365	4.032
6	1.943	2.447	3.143	3.707
7	1.895	2.365	2.998	3.499
8	1.860	2.306	2.896	3.355
9	1.833	2.262	2.821	3.250
10	1.812	2.228	2.764	3.169
11	1.796	2.201	2.718	3.106
12	1.782	2.179	2.681	3.055
13	1.771	2.160	2.650	3.012
14	1.761	2.145	2.624	2.977
15	1.753	2.131	2.602	2.947
16	1.746	2.120	2.583	2.921
17	1.740	2.110	2.567	2.898
18	1.734	2.101	2.552	2.878
19	1.729	2.093	2.539	2.861
20	1.725	2.086	2.528	2.845
21	1.721	2.080	2.518	2.831
22	1.717	2.074	2.508	2.819
23	1.714	2.069	2.500	2.807
24	1.711	2.064	2.492	2.797
25	1.708	2.060	2.485	2.787
26	1.706	2.056	2.479	2.779
27	1.703	2.052	2.473	2.771
28	1.701	2.048	2.467	2.763
29	1.699	2.045	2.462	2.756
30	1.697	2.042	2.457	2.750
60	1.671	2.000	2.390	2.660
∞	1.645	1.960	2.326	2.576

Table B.4 is taken from Table 3 of Fisher & Yates, *Statistical Tables for Biological, Agricultural and Medical Research,* published by Longman Group Ltd., London (previously published by Oliver & Boyd Ltd., Edinburgh). By permission of the authors and publishers.

Table B.5 CRITICAL VALUES OF F (TOP NUMBER IN EACH CELL IS FOR TESTING AT 0.05 LEVEL; BOTTOM NUMBER FOR TESTING AT 0.01 LEVEL.)

DEGREES OF FREEDOM FOR NUMERATOR—df_{bg}

df_{wg}	1	2	3	4	5	6	8	12	24	∞
1	161.45 / 4032.10	199.50 / 4999.03	215.72 / 5403.49	224.57 / 5625.14	230.17 / 5764.08	233.97 / 5859.39	238.89 / 5981.34	243.91 / 6105.83	249.04 / 6234.16	254.32 / 6366.48
2	18.51 / 98.49	19.00 / 99.01	19.16 / 99.17	19.25 / 99.25	19.30 / 99.30	19.33 / 99.33	19.37 / 99.36	19.41 / 99.42	19.45 / 99.46	19.50 / 99.50
3	10.13 / 34.12	9.55 / 30.81	9.28 / 29.46	9.12 / 28.71	9.01 / 28.24	8.94 / 27.91	8.84 / 27.49	8.74 / 27.05	8.64 / 26.60	8.53 / 26.12
4	7.71 / 21.20	6.94 / 18.00	6.59 / 16.69	6.39 / 15.98	6.26 / 15.52	6.16 / 15.21	6.04 / 14.80	5.91 / 14.37	5.77 / 13.93	5.63 / 13.46
5	6.61 / 16.26	5.79 / 13.27	5.41 / 12.06	5.19 / 11.39	5.05 / 10.97	4.95 / 10.67	4.82 / 10.27	4.68 / 9.89	4.53 / 9.47	4.36 / 9.02
6	5.99 / 13.74	5.14 / 10.92	4.76 / 9.78	4.53 / 9.15	4.39 / 8.75	4.28 / 8.47	4.15 / 8.10	4.00 / 7.72	3.84 / 7.31	3.67 / 6.88
7	5.59 / 12.25	4.74 / 9.55	4.35 / 8.45	4.12 / 7.85	3.97 / 7.46	3.87 / 7.19	3.73 / 6.84	3.57 / 6.47	3.41 / 6.07	3.23 / 5.65
8	5.32 / 11.26	4.46 / 8.65	4.07 / 7.59	3.84 / 7.01	3.69 / 6.63	3.58 / 6.37	3.44 / 6.03	3.28 / 5.67	3.12 / 5.28	2.93 / 4.86
9	5.12 / 10.56	4.26 / 8.02	3.86 / 6.99	3.63 / 6.42	3.48 / 6.06	3.37 / 5.80	3.23 / 5.47	3.07 / 5.11	2.90 / 4.73	2.71 / 4.31
10	4.96 / 10.04	4.10 / 7.56	3.71 / 6.55	3.48 / 5.99	3.33 / 5.64	3.22 / 5.39	3.07 / 5.06	2.91 / 4.71	2.74 / 4.33	2.54 / 3.91
11	4.84 / 9.65	3.98 / 7.20	3.59 / 6.22	3.36 / 5.67	3.20 / 5.32	3.09 / 5.07	2.95 / 4.74	2.79 / 4.40	2.61 / 4.02	2.40 / 3.60
12	4.75 / 9.33	3.88 / 6.93	3.49 / 5.93	3.26 / 5.41	3.11 / 5.06	3.00 / 4.82	2.85 / 4.50	2.69 / 4.16	2.50 / 3.78	2.30 / 3.36
14	4.60 / 8.86	3.74 / 6.51	3.34 / 5.56	3.11 / 5.03	2.96 / 4.69	2.85 / 4.46	2.70 / 4.14	2.53 / 3.80	2.35 / 3.43	2.13 / 3.00
16	4.49 / 8.53	3.63 / 6.23	3.24 / 5.29	3.01 / 4.77	2.85 / 4.44	2.74 / 4.20	2.59 / 3.89	2.42 / 3.55	2.24 / 3.18	2.01 / 2.75

DEGREES OF FREEDOM FOR DENOMINATOR—df_{wg}

Table B.5 (*Continued*)

DEGREES OF FREEDOM FOR NUMERATOR—df_{bg}

DEGREES OF FREEDOM FOR DENOMINATOR—df_{wg} (CONTINUED)

df_{wg}	1	2	3	4	5	6	8	12	24	∞
18	4.41	3.55	3.16	2.93	2.77	2.66	2.51	2.34	2.15	1.92
	8.28	6.01	5.09	4.58	4.25	4.01	3.71	3.37	3.01	2.57
20	4.35	3.49	3.10	2.87	2.71	2.60	2.45	2.28	2.08	1.84
	8.10	5.85	4.94	4.43	4.10	3.87	3.56	3.23	2.86	2.42
25	4.24	3.38	2.99	2.76	2.60	2.49	2.34	2.16	1.96	1.71
	7.77	5.57	4.68	4.18	3.86	3.63	3.32	2.99	2.62	2.17
30	4.17	3.32	2.92	2.69	2.53	2.42	2.27	2.09	1.89	1.62
	7.56	5.39	4.51	4.02	3.70	3.47	3.17	2.84	2.47	2.01
40	4.08	3.23	2.84	2.61	2.45	2.34	2.18	2.00	1.79	1.52
	7.31	5.18	4.31	3.83	3.51	3.29	2.99	2.66	2.29	1.82
50	4.03	3.18	2.79	2.56	2.40	2.29	2.13	1.95	1.74	1.44
	7.17	5.06	4.20	3.72	3.41	3.19	2.89	2.56	2.18	1.68
60	4.00	3.15	2.76	2.52	2.37	2.25	2.10	1.92	1.70	1.39
	7.08	4.98	4.13	3.65	3.34	3.12	2.82	2.50	2.12	1.60
70	3.98	3.13	2.74	2.50	2.35	2.23	2.07	1.89	1.67	1.35
	7.01	4.92	4.07	3.60	3.29	3.07	2.78	2.45	2.07	1.53
80	3.96	3.11	2.72	2.49	2.33	2.21	2.06	1.88	1.65	1.32
	6.98	4.88	4.04	3.56	3.26	3.04	2.74	2.42	2.03	1.49
90	3.95	3.10	2.71	2.47	2.32	2.20	2.04	1.86	1.64	1.30
	6.92	4.85	4.01	3.53	3.23	3.01	2.72	2.39	2.00	1.46
100	3.94	3.09	2.70	2.46	2.30	2.19	2.03	1.85	1.63	1.28
	6.90	4.82	3.98	3.51	3.21	2.99	2.69	2.37	1.98	1.43
200	3.89	3.04	2.65	2.42	2.26	2.14	1.98	1.80	1.57	1.19
	6.97	4.71	3.88	3.41	3.11	2.89	2.60	2.28	1.88	1.28
∞	3.84	2.99	2.60	2.37	2.21	2.09	1.94	1.75	1.52	1.00
	6.64	4.60	3.78	3.32	3.02	2.80	2.51	2.18	1.79	1.00

Adapted from Table F of H. E. Garrett, *Statistics in Psychology and Education*, 5th Edition, Copyright 1958, David McKay Co., Inc.

References

American Psychological Association. (1983). *Ethical standards of psychologists* (rev. ed.). Washington, DC: Author.

American Psychological Association. (1983). *Standards for providers of psychological services* (rev. ed.). Washington, DC: Author.

American Psychological Association. (1992). Ethical principles of psychologists and code of conduct. *American Psychologist, 47,* 1597–1611.

Asch, S. (1952). *Social psychology.* Englewood Cliffs, NJ: Prentice-Hall.

Ayllon, T. (1963). Intensive treatment of psychotic behavior by stimulus satiation and food reinforcement. *Behavior Research and Therapy, 1,* 53–61.

Ayllon, T., & Azrin, N. H. (1968). *The token economy: A motivational system for therapy and rehabilitation.* Englewood Cliffs, NJ: Prentice-Hall.

Bayoff, A. G. (1940). The experimental social behavior of animals: II. The effect of early isolation of white rats on their competition in swimming. *Journal of Comparative Psychology, 29,* 293–306.

Benbow, C., & Stanley, J. C. (1980). Sex differences in mathematical ability: Fact or artifact? *Science, 210*(4475), 1262–1264.

Bitterman, M. E. (1969). Thorndike and the problem of animal intelligence. *American Psychologist, 24,* 444–453.

Bower, G. H., Karlin, M. B., & Dueck, A. (1975). Comprehension and memory for pictures. *Memory and Cognition, 3,* 216–220.

Bryson, J. B., & Hamblin, K. (1988). Reporting infidelity: The MUM effect and the double standard. Paper presented at WPA meeting: April 29, 1988, Burlingame, CA.

Campbell, D. T. (1969). Reforms as experiments. *American Psychologist, 24,* 409–429.

Campbell, D. T., & Stanley, J. C. (1966). *Experimental and quasi-experimental designs for research.* Chicago: Rand McNally.

Chi, M. T. (1978). Knowledge structures and memory development. In R. S. Siegler (Ed.), *Children's thinking: What develops?* Hillsdale, NJ: Erlbaum.

Conant, J. B. (1951). *Science and common sense.* New Haven, CT: Yale University Press.

Cook, T. D., & Campbell, D. T. (1979). *Quasi-experimentation: Design & analysis issues for field settings.* Chicago: Rand McNally.

Crespi, L. (1942). Quantitative variation of incentive and performance in the white rat. *American Journal of Psychology, 55,* 467–517.

Cunningham, M. (1988). Reactions to heterosexual opening lines: Female selectivity and male responsiveness. *Personality and Social Psychology Bulletin.*

Czeisler, C. A., Moore, C., Ede, M. C., & Coleman, R. M. (1982). Rotating shift work schedules that disrupt sleep are improved by applying circadian principles. *Science, 217,* 460.

Dempster, F. N. (1981). Memory span: Sources of individual and developmental differences. *Psychological Bulletin, 89,* 66.

Ehrenfreund, D., & Badia, P. (1962). Response strength as a function of drive level and pre- and postshift incentive magnitude. *Journal of Experimental Psychology, 63,* 468–471.

Elliott, M. H. (1928). The effect of change of reward on the maze performance of rats. *University of California Publications in Psychology, 4,* 19–30.

Ferster, C. B., & Perrott, M. C. (1968). *Behavior principles.* Englewood Cliffs, NJ: Prentice-Hall.

Festinger, I. (1957). *A theory of cognitive dissonance.* New York: Harper & Row.

Fisher, J. I., & Harris, M. B. (1973). Effect of note taking and review on recall. *Journal of Educational Psychology, 65,* 321–325.

Gold, D. B., & Wegner, D. M. (1991). Fanning old flames: Arousing romantic obsession through thought suppression. Paper presented at the annual meeting of the American Psychological Association, August 1991, San Francisco, CA.

Goodenough, D. R., Shapiro, A., Holden, M., & Steinschriber, L. (1959). A comparison of "dreamers" and "nondreamers": Eye movements, electroencephalograms, and the recall of dreams. *Journal of Abnormal and Social Psychology, 59,* 295–302.

Hayes, S. (1981). Single case experimental design and empirical clinical practice. *Journal of Consulting and Clinical Psychology, 49,* 193–211.

Hilgard, E. R. (1969). Pain as a puzzle for psychology and physiology. *American Psychologist, 24,* 103–113.

Hirsh, I. J. (1966). Audition. In J. B. Sidowski (Ed.), *Experimental methods and instrumentation in psychology.* New York: McGraw-Hill.

Howes, D. H., & Solomon, R. L. (1950). A note on McGuinnes' "Emotionality and perceptual defense." *Psychological Review, 57,* 229–234.

Howes, D. H., & Solomon, R. L. (1951). Visual duration threshold as a function of word-probability. *Journal of Experimental Psychology, 41,* 401–410.

Humphreys, A. P., & Smith, P. K. (1987). Rough and tumble, friendship, and dominance in school children: Evidence for continuity and change with age. *Child Development, 58,* 201–212.

Johnson, H. H., & Scileppi, J. A. (1969). Effects of ego-involvement conditions on attitude change to high and low credibility communicators. *Journal of Personality and Social Psychology, 13,* 31–36.

Kleinke, C. L., Meeker, F. B., & Staneski, R. A. (1986). Preference for opening lines: Comparing ratings by men and women. *Sex Roles, 15,* 585–600.

Lambert, W. W., & Solomon, R. L. (1952). Extinction of a running response as a function of block point from the goal. *Journal of Comparative and Physiological Psychology, 45,* 269–279.

Larson, C. C. (1982). Animal research: Striking a balance. *APA Monitor, 13*(Jan.), 1 ff.

Linder, D. E., Cooper, J., & Jones, E. E. (1967). Decision freedom as a determinant of the role of incentive magnitude in attitude change. *Journal of Personality and Social Psychology, 6,* 245–254.

Loftus, E. F., & Palmer, J. C. (1974). Reconstruction of automobile destruction: An example of the interaction between language and memory. *Verbal Learning and Verbal Behavior, 13,* 585–589.

Lorge, I. (1930). Influence of regularly interpolated time intervals upon subsequent learning. *Teachers College, Columbia University Contributions to Education* (Whole No. 438).

Mahoney, M. J., Moura, N. G. M., & Wade, T. C. (1973). Relative efficacy of self-reward, self-punishment, and self-monitoring techniques for weight loss. *Consulting and Clinical Psychology, 40,* 404–407.

Marshall, J. (1969). *Law and psychology in conflict.* New York: Anchor Books.

Mazur, A. (1986). U.S. trends in feminine beauty and overadaptation. *The Journal of Sex Research, 22,* 281–303.

More, A. J. (1969). Delay of feedback and the acquisition and retention of verbal materials in the classroom. *Journal of Educational Psychology, 60,* 339–342.

Paul, G. L. (1966). *Insight versus desensitization in psychotherapy.* Stanford, CA: Stanford University Press.

Posner, M. I. (1969). Abstraction and the process of recognition. In J. T. Spence & G. H. Bower (Eds.), *The psychology of learning and motivation: Advances in learning and motivation* (Vol. 3). New York: Academic Press.

Posner, M. I., Boies, S. J., Eichelman, W., & Taylor, R. L. (1969). Retention of visual and name codes of single letters. *Journal of Experimental Psychology, 73,* 28–38.

Posner, M. I., & Keele, S. W. (1968). On the genesis of abstract ideas. *Journal of Experimental Psychology, 77,* 353–363.

Postman, L., Bronson, W. C., & Gropper, G. L. (1952). Is there a mechanism of perceptual defense? *Abnormal and Social Psychology, 48,* 215–224.

Pryor, K. W., Haag, R., & O'Reilly, J. (1969). The creative porpoise: Training for novel behavior. *Experimental Analysis of Behavior, 12,* 653–661.

Rosenthal, R. (1966). *Experimenter effects in behavioral research.* New York: Appleton-Century-Crofts.

Rosenthal, R. (1969). Interpersonal expectations: Effects of the experimenter's hypothesis. In R. Rosenthal & R. L. Rosnow (Eds.), *Artifact in behavioral research.* New York: Academic Press.

Rosenthal, R., & Fode, K. (1963). The effects of experimenter bias on the performance of the albino rat. *Behavioral Science, 8,* 183–189.

Rosenthal, R., & Rosnow, R. L. (Eds.). (1969). *Artifact in behavioral research.* New York: Academic Press.

Sands, S. F., Lincoln, C. E., & Wright, A. A. (1982). Pictorial similarity judgments and the organization of visual memory in the rhesus monkey. *Journal of Experimental Psychology: General, 3,* 369.

Simon, C. W., & Emmons, W. H. (1956). Responses to material presented during various levels of sleep. *Journal of Experimental Psychology, 51,* 89–97.

Smith, A. P., Tyrrell, D. A. J., Coyle, K., & Willman, J. S. (1987). Selective effects of minor illnesses on human performance. *British Journal of Psychology, 78,* 183–188.

Snodgrass, J. G., Levy-Berger, G., & Haydon, M. (1985). *Human experimental psychology.* New York: Oxford.

Solso, R. L. (1987a). The social-political consequences of the organization and dissemination of knowledge. *American Psychologist, 42,* 824–825.

Solso, R. L. (1987b). Recommended readings in psychology over the past 33 years. *American Psychologist, 42,* 1130–1132.

Solso, R. L., & McCarthy, J. E. (1981). Prototype formation of faces: A case of pseudomemory. *British Journal of Psychology, 72,* 499–503.

Solso, R. L., & Short, B. A. (1979). Color recognition. *Bulletin of the Psychonomic Society, 14,* 275–277.

Sprafkin, J. N., Liebert, R. M., & Poulos, R. W. (1975). Effects of a prosocial televised example on children's helping. *Journal of Experimental Child Psychology, 20,* 119–126.

Stevens, S. S., & Davis, H. (1938). *Hearing, it's psychology and physiology.* New York: Wiley.

Stevenson, H. W., & Odom, R. D. (1962). The effectiveness of social reinforcement following two conditions of social deprivation. *Abnormal and Social Psychology, 65,* 429–431.

Suls, J. M., & Miller, R. L. (1976). Humor as an attributional index. *Personality and Social Psychology Bulletin, 2,* 256–259.

Supa, M., Cotzin, M., & Dallenbach, K. M. (1944). "Facial vision": The perception of obstacles by the blind. *American Journal of Psychology, 57,* 133–183.

Terkel, J., & Rosenblatt, J. S. (1968). Maternal behavior induced by maternal blood plasma injected into virgin rats. *Journal of Comparative and Physiological Psychology, 65,* 479–482.

Thumin, F. J. (1962). Identification of cola beverages. *Journal of Applied Psychology, 46,* 358–360.

Underwood, B. J. (1975). *Psychological research.* New York: Appleton-Century-Crofts.

Underwood, B. J. (1983). *Attributes of memory.* Glenview, IL: Scott-Foresman.

Underwood, B. J., Rehula, R., & Keppel, G. (1962). Item selection in paired-associate learning. *American Journal of Psychology, 75,* 353–371.

Walk, R. D. (1969). Two types of depth discrimination by the human infant with five inches of visual depth. *Psychonomic Science, 14,* 253–254.

Walter, R. H., & Parke, R. D. (1964). Emotional arousal, isolation, and discrimination learning in children. *Journal of Experimental Child Psychology, 1,* 163–173.

Webb, E. J., Campbell, D. T., Schwartz, R. D., & Sechrest, L. (1966). *Unobtrusive measures: Nonreactive research in the social sciences.* Chicago: Rand McNally.

Westrup, D. A., Keller, S. R., Nellis, T. A., & Hicks, R. A. (1992). Bruxism and arousal predisposition in college students. Paper presented at the Western Psychological Association meeting, April 1992, Portland, OR.

Wilson, G. D., Nias, D. K. B., & Brazendale, A. H. (1975). Vital statistics, perceived sexual attractiveness, and response to risque humor. *The Journal of Social Psychology, 95,* 201–255.

Wynder, E. L., & Stellman, S. D. (1977). Comparative epidemiology of tobacco-related cancer. *Cancer Research, 37,* 4608–4622.

Name Index

Aristotle, 5–7
Asch, S., 21–22, 82
Ayllon, T., 63, 64

Badia, P., 43–44
Bailey, L., 131
Bartlett, F., 15
Bayoff, 109
Berenbaum, S., 9, 220, 222
Benbow, C., 22
Boies, S., 53
Bower, G., 205, 206
Brazendale, A., 30
Bronson, W., 89
Bryson, J., 39–40

Campbell, D., 56, 57, 182
Chi, M., 49–50
Coleman, R., 148
Conant, J., 4
Cook, T., 182
Cooper, J., 45–46, 47, 101
Cotzin, M., 24
Coyle, K., 123
Curie, M., 15
Czeisler, C., 148

Dallenbach, K., 24
Darwin, C., 12–13, 15
Dean, L., 39, 180, 183
Dempster, F., 150, 152
Dueck, A., 205, 206

Ede, M., 148
Ehrenfreund, D., 43–44
Eichelman, W., 53

Einstein, A., 4
Emmons, W., 84–85

Ferster, C., 61
Festinger, L., 45
Fisher, J., 313, 314, 315
Fode, K., 26

Galileo, 5–7
Gardner, A., 31–32
Gardner, B., 31–32
Gold, D., 192, 194
Gropper, G., 89

Haag, R., 252, 254
Hamblin, K., 39–40
Harlow, H., 235
Harris, M., 313, 314, 315
Hewitt, J., 39, 180, 183
Hicks, R., 201
Hilterbrand, K., 245, 246
Hines, M., 220, 222
Hirsch, I., 149
Howes, D., 89

James, W., 15, 235
Johnson, H., 102
Jones, E., 45–46, 47, 100

Karlin, M., 205, 206
Keele, S., 53
Keller, S., 201
Keppel, G., 91
Krushinski, K., 135

Laird, R., 135
Larson, C., 115

Subject Index